BIOANALYTICAL
CHEMISTRY

BIOANALYTICAL CHEMISTRY

Susan R. Mikkelsen
Eduardo Cortón

WILEY-INTERSCIENCE

A JOHN WILEY & SONS, INC., PUBLICATION

For general information on our other products and services please contact our Customer Care Department
within the U.S. at 877-762-2974, outside the U.S. at 317-572-3993 or fax 317-572-4002.

Wiley also publishes its books in a variety of electronic formats. Some content that appears in print,
however, may not be available in electronic format.

Library of Congress Cataloging-in-Publication Data:

Mikkelsen, Susan R., 1960–
    Bioanalytical chemistry / Susan R. Mikkelsen, Eduardo Cortón.
      p.   cm.
Includes bibliographical references and index.
    ISBN 0-471-54447-7 (cloth)
    1. Analytical biochemistry.
    [DNLM: 1. Chemistry, Analytical.   QY 90 M637b 2004]   I. Cortón,
Eduardo, 1962– II. Title.
    QP519.7.M54 2004
      572'.36–dc22   2003016568

Printed in the United States of America

10 9 8 7 6 5 4 3 2 1

■■■■■■ CONTENTS

▪ PREFACE

The expanding role of bioanalytical chemistry in academic and industrial environments has made it important for students in chemistry and biochemistry to be introduced to this field during their undergraduate training. Upon introducing a bioanalytical chemistry course in 1990, I found that there was no suitable textbook that incorporated the diverse methods and applications in the depth appropriate to an advanced undergraduate course. Many specialized books and monographs exist that cover one or two topics in detail, and some of these are suggested at the end of each chapter as sources of further information.

This book is intended to be used as a textbook by advanced undergraduate chemistry and biochemistry students, as well as bioanalytical chemistry graduate students. These students will have completed standard introductory analytical chemistry and biochemistry courses, as well as instrumental analysis. We have assumed familiarity with basic spectroscopic, electrochemical, and chromatographic methods, as they apply to chemical analysis.

The subject material in each chapter has generally been organized as a progression from basic concepts to applications involving real samples. Mathematical descriptions and derivations have been limited to those that are believed essential for an understanding of each method, and are not intended to be comprehensive reviews. Problems given at the end of each chapter are included to allow students to assess their understanding of each topic; most of these problems have been used as examination questions by the authors.

As research in industrial, government, and academic laboratories moves toward increasingly interdisciplinary programs, the authors hope that this book will be used to facilitate, and to prepare students for, collaborative scientific work.

Waterloo, Canada

SUSAN R. MIKKELSEN
EDUARDO CORTÓN

ACKNOWLEDGMENTS

The support of our colleagues at Concordia University, the University of Waterloo and Universidad de Buenos Aires, during the preparation of this book is gratefully acknowledged.

We also acknowledge the patience and support of our families and friends, who understood the time commitments that were needed for the completion of this project.

Waterloo, Canada   SUSAN R. MIKKELSEN
                                                                       EDUARDO CORTÓN

CHAPTER 1

Spectroscopic Methods for Matrix Characterization

1.1. INTRODUCTION

The objective of many bioassay methods is to selectively quantitate a single biomolecule, such as a particular enzyme or antibody, or to determine the presence or absence of a known DNA sequence in an unknown sample. Methods for these very selective assays will be considered in later chapters.

When faced with a true unknown, for example, during the isolation or purification of a biomolecule, it can be important to characterize the unknown matrix, or the components in the unknown solution that are present along with the species of interest. This involves the estimation of the total quantity of the different types of biomolecules.

Biochemists often estimate the total quantity of protein and nucleic acid in an unknown by the nomograph method.[1] In this method, the absorbance of the unknown solution in a 1-cm cuvette is measured at 260 and 280 nm. The nomograph (Fig. 1.1) is then used to estimate concentrations.

The nomograph method is rapid and involves no chemical derivatization, but suffers from several disadvantages. Interferences result from any species present in the unknown, other than protein or nucleic acid, that absorb at these wavelengths. Phenol, for example, which is used at high concentrations during the purification of nucleic acid, absorbs at both 260 and 280 nm and leads to overestimation of total nucleic acid. Furthermore, the nomograph method does not distinguish between deoxyribonucleic acid (DNA) and ribonucleic acid (RNA), and is useful only for samples containing relatively high concentrations of protein and nucleic acid.

In this chapter, we will consider simple colorimetric and fluorometric methods for the quantitation of total protein, DNA, RNA, carbohydrate, and free fatty acid. All of the methods are based on the generation of a chromophore or fluorophore by a selective chemical reaction.

Bianalytical Chemistry, by Susan R. Mikkelsen and Eduardo Cortón
ISBN 0-471-54447-7   Copyright © 2004 John Wiley & Sons, Inc.

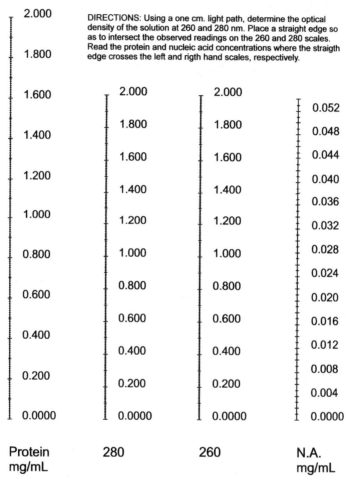

NOMOGRAPH
By E. Adams

Based on the extinction coefficients for enolase and nucleic acid
given by Warbug and Christian, Biochem. Z. 310, 384 (1942).

DIRECTIONS: Using a one cm. light path, determine the optical
density of the solution at 260 and 280 nm. Place a straight edge so
as to intersect the observed readings on the 260 and 280 scales.
Read the protein and nucleic acid concentrations where the straigth
edge crosses the left and rigth hand scales, respectively.

Protein
mg/mL   280   260   N.A.
mg/mL

Figure 1.1. Nomograph used to estimate total protein and nucleic acid concentrations. [Reprinted, with permission, from Calbiochem, San Diego, CA.]

1.2. TOTAL PROTEIN

The three most common assays for total protein[2] are the Lowry (enhanced copper), Smith (bicinchoninic acid, BCA), and Bradford (Coomassie Blue) methods. All are colorimetric methods, and are based on the generation of absorbing species in proportion to the quantity of protein present in the sample. The ninhydrin assay is a recently reported promising method.

1.2.1. Lowry Method

The Lowry, or enhanced alkaline copper method, begins with the addition of an alkaline solution of Cu^{2+} to the sample. The copper forms a complex with nitrogen atoms in the peptide bonds of proteins under these conditions, and is reduced to Cu^+. The Cu^+, along with the R groups of tyrosine, tryptophan, and cysteine residues of the protein then react with the added Folin–Ciocalteau reagent, which contains sodium tungstate, sodium molybdate, phosphoric acid, and HCl (W^{6+}/Mo^{6+}). During this reaction, the Cu^+ is oxidized to Cu^{3+}, which then reacts with the peptide backbone to produce imino peptide ($R_1-CO-N=C(R_2)-Co-R_3$) plus Cu^{2+}, and the Folin reagent is reduced to become molybdenum–tungsten blue. Absorbance is measured in glass or polystyrene cuvettes at 720 nm, or if this value is too high (>2), at 500 nm.

A calibration curve obtained with standard protein solutions [e.g., bovine serum albumin (BSA)] is used to obtain total protein in the unknown. Under optimum conditions, and in the absence of reactive side chains, it has been shown that two electrons are transferred per tetrapeptide unit; however, proteins, with significant proline or hydroxyproline content, or with side chains that can complex copper (such as glutamate) yield less color. The side chains of cysteine, tyrosine and tryptophan contribute one, four and four electrons, respectively.[3] Note that different proteins will produce different color intensities, primarily as a result of different tyrosine and tryptophan contents.

With reagents prepared in advance, the Lowry assay requires $\sim$ 1 h. A 400-µL sample is required, containing 2–100-µg protein (5–250 µg/mL). Nonlinear calibration curves are obtained, due to decomposition of the Folin reagent at alkaline pH following addition to the sample that results in incomplete reaction. Interferences include agents that acidify the solution, chelate copper, or cause reduction of copper(II).

1.2.2. Smith (BCA) Method

The Smith total protein assay is also based on the initial complexation of copper (II) with peptides under alkaline conditions, with reduction to copper(I). The ligand BCA is then added in excess, and the purple color (562-nm peak absorbance) develops upon 2:1 binding of BCA with Cu^+ (Fig. 1.2).

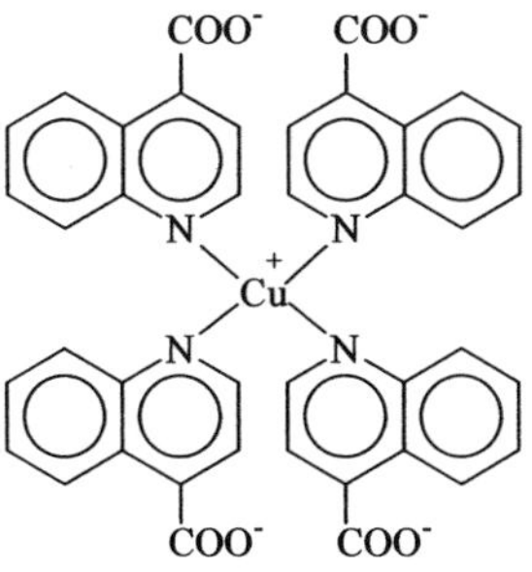

Figure 1.2. Complex formed upon reaction of bicinchoninic acid with Cu^+.

The Smith assay takes ~ 1 h, and requires a minimum volume of 500 μL with 0.2–50-μg protein (0.4–100 μg/mL). As with the Lowry assay, calibration curves are nonlinear, with negative deviation at high protein concentration. Interferences include copper reducing agents (such as reducing sugars) and complexing agents as well as acidifying agents in the unknown sample. Common membrane lipids and phospholipids have also been shown to interfere, and an apparent protein concentration of 17 μg of protein (as BSA) was found for 100 μg of 1,3-dilinoleoyglycerol.[4] Since BCA is a stable reagent at alkaline pH, it can be added to the alkaline copper reagent, so that only one reagent addition is required. The Smith assay has been adapted for 96-well microtiter plates,[5] and a linear dependence of absorption (570 nm) on protein concentration was observed over the 1–10-μg/mL concentration range, with 10-μL sample volumes.

1.2.3. Bradford Method

The Bradford method is based on the noncovalent binding of the anionic form of the dye Coomassie Blue G-250 with protein.[6] The dye reacts chiefly with arginine residues, which have a positively charged side chain, and slight interactions have also been observed with basic residues (histidine and lysine) and aromatic residues (tyrosine, tryptophan, and phenylalanine). In the absence of protein, the dye reagent is a pale red, and upon binding to protein, a blue color is generated with an absorbance maximum at 590 nm. The structure of the dye is shown in Figure 1.3.

The Bradford assay is very popular because it is rapid (5 min) and involves a single addition of the dye reagent to the sample. It provides a nonlinear calibration curve of A_{590} against concentration over the 0.2–20-μg range of total protein contained in a 20-μL sample volume (10–1000 μg/mL). Negative deviation from linearity occurs with this method as with the Lowry and Smith methods. In the Bradford assay, curvature is due to depletion of free dye at high protein concentration, and a better approximation of linearity can be achieved by plotting $A_{595}-A_{465}$ against concentration, to take dye depletion into account.

The Lowry and Bradford methods have been compared for protein quantitation in samples containing membrane fractions.[7] While the Lowry method yielded reproducible results over 2 weeks of sample storage at -20 °C, the Bradford method was shown to significantly underestimate membrane proteins, even after

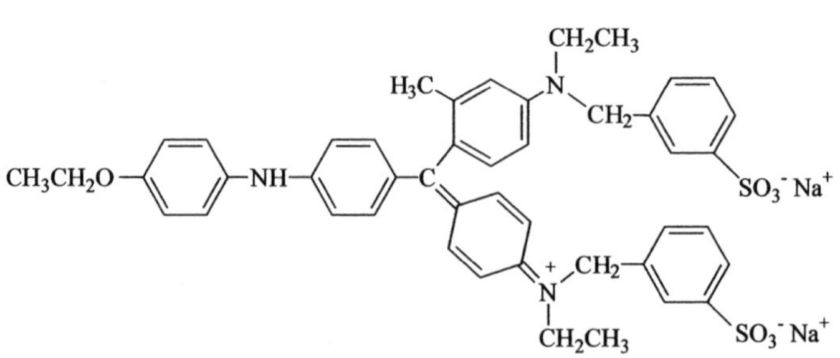

Figure 1.3. Coomassie Blue G-250, used in the Bradford total protein assay.

treatment with base or surfactants, with lower estimates occurring after longer periods of sample storage. Tissue homogenates from rat brain assayed by Bradford method initially showed 52% of the protein estimated by the Lowry method; after 14-days storage, this value had decreased to $\sim$ 30%. Since membrane lipids and phospholipids interfere with the Smith total protein assay, membrane proteins should be quantitated by the Lowry method, or by the new ninhydrin method described below.

1.2.4. Ninhydrin-Based Assay[8]

This method is based on the quantitation of total amino acids following acid hydrolysis of proteins present in tissue samples. Microtiter plates are used in this assay. Tissue samples (10 µg) are first hydrolyzed in 500 µL of 6 M HCl at 100 °C for 24 h to liberate ammonium. The samples are then lyophilized (chilled and evaporated), and the residue, containing ammonium chloride, is dissolved in a known volume of water.

Ninhydrin reagent, containing ninhydrin, ethylene glycol, acetate buffer, and stannous chloride suspension, initially a pale red color, is added to 1–10 µg of protein hydrolysate in a flat-bottom microtiter plate. During the 10-min incubation at 100 °C, ammonia reacts with the ninhydrin reagent to produce diketohydrindylidene-diketohydrindamine (Eq. 1.1).

$$ NH_3 \ + \ 2 \qquad\qquad\qquad\qquad \longrightarrow \qquad\qquad\qquad\qquad\qquad\qquad \tag{1.1} $$

Under these conditions, the product exhibits a broad absorption band between 560 and 580 nm, and a microtiter plate reader set at 575 nm may be used for measurement. Correlation of absorbance with protein concentration has been performed using several protein standards, as shown in Figure 1.4, which also shows results obtained with the same protein standards by the Bradford method. These data demonstrate much better sensitivity with the ninhydrin method, and suggest an $\sim$ 10-fold improvement in detection limit. An important advantage of this method is that the differences in calibration curves obtained using different protein standards are relatively small. Interferences include free amino acids as well as other compounds containing amine groups.

1.2.5. Other Protein Quantitation Methods

The method of nitration has been reported for protein quantitation. In this method, aromatic amino acid residues are treated with nitric acid, and the products of nitration absorb at 358 nm. In the case of tyrosine, nitration occurs at the 3-position of the aromatic ring, producing 3-nitrotyrosine. This method has been compared to the Smith and Bradford assays using BSA standards, and practically identical results were obtained, as shown in Figure 1.5. The detection limit of this assay is 5 µg, and known interferences included Triton X-100 and phenolic compounds.[9]

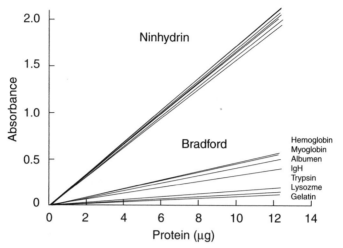

Figure 1.4. Comparison of results of ninhydrin and Bradford methods for total protein concentration in tissue samples. Proteins in both assays are the same.[8] [Reprinted, with permission, from B. Starcher, *Anal. Biochem.* **292**, 2001, 125–129. "A Nirhydrin-Based Assay to Quantitate the Total Protein Content of Tissue Samples." Copyright © 2001 by Academic Press.]

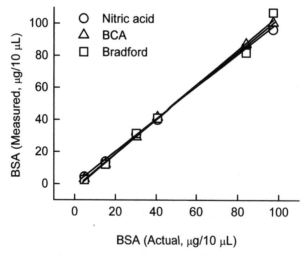

Figure 1.5. Relationship between actual and measured BSA concentration in six samples, showing that this method produces comparable results to the Bradford and BCA total protein assays. [Reprinted, with permission, from K. C. Bible, S. A. Boerner, and S. H. Kaufmann, *Anal. Biochem.* **267**, 1999, 217–221. "A One-Step Method for Protein Estimation in Biological Samples: Nitration of Tyrosine in Nitric Acid." Copyright © 1999 by Academic Press.]

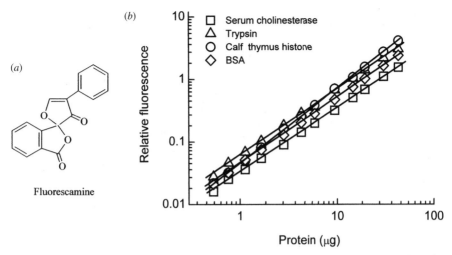

Figure 1.6. (*a*) Chemical structure of fluorescamine, and (*b*) calibration curves obtained for different proteins. [Reprinted, with permission, from P. Böhlen, S. Stein, W. Dairman, and S. Undenfriend, *Archives of Biochemistry and Biophysics* **155**, 1973, 213–220. "Fluorometric Assay of Proteins in the Nanogram range." Copyright © 1973 by Academic Press, Inc.]

An older method employing fluorescamine may be used if improved detection limits are required. In this method, fluorescamine (Fig. 1.6) reacts with the primary amine groups of the protein, generating flourophores that can be excited at 390 nm to generate emission at 475 nm. Figure 1.6 also shows calibration curves for a variety of proteins; this method is capable of detecting as little as 10-ng total protein.[10]

Table 1.1 summarizes detection limits and interferences for the five protein assay methods described here.

1.3. TOTAL DNA

Several methods are available for the colorimetric or flourometric determination of total DNA. The first two that will be considered involve depurination of DNA under acidic conditions, and are followed by chemical reactions with 3,5-diamino-benzoic acid (DABA) or diphenylamine (DPA). Two fluorescence methods based on selective, noncovalent interactions of a probe molecule with DNA will then be considered.

1.3.1. Diaminobenzoic Acid Method

Samples containing DNA are incubated in 1 *N* perchloric acid for 10 min. Under these conditions, DNA is depurinated, deoxyribose is released, and converted to ω-hydroxylevulinylaldehyde. An equal volume of 1.3 *M* DABA dihydrochloride is

TABLE 1.1. Detection Limits and Interferences for Protein Assay Methods

Assay	Detection Limit	Interferences[a]
Lowry	2 µg	Acidifiers, Cu chelating agents (e.g., high phosphate concentration), Cu^{2+} reducing agents (e.g., reducing sugars, thiols), Tris, HEPES, EDTA, neutral detergents
Smith	0.2 µg	Acidifiers, Cu chelating, reducing agents, Tris, ammonium sulfate, EDTA, lipids + phospholipids. (Neutral detergents and SDS do not interfere)
Bradford	0.2 µg	Lipids + phospholipids, neutral detergents, SDS
Ninhydrin	0.02 µg	Amino acids, amine-containing compounds
Nitration	5 µg	Triton X-100, phenolic compounds. (Tris, SDS, urea, thiols, reducing agents, glycerol, ammonium sulfate, neutral detergents do not interfere)
Fluorescamine	10 ng	Primary amine groups, secondary amines at high concentration, TRIS, ammonium sulfate (Phosphate does not interfere)
Ninhydrin	0.02 µg	Free amino acids and other compounds containing amine groups

[a]Tris = tris(hydroxymethyl)aminomethane;   HEPES = N-(2-hydroxyethyl)piperazine-N'-ethanesulfonic acid; SDS = sodium dodecyl sulfate; EDTA = ethylenediaminetetraacetic acid.

added and the solution is heated at 60 °C for 30 min. Under these conditions, the following reactions (Eqs. 1.2 and 1.3) occur:[11]

$$R-CH_2-CHO \longrightarrow R-CH_2-CH{=}\overset{\overset{\displaystyle R}{|}}{C}-CHO \qquad (1.2)$$

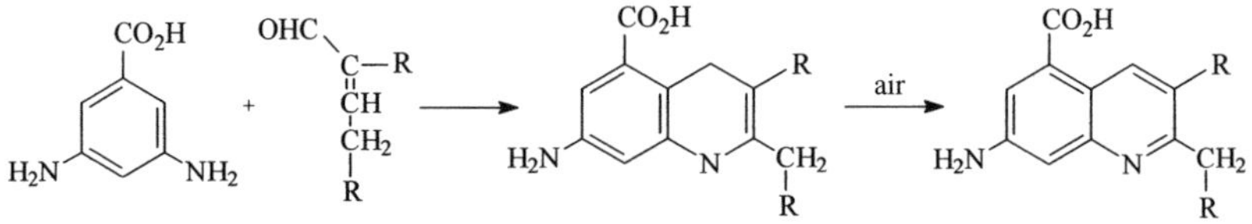

$$(1.3)$$

Following a 10-fold dilution of the sample with 0.6 M perchloric acid, the product of this reaction absorbs maximally at 420 nm. The original assay, developed by Kissane and Robins,[12] uses DABA dihydrochloride for depurination (no perchloric acid was used) and employs fluorescence measurements at 520 nm after excitation at 420 nm. More recent work shows that the assay with perchloric acid can be conducted with 10-µL DNA sample volumes and provides a linear fluorescence calibration curve over the 10–500-ng total DNA range (1–50 µg/mL).[13] The DABA method has since been modified for colorimetric measurements at 420 nm, and has a detection limit of 25 µg.[14] The DABA reaction is specific for DNA, but measurements are affected by the presence of lipids, saccharides, salts, and the surfactant Triton X-100, these species can be removed form DNA samples by ethanol precipitation.[15]

1.3.2. Diphenylamine Method

The DPA method is based on the colorimetric measurement of products formed by reaction of ω-hydroxylevulinylaldehyde (from the deoxyribose released after depurination of DNA) with diphenylamine in 1 M perchloric acid.[14–16] The reaction yields a mixture of products, such as that shown in Eq. 1.4,[17] which together have an absorbance maximum at 595 nm. This method has a similar dynamic range to the DABA assay with colorimetric measurement. Interferences include proteins, lipids, saccharides, and RNA.

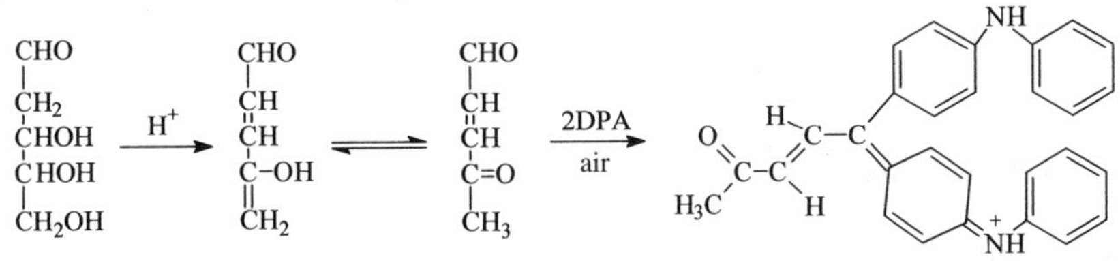

$$(1.4)$$

1.3.3. Other Fluorometric Methods

Sensitive fluorometric assays for double-stranded (native) DNA in tissue extracts are based on the noncovalent interaction of intercalating dyes (Fig. 1.7 such as ethidium bromide, 4′,6′-diamidino-2-phenylindole (DAPI), or 2-(2-[4-hydroxyphenyl]-6-benzimidazolyl)-6- (1-methyl-4-piperazyl)benzimidazol trihydrochloride (Hoechst 33258), with double-stranded DNA in neutral, aqueous solutions.

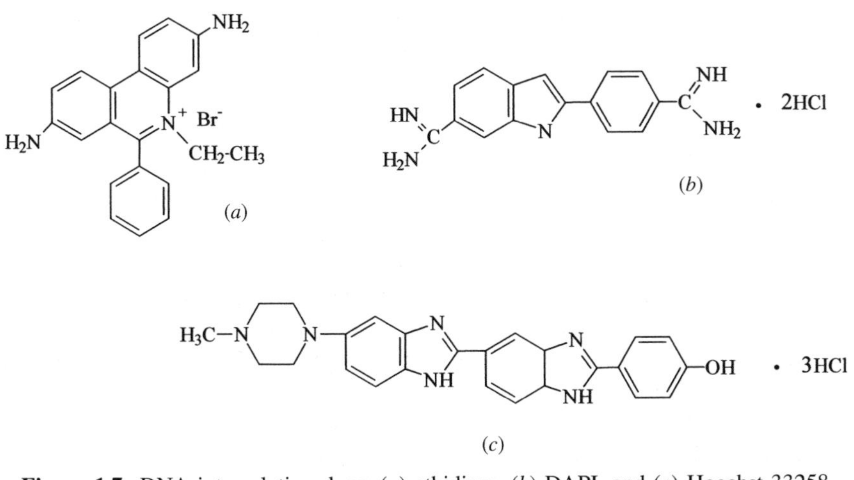

Figure 1.7. DNA-intercalating dyes: (*a*) ethidium, (*b*) DAPI, and (*c*) Hoechst 33258.

When free in aqueous solution, these dyes exhibit limited fluorescence, but upon binding DNA, their fluorescence increases markedly. Ethidium binds to RNA and DNA while DAPI and Hoechst 33258 selectively interact with DNA. Hoechst 33258 at a final concentration of 1 µg/mL in 0.05 M phosphate, pH 7.4, with 2 M NaCl has been used to quantitate as little as 10-ng double-stranded DNA.[18] Because these dye-binding methods involve intercalation between the bases of double-stranded DNA they are not useful for the quantitation of single-stranded DNA.

Single-stranded DNA containing deoxyguanosine residues can be quantitated in the presence of terbium(III). Ten micromolar Tb^{3+} in a pH 6 cacodylate buffer shows no detectable fluorescence, with excitation at 290 nm and emission measured at 488 nm. In the presence of single-stranded DNA, Tb^{3+} coordinates with deoxyguanosine-5-phosphate nucleotides, and a linear dependence of fluorescence on concentration has been observed over the 1–10-µg/mL range of thermally denatured rat liver DNA.[19]

1.4. TOTAL RNA

The only method used to quantitate ribonucleic acid is the orcinol assay.[20] With this method, RNA is depurinated in concentrated HCl and the resulting ribosephosphates are dephosphorylated and dehydrated to produce furfural (Eq. 1.5). Furfural then reacts with orcinol in the presence of Fe^{3+} to yield colored condensation products, as shown in Eq. 1.4, which together possess an absorption maximum at 660 nm.

$$\text{(1.5)}$$

The DNA also undergoes a limited reaction under these conditions, and yields ~ 10% of the color of a similar concentration of RNA; the dehydration step to form furfural is not readily accomplished by 2-deoxyribose.

The original orcinol assay has been modified to improve its selectivity toward RNA over DNA and sugars. The improved method[21] uses a 1.0-mL RNA sample incubated 24 h at 40 °C with 4.0 mL of 85% sulfuric acid prior to addition of the orcinol reagent (containing no Fe^{3+}). Under these conditions, colorimetric measurements are made at 500 nm, where products formed by the reaction of levulinic acid (from DNA) with orcinol do not absorb. Proteins (e.g., BSA) do not interfere, and the method is sixfold more selective to RNA over DNA than the original orcinol assay.

1.5. TOTAL CARBOHYDRATE

Methods reported for the quantitation of reducing sugars, that is, those that possess a terminal aldehyde group, involve redox reactions with oxidizing agents such as Cu^{2+} (in the presence of BCA) or $Fe(CN)_6^{3-}$. Nelson reagents (alkaline copper sulfate and solutions of molybdate and arsenate in dilute sulfuric acid that reduce to molybdenum blue),[22] or 2-cyanoacetamide.[23] The Cu^{2+}-BCA assay has been described earlier for protein quantitation, and the method used for quantitation of reducing carbohydrates is almost identical to the total protein method.[24] Methods for total carbohydrate determination employ strong acids to affect hydrolysis of oligosaccharides and polysaccharides, and dehydration to reactive species such as furfural. These reactive intermediates form chromophores in the presence of phenol, 2-aminotoluene[25] or anthrone[26] and fluorophores with 2-aminothiophenol. We will consider the ferricyanide method for reducing sugars, the phenol-sulfuric acid and 2-aminothiophenol methods for total carbohydrate, and the purpald assay for bacterial polysaccharides.

1.5.1. Ferricyanide Method

This assay employs an alkaline solution of potassium ferricyanide (2.5 mg/mL in phosphate buffer) as the only reagent. A reducing sugar solution of 80 μL is combined with 20 μL of this reagent. The mixture is heated in a sealed tube to 100 °C for 10 min. Following the addition of 300 μL of cold water, absorbance is measured at 237 nm, which is the wavelength of maximum absorbance for ferrocyanide. This method allows the quantitation of mannose over the 1–25-nmol ($\sim$ 1.3–30-μg/mL) range of reducing monosaccharides. Tryptophan and tyrosine residues of proteins were found to interfere significantly with the assay, but the extent of interference can be quantitated by measuring A_{237} following an initial incubation at 40 °C, where these residues reduce ferricyanide, but reducing sugars remain inert. This method was developed to determine the carbohydrate content of glycoproteins.[27]

1.5.2. Phenol–Sulfuric Acid Method

In concentrated sulfuric acid, polysaccharides are hydrolyzed to their constituent monosaccharides, which are dehydrated to reactive intermediates. In the presence of phenol, these intermediates form yellow products, such as that shown in Eq. 1.6,[28] with a combined maximal absorbance at 492 nm.

$$\text{R}-\underset{O}{\text{furyl}}-\text{CHO} \; + \; 2 \; \text{phenol} \; \xrightarrow[\text{air}]{\text{H}_2\text{SO}_4} \; \text{product} \quad (1.6)$$

The phenol–sulfuric acid method has been adapted for use with 96-well microtiter plates. Carbohydrate samples are combined with an equal volume of 5% (w/v) phenol in the wells of a microtiter plate, the plate is placed on a bed of ice and 0.125 mL of concentrated sulfuric acid is added to each well. Following incubation of the covered plate at 80 °C for 30 min and cooling to room temperature, the absorbance of each well measured at 492 nm. This method shows good linearity for maltose over the 0.2–25-µg range, using an initial sample volume of 25 µL (10–200 µg/mL).[24]

1.5.3. 2-Aminothiophenol Method

In this assay, furfural and related monosaccharide dehydration products formed in strong acid are reacted with 2-aminothiophenol to form highly fluorescent 2-(2-furyl)benzothiazole and related species (Eq. 1.7).

$$\tag{1.7}$$

The reagent solution contains 0.4% (w/v) 2-aminothiophenol, and is prepared by combining equal volumes (0.34 mL) of aminothiophenol and ethanol, and diluting to 100 mL with 120 mM HCl. Samples (0.5 mL) are combined with reagent (0.2 mL) and 30% (w/v) H_2SO_4 (0.5 mL) in sealed vials, heated to 150 °C for 15 min, cooled, and diluted by the addition of 0.8-mL water.

Fluorescence is measured at 411 nm with excitation at 361 nm. The assay can easily detect 50 ng of carbohydrates such as galactose, glucose and xylose, and pentoses such as arabinose are detectable at the 10-ng level. Calibration curves show slight curvature, with negative deviations from linearity. Most of the carbohydrates studied yielded useful curves over the 50–600-ng range (0.5–6 µg/mL).[29] Interferences from surfactants (0.5% w/v) such as SDS and Triton X-100 are significant and are thought to alter the polarity of the environment of the fluorophores. Tris buffers and high protein concentration (e.g., 0.1-mg/mL BSA) should also be avoided, unless blanks containing similar concentrations of these species can be prepared.

1.5.4. Purpald Assay for Bacterial Polysaccharides

Polysaccharides containing substituted or unsubstituted glycols in residues such as glycerol, ribitol, arabinitol, furanosyl galactose, and sialic acid may be quantitated by reaction with sodium periodate, follow by reaction with the purpald reagent, as shown in Figure 1.8. The formaldehyde produced in the first step reacts with the purpald reagent, and the product of this reaction is further oxidized by periodate to form a purple product with an absorbance maximum at 550 nm. The limit of detection has been reported as 15 µg/mL using the native polysaccharide from *Salmonella pneumoniae*.[30]

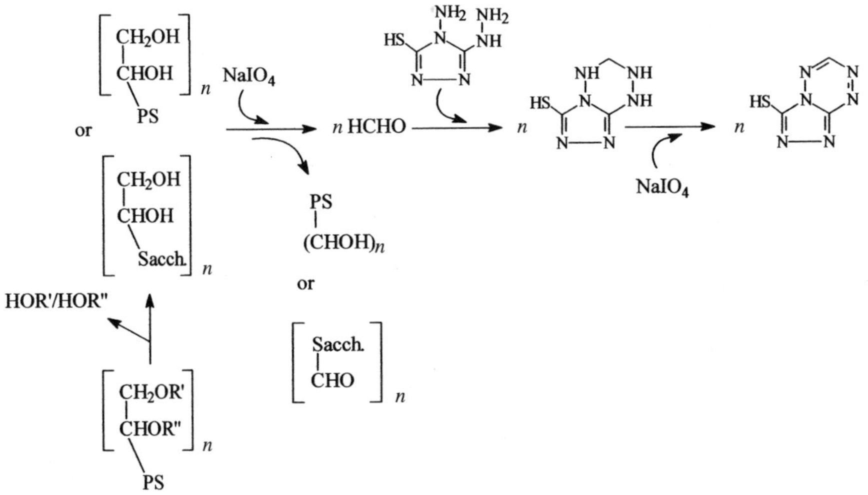

Figure 1.8. Chemistry of the purpald assay for bacterial polysaccharides (PS). The substituent groups R′ and R″ must be released during treatment with NaOH and H_2SO_4, before the periodate reaction.

1.6. FREE FATTY ACIDS

Most methods for the quantitation of free fatty acids involve chemical derivatization followed by gas chromatography–mass spectrometry. However, a colorimetric method is available for the quantitation of long-chain (C > 10) free fatty acids in plasma, and is based on the color developed by cobalt soaps of free fatty acids dissolved in chloroform.[31]

The assay involves four solutions: (1) chloroform:heptane:methanol (4:3:2 by volume); (2) 0.035 M HCl; (3) the salt reagent, consisting of 8 mL of triethanolamine added to 100 mL of an aqueous salt solutions containing 20 g Na_2SO_4, 10 g Li_2SO_4, and 4 g $Co(NO_3)_2.6H_2O$; and (4) the indicator solution consisting of 20-mg 1-nitroso-2-naphthol in 100 mL of 95% ethanol.

The procedure uses 100-μL samples (heptane blank, standard, or plasma sample) combined with 4.0 mL of chloroform/heptane/methanol and 1.0 mL 0.035 M HCl in sealed vials. After mixing and centrifugation steps, the upper aqueous methanol phase is discarded, and 2.0 mL of the salt reagent is added and mixed. A second centrifugation step is followed by combining 2.0 mL of the upper phase with 1.0 mL of the indicator solution. After 20-min absorbance is measured at 435 nm.

Linear calibration curves were obtained for palmitic acid over the 0.10–1.2-μM concentration range, with few interferences. Protein (2% BSA) and many major metabolites such as lactic acid or β-hydroxybutyrate do not interfere, but some color development is observed for high concentrations of phospholipids such as lecithin.

REFERENCES

1. R. R. Alexander and J. M. Griffiths, *Basic Biochemical Methods*, Wiley-Liss, New York, 1993, pp. 29–31.

2. C. M. Stoscheck, *Methods Enzymol.* **182**, 1990, 50–68, and references cited therein.

3. G. Legler, C. M. Muller-Platz, M. Mentges-Hettkamp, G. Pflieger, and E. Julich, *Anal. Biochem.* **150**, 1985, 278–287.

4. R. J. Kessler and D. D. Fanestil, *Anal. Biochem.* **159**, 1986, 138–142.

5. M. G. Redinbaugh and R. B. Turley, *Anal. Biochem.* **153**, 1986, 267–271.

6. S. J. Compton and C. G. Jones, *Anal. Biochem.* **151**, 1985, 369–374.

7. L. P. Kirazov, L. G. Venkov, and E. P. Kirazov, *Anal. Biochem.* **208**, 1993, 44–48.

8. B. Starcher, *Anal. Biochem.* **292**, 2001, 125–129.

9. K. C. Bible, S. A. Boerner, and S. H. Kaufmann, *Anal. Biochem.* **267**, 1999, 217–221.

10. P. Böhlen, S. Stein, W. Dairman, and S. Undenfriend, *Arch. Biochem. Biophys.* **155**, 1973, 213–220.

11. M. Pesez and J.Bartos, *"Colorimetric and Fluorimetric Analysis of Organic Compounds and Drugs,"* Marcel Dekker, New York, 1974, pp. 284–285.

12. J. M. Kissane and E. Robins, *J. Biol. Chem.* **233**, 1958, 184–188.

13. R. Vytasek, *Anal. Biochem.* **120**, 1982, 243–248.

14. F. Setaro and C. D. G. Morley, *Anal. Biochem.* **81**, 1977, 467–471.

15. P. S. Thomas and M. N. Farquhar, *Anal. Biochem.* **89**, 1978, 35–44.

16. K. W. Giles and A. Meyer, *Nature (London)* **206**, 1965, 93.

17. M. Pesez and J. Bartos, *Colorimetric and Fluorimetric Analysis of Organic Compounds and Drugs*, Marcel Dekker, New York, 1974, pp. 436–437.

18. C. Labarca and K. Paigen, *Anal. Biochem.* **102**, 1980, 344–352.

19. D. P. Ringer, B. A. Howell, and D. E. Kizer, *Anal. Biochem.* **103**, 1980, 337–342.

20. Z. Dische and K. Schwartz, *Mikrochim. Acta* **2**, 1937, 13–19.

21. R. Almog and T. L. Shirey, *Anal. Biochem.* **91**, 1978, 130–137.

22. X. Boyer, *Modern Experimental Biochemistry*, 2nd ed., Benjamin/Cummings, New York, 1993.

23. S. Honda, Y. Nishimura, M. Takahashi, H. Chiba, and K. Kakehi, *Anal. Biochem.* **119**, 1982, 194–199.

24. J. D. Fox and J. F. Robyt, *Anal. Biochem.* **195**, 1991, 93–96.

25. T. Morcol and W. H. Velander, *Anal. Biochem.* **195**, 1991, 153–159.

26. M. A. Jermyn, *Anal. Biochem.* **69**, 1975, 332–335.

27. G. Krystal and A. F. Graham, *Anal. Biochem.* **70**, 1976, 336–345.

28. M. Pesez and J. Bartos, *Colorimetric and Fluorimetric Analysis of Organic Compounds and Drugs*, Marcel Dekker, New York, 1974, pp. 411–412.

29. J. K. Zhu and E. A. Nothnagel, *Anal. Biochem.* **195**, 1991, 101–104.

30. Che-Hung Lee and C. E. Frasch, *Anal. Biochem.* **296**, 2001, 73–82.

31. S. W. Smith, *Anal. Biochem.* **67**, 1975, 531–539.

PROBLEMS

1. Total protein in an unknown sample was estimated using the Lowry and Bradford assays. Results were 33 ± 2 µg/mL from the Lowry assay and 21 ± 1 µg/mL from the Bradford assay, using stock solutions of BSA as standards in each assay. The unknown sample was then thoroughly oxygenated, and the assays were repeated. The Lowry method yielded 22 ± 1 µg/mL, while the Bradford results did not change. Why did the results of the Lowry assay decrease by 33%?

2. The Lowry, Smith, and Bradford assays provide an estimate, rather than an exact measure of total protein. Why? What is the importance of the standard protein chosen for the construction of calibration curves?

3. An analyst suspects that an unknown sample containing double-stranded DNA also contains a significant quantity of carbohydrate. The DABA assay for total DNA yields a concentration of 44 ± 3 µg/mL, while the ferricyanide assay for reducing carbohydrate, conducted on an 10-fold dilution of the unknown sample, yielded a result of 1.8 ± 0.3 µg/mL. The formula weight of one deoxynucleotide monophosphate residue of DNA is ~ 330 g/mol, while monosaccharides have a molecular weight of ~ 180 g/mol. Assuming that interference in the DABA assay is due solely to reducing monosaccharides, and that one monosaccharide molecule yields the same absorbance as one deoxynucleotide monophosphate in the DABA assay, calculate the actual quantity of DNA in the sample. Suggest a more selective assay for double-stranded DNA that could be used to confirm this value.

CHAPTER 2

Enzymes

2.1. INTRODUCTION

Enzymes are biological catalysts that facilitate the conversion of substrates into products by providing favorable conditions that lower the activation energy of the reaction.

$$\text{Substrate(s)} \xrightarrow{\text{enzyme}} \text{Product(s)} \tag{2.1}$$

An enzyme may be a protein or a glycoprotein, and consists of at least one polypeptide moiety. The regions of the enzyme that are directly involved in the catalytic process are called the *active sites*. An enzyme may have one or more groups that are essential for catalytic activity associated with the active sites through either covalent or noncovalent bonds; the protein or glycoprotein moiety in such an enzyme is called the *apoenzyme*, while the nonprotein moiety is called the *prosthetic group*. The combination of the apoenzyme with the prosthetic group yields the *holoenzyme*.

The enzyme D-amino acid oxidase, for example, catalyzes the conversion of D-amino acids to 2-keto acids according to Eq. 2.2:

$$^{+}H_3N-CH(R)-COO^{-}+O_2+H_2O \xrightarrow{\text{D-Amino acid oxidase}} R-CO-COO^{-}+H_2O_2+NH_4^{+} \tag{2.2}$$

This enzyme is monomeric, consisting of one polypeptide apoenzyme and one prosthetic group, a flavin adenine dinucleotide (FAD) moiety (Fig. 2.1), that is noncovalently bound, with an association constant[1] of $3.6 \times 10^6\ M^{-1}$. The active holoenzyme can be prepared, or reconstituted, by the addition of FAD to a solution containing the apoenzyme.

FAD is a redox-active prosthetic group commonly found at the active site of oxidase enzymes. A related species, flavin mononucleotide (FMN) is present at the active sites of many dehydrogenase enzymes. What these two prosthetic groups

Bianalytical Chemistry, by Susan R. Mikkelsen and Eduardo Cortón
ISBN 0-471-54447-7 Copyright © 2004 John Wiley & Sons, Inc.

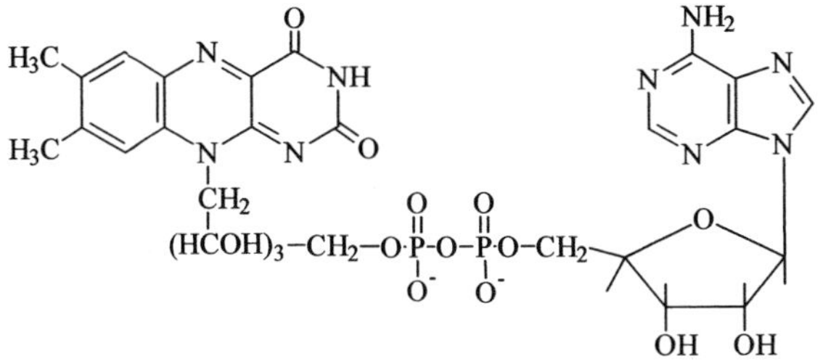

Figure 2.1. Structure of flavin adenine dinucleotide.

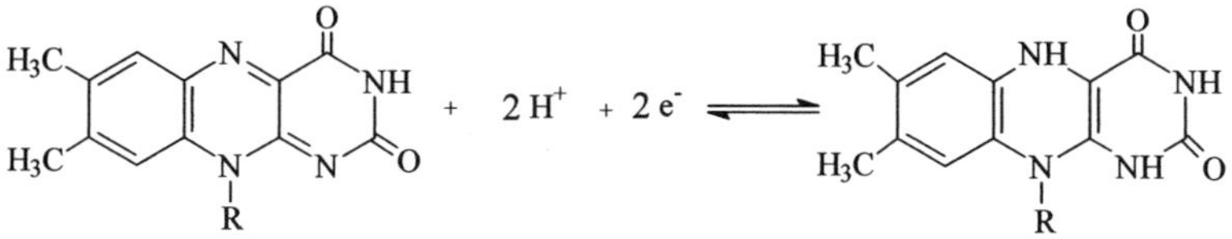

Figure 2.2. Redox half-reaction of flavin species.

have in common is the flavin moiety, which is capable of undergoing a reversible two-proton, two-electron redox cycle, shown in Figure 2.2.

The prosthetic group accepts electrons from the substrate species to generate reduced flavin from the oxidized form in the active site; the primary product diffuses away from the active site, and a secondary substrate, an electron acceptor such as molecular oxygen, then regenerates the oxidized form of the flavin in the active site to complete the catalytic cycle. This is represented by Eqs. 2.3–2.5:

$$\text{Enzyme}(\text{FAD}) + \text{Substrate} \leftrightarrow \text{Enzyme}(\text{FADH}_2) + \text{Product} \tag{2.3}$$

$$\text{Enzyme}(\text{FADH})_2 + O_2 \leftrightarrow \text{Enzyme}(\text{FAD}) + H_2O_2 \tag{2.4}$$

Eq. 2.5 is the sum of Eqs. 2.3 and 2.4

$$\text{Substrate} + O_2 \leftrightarrow \text{Product} + H_2O_2 \tag{2.5}$$

This cycle is typical of many enzymatic reactions, in that the initial substrate conversion is followed by an enzyme regeneration step.

2.2. ENZYME NOMENCLATURE

Enzyme names apply to a single catalytic entity, rather than to a series of individually catalyzed reactions. Names are related to the function of the enzyme, in particular, to the type of reaction catalyzed. This convention implies that one name may, in fact, designate a family of enzymes that are slightly different from each other yet still catalyze the same reaction. For example, lactate dehydrogenase (LDH) has five such *isoenzymes* in humans. LDH catalyzes the oxidation of L-lactic acid by

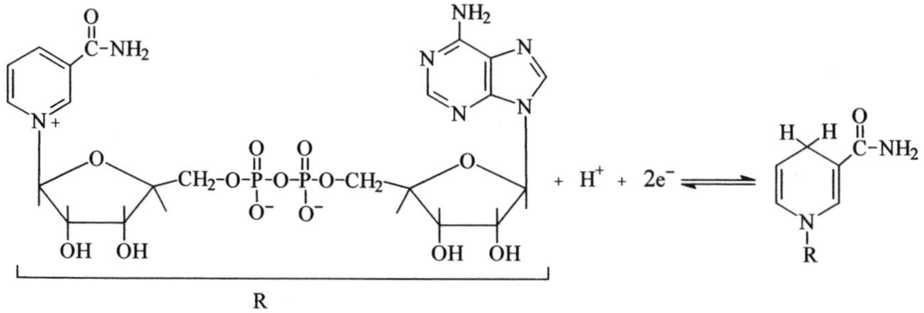

Figure 2.3. Redox half-reaction of nicotinamide adenine dinucleotide.

nicotinamide adenine dinucleotide (NAD^+), a common *cofactor* for dehydrogenase reactions, as shown in Eq. 2.6 and Figure 2.3.

$$\text{L-Lactate} + NAD^+ \xrightarrow{\text{LDH}} \text{Pyruvate} + NADH + H^+ \tag{2.6}$$

where NADH = reduced NAD.

While each of the five LDH isoenzymes catalyzes the conversion of lactic acid to pyruvic acid, the isoenzymes are produced in different organs. Because of this, the polypeptide moieties and the rates at which lactate can be converted to pyruvate are slightly different for each isoenzyme. Similarly, different species often possess identical metabolic pathways, and have equivalent but slightly different enzymes that catalyze identical reactions. The differences that occur within such a family of enzymes usually occur in noncritical regions of the polypeptide moiety, by the substitution of one amino acid residue for another, or by the deletion of amino acid residues.

2.3. ENZYME COMMISSION NUMBERS

The ultimate identification of a particular enzyme is possible through its Enzyme Commission (E.C.) number.[2] The assignment of E.C. numbers is described in guidelines set out by the International Union of Biochemistry, and follows the format E.C. w.x.y.z, where numerical values are substituted for w, x, y and z. The value of w is always between 1 and 6, and indicates one of six main divisions; values of x indicate the subclassification, and are often related to either the prosthetic group or the cofactor required for the reaction; values of y indicate a subsubclassification, related to a substrate or product family; and the value of z indicates the serial number of the enzyme.

The six main divisions are (1) the oxidoreductases, (2) the transferases, (3) the hydrolases, (4) the lyases, (5) the isomerases, and (6) the ligases (synthetases). All known enzymes fall into one of these six categories. Oxidoreductases catalyze the transfer electrons and protons from a donor to an acceptor. Transferases catalyze the transfer of a functional group from a donor to an acceptor. Hydrolases

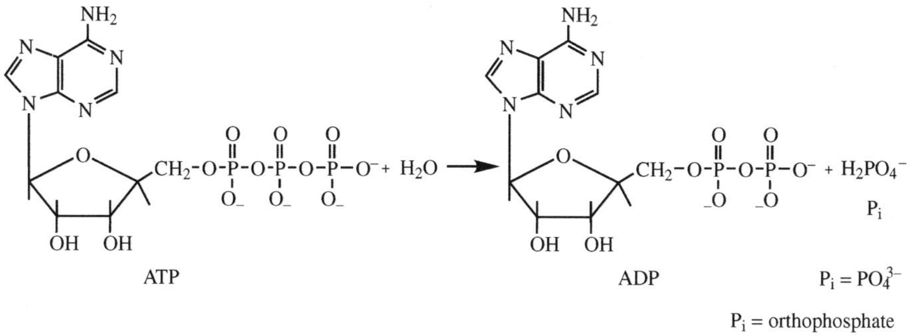

Figure 2.4. Hydrolysis of ATP to adenosine diphosphate (ADP) and inorganic phosphate (P_i).

facilitate the cleavage of C–C, C–O, C–N and other bonds by water. Lyases catalyze the cleavage of these same bonds by elimination, leaving double bonds (or, in the reverse mode, catalyze the addition of groups across double bonds). Isomerases facilitate geometric or structural rearrangements or isomerizations. Finally, ligases catalyze the joining of two molecules, and often require the hydrolysis of a pyrophosphate bond in the cofactor adenosine triphosphate (ATP, Fig. 2.4) to provide the energy required for the synthetic step.

The enzymes D-amino acid oxidase and lactate dehydrogenase (Eqs. 2.2 and 2.6) have the numbers E.C. 1.4.3.3 and E.C. 1.1.1.28, respectively; both are oxidoreductases, and therefore fall into the first of the six main divisions. The enzyme cholesterol esterase catalyzes the hydrolysis of cholesterol esters into cholesterol and free fatty acids (Eq. 2.7), and has been assigned E.C. 3.1.1.13.

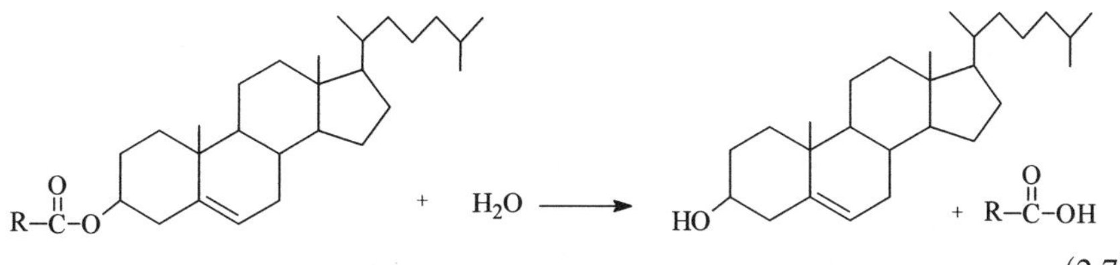

$$(2.7)$$

In addition to the E.C. number, the source of a particular enzyme is usually given, listing the species and the organ or tissue from which it was isolated.

2.4. ENZYMES IN BIOANALYTICAL CHEMISTRY

Enzymes can be employed to measure substrate concentrations as well as the concentrations of species that affect the catalytic activity of the enzyme toward its substrate, such as activators and inhibitors. The first known enzymatic assay was reported by Osann in 1845: hydrogen peroxide (H_2O_2) was quantitated using the enzyme peroxidase. In 1851, Schönbein reported a detection limit of 1 part H_2O_2 in 2×10^6 [i.e., 500 parts per billion] using this method! Enzymatic methods

TABLE 2.1. **Substrate Selectivity of Glucose Oxidase**[a]

Substrate	Relative Rate of Oxidation (%)
β-D-Glucose	100
2-Deoxy-D-glucose	25
6-Deoxy-6-fluoro-D-glucose	3
6-Methyl-D-glucose	1.85
4,6-Dimethyl-D-glucose	1.22
D-Mannose	0.98
D-Xylose	0.98
α-D-Glucose	0.22

[a] See Ref. 3. [Reprinted, with permission, from M. Dixon and E. C. Webb, *Enzymes*, 3rd ed., Academic Press, New York, 1979, p. 243. Third edition © Longman Group Ltd. 1979.]

are popular because they are relatively simple, require little or no sample pretreatment, and do not require expensive instrumentation. The single critical advantage, however, is the lack of interferences due to the selectivity of enzymes for their natural substrates.

The selectivity of glucose oxidase (E.C. 1.1.3.4) from *Aspergillus niger* has been studied through a comparison of the maximum rates of product formation from a variety of structurally related sugars, shown in Table 2.1.[3] Glucose oxidase is most reactive toward its natural substrate, β-D-glucose (Eq. 2.8), so that this substrate has been assigned a relative oxidation rate of 100%.

$$+ \ O_2 \ \longrightarrow \ + \ H_2O_2 \qquad (2.8)$$

The data in Table 2.1 show that the only substrate that would represent a significant interference in an enzymatic assay for glucose using glucose oxidase is 2-deoxy-D-glucose. It is of particular interest that the anomeric form of the enzyme's natural substrate, α-D-glucose, cannot be oxidized at a significant rate, even though the two compounds differ only in the position of the hydroxyl group at the C1 position of the sugar (Eq. 2.9). It is this exquisite selectivity that is exploited in enzymatic assays, enabling their use for substrate quantitation in matrices that may be as complex as blood or fermentation broths.

$$\beta \qquad \rightleftharpoons \qquad \rightleftharpoons \qquad \alpha \qquad (2.9)$$

To understand how enzymes are used in bioanalytical methods, and how their concentrations are represented and determined, it is first necessary to examine the kinetics of one- and two-substrate enzymatic reactions.

2.5. ENZYME KINETICS

The dramatic increases in reaction rates that occur in enzyme-catalyzed reactions can be seen for representative systems in the data given in Table 2.2.[4] The hydrolysis of the representative amide benzamide by acid or base yields second-order rate constants that are over six orders of magnitude lower than that measured for benzoyl-L-tyrosinamide in the presence of the enzyme α-chymotrypsin. An even more dramatic rate enhancement is observed for the hydrolysis of urea: The acid-catalyzed hydrolysis is nearly *13 orders of magnitude* slower than hydrolysis with the enzyme urease. The disprotionation of hydrogen peroxide into water and molecular oxygen is enhanced by a factor of ~1 million in the presence of catalase.

Enzymes derive both their selectivities and their reaction rate enhancements by the formation of enzyme–substrate complexes. This complex formation results in a transition state for the reaction, that lowers $\Delta G^{\ddagger}$, the activation energy, but does not affect the net free energy for the conversion of substrate to product. This finding is represented in Figure 2.5, for the simple one-substrate enzyme reaction shown in Eq. 2.10:

$$E + S \leftrightarrow E \cdot S \longrightarrow E + P \qquad (2.10)$$

where E is the enzyme, S is the substrate, $E \cdot S$ is the enzyme–substrate complex, and P is the product of the reaction.

The critical first step in enzyme-catalyzed reactions is the formation of $E \cdot S$, and this is usually represented as a simple association reaction, as in Eq. 2.10. Because it is the $E \cdot S$ complex that is the reactant in the substrate conversion step, its concentration determines the rate of the reaction; it follows, then, that the reaction rate

TABLE 2.2. Examples of the Catalytic Power of Enzymes[a]

Substrate	Catalyst	$T(K)$	$k(M^{-1}\,s^{-1})$
Amide (hydrolysis)			
-Benzamide	H^+	325	2.4×10^{-6}
-Benzamide	OH^-	326	8.5×10^{-6}
-Benzoyl-L-tyrosinamide	α-Chymotrypsin	298	14.9
Urea (hydrolysis)	H^+	335	7.4×10^{-7}
	Urease	294	5.0×10^6
Hydrogen peroxide	Fe^{2+}	295	56
	Catalase	295	3.5×10^7

[a] See Ref. 4. [Reprinted, with permission, from N. C. Price and L. Stevens, *Fundamentals of Enzymology*, 2nd ed., Oxford University Press, 1989 © N. C. Price and L. Stevens, 1989.]

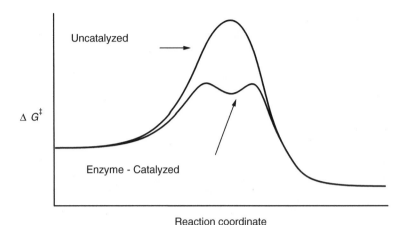

Figure 2.5. Free energy profiles during uncatalyzed and enzyme–catalyzed reactions.

will reach a maximum when all available enzyme effectively exists in the form of the E • S complex. This situation occurs at high substrate concentrations, where the enzyme is said to be *saturated* with substrate. If enzyme concentration is held constant, a plot of initial reaction rate against initial substrate concentration (Fig. 2.6) will yield a curve typical of saturation kinetics, where a plateau is observed at high [S].

The plot shown in Figure 2.6 has two regions that are important in analytical methods. First, when [S] is very low, the reaction rate is linearly related to [S]; under these conditions of first-order kinetics, reaction rates can be used to quantitate substrate. The second region of interest occurs at very high [S], where the reaction

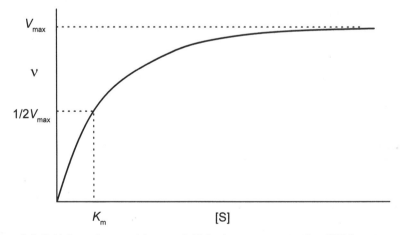

Figure 2.6. Initial reaction rate (ν) versus initial substrate concentration ([S]) for an enzyme-catalyzed reaction at constant enzyme concentration.

rate is independent of [S]; under these conditions of zero-order kinetics, all of the available enzyme exists as $E \cdot S$, and the reaction rate can be used to determine the total amount of enzyme present in the sample.

2.5.1. Simple One-Substrate Enzyme Kinetics

One-substrate enzyme kinetics are applied to many reactions that require water as a cosubstrate, that is, the hydrolases (such as esterases and proteases), since aqueous solutions have a water concentration of 55.6 M. The kinetic model is based on the initial formation of the enzyme–substrate complex, with the rate constants as shown in Eq. 2.11.

$$E + S \underset{k_{-1}}{\overset{k_1}{\longleftrightarrow}} E \cdot S \overset{k_2}{\rightarrow} E + P \tag{2.11}$$

From this model, the dissociation constant of the enzyme substrate complex, K_d, will be equal to k_{-1}/k_1, which is equal to $[E][S]/[E \cdot S]$.

The mathematical derivation of the expression for the reaction rate is based upon the initial rate of the reaction. We will assume that, almost immediately after the reaction begins, the rate of change of concentration of the enzyme–substrate complex is zero; that is, a steady-state $[E \cdot S]$ value is achieved. The rate of change of $[E \cdot S]$ with time, $d[E \cdot S]/dt$, may be expressed as in Eq. 2.12:

$$d[E \cdot S]/dt = k_1[E][S] - k_{-1}[E \cdot S] - k_2[E \cdot S] = 0 \tag{2.12}$$

where $[E \cdot S]$ is formed only from the association reaction, but is removed either by dissociation into E and S, or by conversion to product.

The enzyme exists in solution as either E or $E \cdot S$, so that at any time, the total enzyme concentration is equal to the sum of these concentrations, or $[E]_0 = [E] + [E \cdot S]$. Using this expression to substitute for [E] in Eq. 2.12 yields

$$d[E \cdot S]/dt = k_1([E]_0 - [E \cdot S]) \times [S] - k_{-1}[E \cdot S] - k_2[E \cdot S] = 0 \tag{2.13}$$

Rearranging to collect terms in $[E \cdot S]$ yields Eqs. 2.14 and 2.15:

$$k_1[E]_0[S] - (k_1[S] + k_{-1} + k_2) \times [E \cdot S] = 0 \tag{2.14}$$

$$[E \cdot S] = [E]_0[S]/\{[S] + (k_{-1} + k_2)/k_1\} \tag{2.15}$$

Equation 2.15 can now be used to find the initial rate of formation of product, as given in Eq. 2.16:

$$v = k_2[E \cdot S] = k_2[E]_0[S]/\{[S] + (k_{-1} + k_2)/k_1\} \tag{2.16}$$

Recalling that the maximum initial rate of reaction V_{max} occurs when all available enzyme exists in the form of the enzyme–substrate complex (i.e., $[E \cdot S] = [E]_0$), the term V_{max} may be substituted for $k_2[E]_0$; in addition, the constant K_m is used to represent $(k_{-1} + k_2)/k_1$, so that Eq. 2.16 simplifies to the standard form of the *Michaelis–Menten Equation* (Eq. 2.17):

$$v = V_{max}[S]/(K_m + [S]) \tag{2.17}$$

Note that the Michaelis–Menten constant, K_m, is related to the dissociation constant for the enzyme–substrate complex, K_d; in fact, K_m will always be $> K_d$, but will approach K_d as k_2 approaches 0. It is also noteworthy that when $[S] = K_m$, $v = V_{max}/2$; in other words, the reaction rate is half-maximal when the substrate concentration is equal to K_m.

There are two simplifications of the Michaelis–Menten equation that are of tremendous analytical importance. The first occurs at low substrate concentration: if $[S] \ll K_m$, then $K_m + [S] \approx K_m$. Under these conditions, Eq. 2.17 simplifies to Eq. 2.18:

$$v = V_{max}[S]/K_m = \text{Constant} \times [S] \tag{2.18}$$

The initial rate of reaction is therefore directly proportional to initial substrate concentration at low [S], and can be used to quantitate substrate. This is shown in the initial linear region of the plot of reaction rate against substrate concentration in Figure 2.5, where the slope in this region is equal to V_{max}/K_m.

The second simplification of Eq. 2.17 occurs at high substrate concentration. If the substrate concentration greatly exceeds K_m, then $[S]/(K_m + [S]) \approx 1$ and $v \approx V_{max}$. Under these conditions,

$$v = V_{max} = k_2[E]_0 \tag{2.19}$$

The initial reaction rate is now independent of [S], as shown in the plateau region of Figure 1.5. The rate now depends linearly on enzyme concentration, and a plot of initial rate against total enzyme concentration will have a slope of k_2. Thus, at high substrate concentrations, initial reaction rates can be used to determine enzyme concentrations.

In practice, Eq. 2.18 may be used if $[S] < 0.1K_m$, while Eq. 2.19 is useful if $[S] > 10\,K_m$. Remember that the steady-state assumption underlying these expressions is only valid if $[S] \gg [E]_0$, and that, in practice, substrate concentrations in excess of $10^3[E]_0$ are used.

2.5.2. Experimental Determination of Michaelis–Menten Parameters

Equations 2.18 and 2.19 show that the two regions of analytical utility in the rate versus substrate concentration profile of an enzyme occur at $[S] < 0.1\,K_m$ (for substrate quantitation) and $[S] > 10\,K_m$ (for enzyme quantitation). In developing a new assay, or in adapting an established assay to new conditions, it is thus important to establish the K_m value of the enzyme. It is of practical importance to also establish

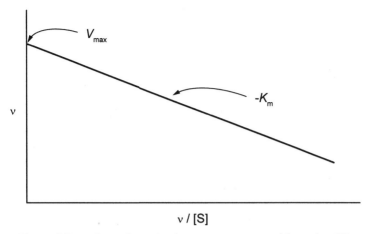

Figure 2.7. Eadie–Hofstee plot for v measurements with varying [S].

and adjust V_{max}, so that an assay may be accomplished in the minimum time required to yield a given precision.

Four graphical methods may be used to establish K_m and V_{max} values under given experimental conditions, called the *Eadie–Hofstee, Hanes, Lineweaver–Burk*, and *Cornish–Bowden–Eisenthal* methods. All four involve the measurement of initial rates of reaction as a function of initial substrate concentration, at constant enzyme concentration.

2.5.2.1. Eadie–Hofstee Method.[5,6] The Michaelis–Menten equation (Eq. 2.17) can be algebraically rearranged to yield Eq. 2.20.

$$v = V_{max} - K_m(v/[S]) \tag{2.20}$$

A plot of v (as y) against $v/[S]$ (as x) will yield, after linear regression, a y intercept of V_{max} and a slope of $-K_m$ (Fig. 2.7). This plot is the preferred "linear regression" method for determining K_m and V_{max}, since precision and accuracy are somewhat better than those obtained using the Hanes plot, and much better than those found using the Lineweaver–Burk method.

2.5.2.2. Hanes Method.[7] A different rearrangement of the Michaelis–Menten equation yields the form that is used in the Hanes plot:

$$[S]/v = K_m/V_{max} + [S]/V_{max} \tag{2.21}$$

In the Hanes plot (Fig. 2.8), $[S]/v$ (as y) is plotted against [S], yielding a slope of $1/V_{max}$ and a y intercept of K_m/V_{max} following linear regression. The parameter K_m is then calculated as the y intercept divided by the slope.

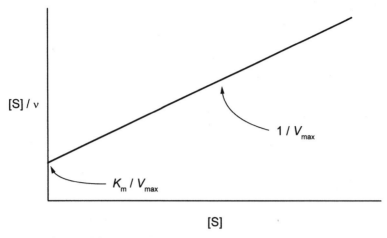

Figure 2.8. Hanes plot for v measured as a function of [S].

2.5.2.3. Lineweaver–Burk Method.[8] The Michaelis–Menten equation may be algebraically rearranged to Eq. 2.22, yielding a third linear plot, called the Lineweaver–Burk or double-reciprocal plot:

$$1/v = 1/V_{max} + K_m/(V_{max} \times [S]) \tag{2.22}$$

The reciprocal rate $1/v$ is plotted against the reciprocal substrate concentration $1/[S]$, yielding a straight line with a positive slope of K_m/V_{max} and a y intercept of $1/V_{max}$. The parameter K_m is then calculated from the linear regression values of the slope divided by the y intercept (Fig. 2.9). While the Lineweaver–Burk plot is

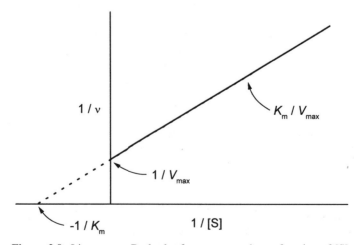

Figure 2.9. Lineweaver–Burk plot for v measured as a function of [S].

the most commonly used graphical method for the determination of Michaelis–Menten parameters, it yields the poorest accuracy and precision in these values. The main reason for this is that, since it is a reciprocal plot, the smallest [S] and v values yield the largest values on the plot. These small values have the greatest relative imprecision, yet standard linear regression programs assume constant uncertainty in all y values $(1/v)$ plotted. One approach to minimizing this problem uses weighted linear regression, where these imprecise points obtained at lower concentrations are given less significance in the calculation of slope and intercept values. The popularity of the Lineweaver–Burk plot results from its use as a diagnostic plot for enzyme inhibitors, a topic discussed in a Section 2.7.

2.5.2.4. Cornish–Bowden–Eisenthal Method.[9] This method is distinct from the previous three linear regression methods, in that each pair of $(v, [S])$ values is used to construct a separate line on a plot in which V_{max} and K_m form the y and x axes, respectively. Beginning with another version of the Michaelis–Menten equation, in which V_{max} is the y value and K_m is the x value as shown in Eq. 2.23,

$$V_{max} = v + (v/[S]) \times K_m, \qquad (2.23)$$

it can be seen that each $(v, [S])$ data point will yield a unique y intercept (v) and slope $(v/[S])$. These values define a unique line on the V_{max} versus K_m plot. Each $(v/[S])$ pair will define a different line, but in the absence of experimental uncertainty, *all of these lines will intersect at an identical point that defines the V_{max} and K_m for the enzyme studied.* This type of plot is shown in Figure 2.10.

In practice, the lines do not pass through a single unique point as shown in Figure 2.10, but instead a cluster of intersection points is observed. All of the intersection points are then used to calculate average K_m and V_{max} values. This unusual method has been shown to yield the best accuracy and precision for statistically treated model data sets.

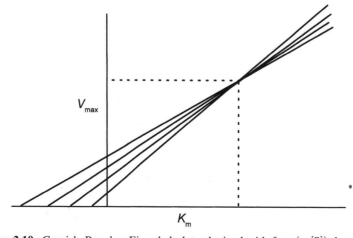

Figure 2.10. Cornish–Bowden–Eisenthal plots obtained with four $(v, [S])$ data pairs.

TABLE 2.3. Error-Free Michaelis–Menten Data for Method of Comparison

[S]	v
0.25	0.200
0.50	0.333
0.75	0.428
1.00	0.500
1.25	0.556
1.50	0.600
1.75	0.636

2.5.3. Comparison of Methods for the Determination of K_m Values[10]

The precisions and accuracies of the four methods described above for the determination of K_m values were compared statistically by first generating an "error-free" data set of v and [S] data, and then introducing different types of error to the v values to generate a total of 100 individual data sets. The error-free data set was generated using Eq. 2.17 with $V_{max} = K_m = 1$, and the seven individual values are given below in Table 2.3.

Fifty data sets containing absolute errors were then generated by adding random numbers of mean zero and standard deviation 0.05 to each v value. Another 50 data sets were generated to contain relative error, by multiplying each v value by a random number of mean one and standard deviation 0.10. These 100 data sets were then analyzed individually by each of the four methods, to generate 100 K_m values for each method. The average K_m value and its uncertainty for each method and each error type are summarized in Table 2.4.

The values summarized in Table 2.4 show that, while all four methods give accurate K_m values (in all cases, $K_m = 1.00$ is within the uncertainties reported), the

TABLE 2.4. Mean Estimates of K_m (True Value = 1.00) for Data with Different Error Types

Method	Error Type	
	Absolute	Relative
Lineweaver–Burk (1/v vs. 1/[S])	1.13 ± 0.82	1.09 ± 0.43
Hanes ([S]/v vs. [S])	1.05 ± 0.46	1.05 ± 0.33
Eadie–Hofstee (v vs. v/[S])	0.93 ± 0.46	0.88 ± 0.31
Cornish–Bowden–Eisenthal (V_{max} vs. K_m)	0.79 ± 0.31	0.94 ± 0.29

Lineweaver–Burk method yields significantly poorer precision than any of the other three methods. The Cornish–Bowden–Eisenthal method clearly yields the most precise values for K_m, in the presence of either absolute or relative error.

2.5.4. One-Substrate, Two-Product Enzyme Kinetics

Hydrolytic enzymes that release two products in a defined sequence require a slightly more complicated model for their kinetic behavior. These enzymes pass through two intermediate stages in the catalytic cycle, $E \bullet S$ and $E \bullet S'$, and the formation of products involves two steps with associated rate constants k_2 and k_3:

$$E + S \underset{k_{-1}}{\overset{k_1}{\longleftrightarrow}} E \bullet S \overset{k_2}{\longrightarrow} E \bullet S' + P_1 \overset{k_3}{\longrightarrow} E + P_2 + P_1 \tag{2.24}$$

The rate-determining step is assumed to involve the release of the first product P_1 and the conversion of $E \bullet S$ into $E \bullet S'$. Application of the steady-state approximation to both $[E \bullet S]$ and $[E \bullet S']$ (i.e., assuming that $d[E \bullet S]/dt = 0$ and $d[E \bullet S']/dt = 0$) yields Eq. 2.25 for the reaction rate, after steps analogous to Eqs. 2.12–2.16:

$$v = k_2[E \bullet S] = \frac{\{k_2 k_3/(k_2 + k_3)\} \times [E]_0 \times [S]}{\{(k_{-1} + k_2)/k_2\} \times \{k_3/(k_2 + k_3)\} + [S]} \tag{2.25}$$

While Eq. 2.25 looks much more complicated, it has the same form as Eq. 2.17. In fact, if the substitutions $V_{max} = k_{cat}[E]_0 = \{k_2 k_3/(k_2 + k_3)\}[E]_0$, and $K_m = \{(k_{-1} + k_2)/k_2\}\ \{k_3/(k_2 + k_3)\}$ are made, Eq. 2.17 is obtained, and Michaelis–Menten kinetics are observed with these types of enzymes.

2.5.5. Two-Substrate Enzyme Kinetics

The kinetic behavior of many enzymes cannot be described by the simple Michaelis–Menten equation, because more than one substrate is involved in the reaction. Such is the case for the oxidoreductases, the transferases, and the ligases, three of the six major divisions of enzymes. The kinetic behavior of these systems is necessarily more complicated, and depend on whether or not a ternary enzyme–primary substrate–secondary substrate complex is formed.

If a ternary complex is formed, the mechanism may be considered ordered or random. The ordered mechanism requires that the first substrate (S_1) must bind to the enzyme before the second substrate (S_2) will bind, and is represented by Eq. 2.26:

$$E + S_1 + S_2 \leftrightarrow E \bullet S_1 + S_2 \longrightarrow E \bullet S_1 \bullet S_2 \longrightarrow E + P \tag{2.26}$$

Dehydrogenase enzymes that use NAD^+ as a cofactor follow this mechanism, since an enzyme-NAD^+ complex forms initially, and changes the local structure at the enzyme's active site to allow substrate binding.

If no particular order is required for substrate binding, the model shown in Eq. 2.27 is used.

$$E + S_1 + S_2 \leftrightarrow E \cdot S_1 + S_2 \longrightarrow E \cdot S_1 \cdot S_2 \longrightarrow E + P$$
$$\leftrightarrow E \cdot S_2 + S_1 \longrightarrow \qquad (2.27)$$

In this case, either substrate will bind to the enzyme initially, and a ternary complex is then formed that will decompose to products. By assuming that all three complexes, $E \cdot S_1$, $E \cdot S_2$, and $E \cdot S_1 \cdot S_2$ form rapidly, and that the transformation of $E \cdot S_1 \cdot S_2$ into products is relatively slow (and rate determining), we obtain Eq. 2.28 for the initial reaction rate:

$$v = V_{max}/\{1 + K_{m1}/[S_1] + K_{m2}/[S_2] + K_{m12}/([S_1] \times [S_2])\} \qquad (2.28)$$

Equation 2.28 contains a new term, K_{m12}, that represents a change in affinity of the enzyme for one substrate once the other substrate is bound. If the mechanism is ordered, the simple relationship $K_{m12} = K_{m1} \times K_{m2}$ may be applied. For a random mechanism, the value of K_{m12} is determined experimentally. Creatine kinase (CK) is an example of this type of enzyme. Creatinine and ATP bind to the enzyme randomly in nearby, but independent binding sites.

$$(2.29)$$

Creatine   Phosphocreatine

If the two-substrate mechanism does not involve the formation of a ternary complex, then sequential substrate binding and product release occurs in the so-called "ping–pong" mechanism, Eqs. 2.30 and 2.31:

$$E + S_1 \leftrightarrow E \cdot S_1 \longrightarrow E' + P_1 \qquad (2.30)$$

$$E' + S_2 \leftrightarrow E' \cdot S_2 \longrightarrow E + P_2 \qquad (2.31)$$

In this mechanism, two independent steps occur in which the enzyme is initially converted into an intermediate state E', and this altered enzyme, independent of S_1 and P_1, reacts with the second substrate. Aspartate aminotransferase is such an enzyme; it catalyzes the transfer of an amino group from aspartate to 2-oxoglutarate by first removing the amino group from aspartic acid (S_1), to release oxaloacetate (P_1), and then adding the amino group to 2-oxoglutarate (S_2) to produce glutamic acid (P_2). The enzyme's prosthetic group is pyridoxal phosphate, which is readily converted to pyridoxamine phosphate by the removal of an amino group from aspartate, as shown in Figure 2.11.

Aspartate aminotransferase, and all other enzymes possessing a ping–pong mechanism, are described by Eq. 2.32:

$$v = V_{max}/\{1 + K_{m1}/[S_1] + K_{m2}/[S_2]\} \qquad (2.32)$$

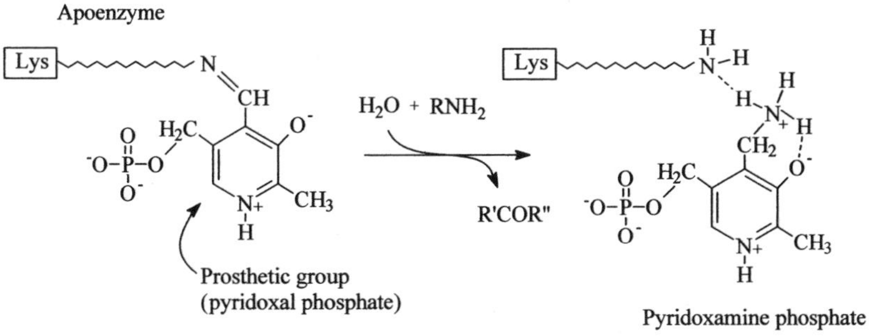

Apoenzyme

H$_2$O + RNH$_2$

R'COR"

Prosthetic group
(pyridoxal phosphate)

Pyridoxamine phosphate

Figure 2.11. Pyridoxal phosphate at the active site of aspartate aminotransferase.

In these cases, the third term in the denominator ($K_{m12}/[S_1] \times [S_2]$ in Eq. 2.36) is absent because no ternary $E \bullet S_1 \bullet S_2$ complex is formed. The individual values of K_{m1} and K_{m2} may be found in two separate experiments in which S_1 and S_2 are varied under saturating conditions of S_2 and S_1, respectively, so that the second and first terms in the denominator become vanishingly small.

2.5.6. Examples of Enzyme-Catalyzed Reactions and Their Treatment

It is not always immediately apparent from an enzymatic reaction whether it should be treated with single- or dual-substrate enzyme kinetic expressions. Lactose hydrolase, which catalyzes the hydrolysis of lactose according to

$$\text{Lactose} + \text{H}_2\text{O} \longrightarrow \text{D-Galactose} + \text{D-Glucose} \qquad (2.33)$$

is best (and perhaps obviously) treated by simple one-substrate Michaelis–Menten kinetics, since H$_2$O may be safely ignored as a second substrate. Glucose oxidase catalyzes a two- substrate reaction, however,

$$\text{D-Glucose} + \text{O}_2 \longrightarrow \delta\text{-D-Gluconolactone} + \text{H}_2\text{O}_2 \qquad (2.34)$$

and must be treated with two-substrate kinetics; it has been shown that the ping–pong mechanism applies in this case.

Lactate dehydrogenase (misnamed since it usually functions in the reverse direction and so would be more aptly named pyruvate hydrogenase) represents a yet more complicated system:

$$\text{Pyruvate} + \text{NADH} + \text{H}^+ \longrightarrow \text{L-Lactate} + \text{NAD}^+ \qquad (2.35)$$

In this case, the question is whether H$^+$ should be considered a third substrate. It has, in fact, been treated in this manner by researchers, who have shown that if both pyruvate and NADH concentrations are maintained at saturating levels, the initial rate of the enzymatic reaction is pH dependent, showing a Gaussian-like curve over the pH range 6–8. Data taken on the alkaline side of this curve may be used to obtain a K_{mH^+} value, or, alternately, the alkaline pH value at $V_{max}/2$ value may be read directly from the v versus pH curve. Research has shown that, in general, $V_{max}/2$ occurs at a pH value that is at least two orders of magnitude (two pH units) alkaline of the optimum pH for the reaction, where $v = V_{max}$. This finding means that, at the pH optimum, $[H^+] \approx 100 \times K_{mH^+}$, so that the deviation of v from V_{max} is $<1\%$. Furthermore, in a buffered solution, $[H^+]$ is constant, so that deviations of v from V_{max}, even at an unusually narrow pH optimum, appear as constants in the kinetic expressions. It has therefore been concluded that, for all practical purposes, H^+ may be safely ignored as a second or third substrate.

It must also be remembered that enzymes have structures that depend on pH. Under excessively acid or alkaline conditions, denaturation of the tertiary protein structure will occur due to disruption of the normal hydrogen bonding modes, and this denaturation will have dramatic effects on enzyme activity.

2.6. ENZYME ACTIVATORS

A large variety of chemical and biochemical compounds are known to affect the activity of enzymes without themselves being involved in the enzyme-catalyzed reaction. Activators are species that increase enzyme activity; they may be necessary for the enzyme to possess any catalytic activity (such as a prosthetic group), or they may increase the specific activity of an already active enzyme (e.g., ions such as Ca^{2+} and Mg^{2+} that interact with phosphate-containing substrates). At constant, saturating levels of substrate(s), increasing concentrations of activator yield increasing initial reaction rates, with a limiting rate reached at high activator concentrations. In some cases, such as for prosthetic group activation, the following sequence of reactions may be used as a model for kinetic descriptions:

$$A + E_{inactive} \longrightarrow E_{active} \tag{2.36}$$

$$S + E_{active} \longrightarrow S \cdot E_{active} \longrightarrow E_{active} + P \tag{2.37}$$

where A represents the activator species. Parallel reaction sequences must be considered if the enzyme is catalytically active in the absence of activator. With the exception of prosthetic group activation, activators are not usually specific, and several species may have similar activating effects on an enzyme. For example, isocitrate dehydrogenase is activated by both Mn^{2+} and Mg^{2+}; it has been shown that, in the absence of Mg^{2+}, Mn^{2+} levels as low as 5 ppb may be determined via measurement of isocitrate dehydrogenase activity. Anions are also relatively nonspecific activators, and the activation of α-amylase by buffer anions (especially Cl^-) has been studied in detail. An exception to this nonspecific ion activation occurs with pyruvate kinase: This enzyme is activated by K^+, but is inhibited by Na^+.

While activator concentrations may be quantitated through activity assays, the presence of unknown concentrations of activators in samples requiring substrate or enzyme quantitation represents a significant source of error. If activators are present or suspected in such samples, activators are added in excess to both samples and calibration standards.

2.7. ENZYME INHIBITORS

Enzyme inhibitors are species that cause a decrease in the activity of an enzyme. Inhibitors usually interact with the enzyme itself, forming enzyme–inhibitor (E•I) complexes, but in a few cases, the inhibition mechanism involves reaction with one of the substrates. Inhibition is considered to be *reversible* if the enzyme recovers its activity when the inhibitor is removed, and *irreversible* if the inhibitor causes a permanent loss of activity. Reversible inhibition affects the specific activity and apparent Michaelis–Menten parameters for the enzyme, while irreversible inhibition (where the E•I complex formation is irreversible) simply decreases the concentration of active enzyme present in the sample. A well-known example of irreversible inhibition is the effect of nerve gas on the enzyme cholinesterase.

Inhibitors may be quantitated using enzyme activity assays if the unknown inhibitor sample does not contain the enzyme employed in the inhibition assay. Standards containing constant enzyme and saturating substrate concentrations are prepared with varying concentrations of inhibitor. The resulting activity versus [inhibitor] calibration curves, for reversible and irreversible inhibitors, are shown below in Figure 2.12.

The curvature seen for reversible inhibition [Fig. 2.12 (*a*)] indicates that an inhibitor-binding equilibrium precedes the conversion of substrate to product. Three types of reversible inhibition may be distinguished. (1) Competitive inhibition occurs when the degree of inhibition decreases as substrate concentration increases and V_{max} is unaffected. (2) Noncompetitive inhibition exists when the degree of inhibition does not vary with substrate concentration, and K_m is unaffected. (3) Uncompetitive inhibition exists if the degree of inhibition increases as substrate concentration increases; both V_{max} and K_m are affected. Uncompetitive inhibition is often thought of as a mixture of competitive and noncompetitive behavior.

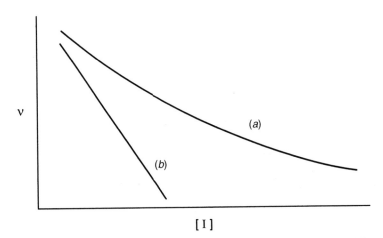

Figure 2.12. Calibration curves for (*a*) reversible and (*b*) irreversible inhibitors.

2.7.1. Competitive Inhibition

Competitive inhibitors compete with the substrate for the enzyme's active site, but are not converted to products after they are bound. They block the active site from substrate, and their effectiveness is described by their inhibition constant, K_I, which is the dissociation constant of the enzyme–inhibitor complex (k_{-3}/k_3):

$$E + S \underset{k_{-1}}{\overset{k_1}{\leftrightarrow}} E \bullet S \overset{k_2}{\longrightarrow} E + P \tag{2.38}$$

$$E + I \underset{k_{-3}}{\overset{k_3}{\leftrightarrow}} E \bullet I \tag{2.39}$$

In this model, the enzyme–inhibitor complex is completely inactive, but is in equilibrium with the active form of the enzyme. For a simple one-substrate reaction, the effect of a competitive inhibitor on the initial rate of the reaction is described by Eq. 2.40.

$$v = V_{max}/\{1 + (K_m/[S]) \times (1 + [I]/K_I)\} \tag{2.40}$$

It can be seen from this equation that competitive inhibitors have no effect on the V_{max} of the enzyme, but alter the apparent K_m. In the presence of inhibitor, K_m will be increased by a factor of $(1 + [I]/K_I)$. Lineweaver–Burk plots constructed at various inhibitor concentrations provide a useful diagnostic for this type of inhibition. Figure 2.13 shows that identical y intercepts ($1/V_{max}$) are obtained at different inhibitor concentrations, while x intercepts (reciprocal of apparent K_m) decrease with increasing [I], and are equal to $-1/\{K_m(1 + [I]/K_i)\}$.

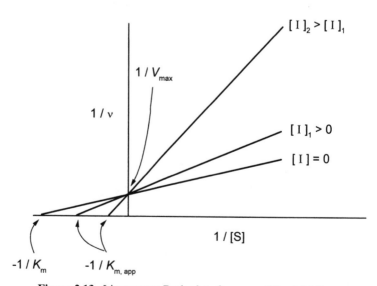

Figure 2.13. Lineweaver–Burk plots for competitive inhibitors.

2.7.2. Noncompetitive Inhibition

Noncompetitive inhibitors interact reversibly with enzymes to form an inactive species, effectively "removing" active enzyme and thus interfering with the rate of conversion of substrate to product. The inhibitor may interact with free enzyme, or with the enzyme–substrate complex. The key feature of noncompetitive inhibition that distinguishes it from competitive inhibition is that inhibition does not affect the apparent affinity of the enzyme for its substrate (i.e., the apparent K_m). For example, a noncompetitive inhibitor may bind in a region remote from the active site to cause a reversible change in enzyme tertiary structure that completely prevents substrate binding and product formation. In this type of inhibition, the quantity of active enzyme appears to decrease as inhibitor concentration increases, so that the apparent V_{max} for the reaction decreases.

$$v = V_{max}/\{(1 + [I]/K_i) \times (1 + K_m/[S])\} \tag{2.41}$$

where $V_{max,app} = V_{max}/\{1 + [I]/K_i\}$.

Figure 2.14 shows Lineweaver–Burk plots that are typical of noncompetitive inhibition.

2.7.3. Uncompetitive Inhibition

Uncompetitive inhibitors bind with the ES complex, affecting both the apparent K_m and the apparent V_{max} of an enzymatic reaction. Their behavior is approximated by Eq. 2.42:

$$v = V_{max}/\{(K_m/[S]) + ([I]/K_i) + 1\} \tag{2.42}$$

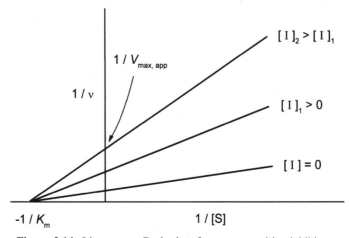

Figure 2.14. Lineweaver–Burk plots for noncompetitive inhibitors.

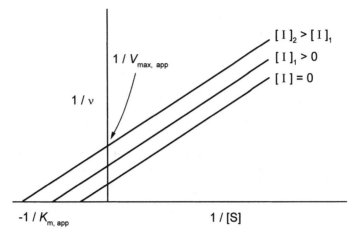

Figure 2.15. Lineweaver–Burk plots for uncompetitive inhibitors.

In this case, Lineweaver–Burk plots (Fig. 2.15) yield a series of parallel lines (in the case of one-substrate enzymes), in which the y intercept is equal to the reciprocal of the apparent V_{max}, $1/V_{max,app} = \{1 + [I]/K_i\}/V_{max}$, while the x intercept is the negative reciprocal of the apparent K_m, $1/K_{m,app} = -\{1 + [I]/K_i\}/K_m$. The model used to describe uncompetitive inhibition predicts an apparent equivalent change in K_m and V_{max}, resulting in lines of identical slope.[11]

2.8. ENZYME UNITS AND CONCENTRATIONS

Enzyme concentrations may be represented in an identical manner as with noncatalytic species, such as is in molar units or as mass per unit volume. The information required for this calculation is minimal: the percentage purity of the enzyme preparation, the molecular weight of the enzyme, and the mass and volume of the solution prepared. Enzyme concentrations are rarely represented in this way, however, because such values say nothing about the catalytic power of the solution prepared.

The concentration of an enzyme solution is most commonly given as the number of *International Units* (I.U.) per unit volume. The I.U. is defined as that quantity of enzyme required to consume one micromole of substrate per minute at a given temperature and pH, under conditions of substrate saturation:

$$\text{International Unit (I.U.)} = 1\text{-}\mu\text{mol substrate consumed/min} \qquad (2.43)$$

Since 1 I.U. is usually a very large quantity of enzyme, concentrations are often given as milliunits (mU) or microunits (μU) per liter (L) or milliliter (mL).

To convert these units to molar concentrations, it is necessary to know the turnover number, k_{cat}, of the particular enzyme used. For a simple one-substrate enzyme, $k_{cat} = k_2$. This value is a fundamental characteristic of an enzyme, indicating the maximum rate at which substrate can be consumed, and is generally given in units of reciprocal seconds (s^{-1}). Under conditions of substrate saturation,

$$v = V_{max} = k_{cat}[E]_0 (\text{mol S converted} \times s^{-1} \times L^{-1}) \qquad (2.44)$$

V_{max} is obtained from the I.U. of enzyme activity present in a given volume (V_T) of enzyme solution:

$$V_{max} = (10^{-6} \text{ mol}/\mu \text{ mol}) \times (1 \text{ min}/60 \text{ s}) \times (\text{I.U.}/V_T)$$

or

$$= (1.67 \times 10^{-8}) \times (\text{I.U.}/V_T) \qquad (2.45)$$

Eq. 2.45 has units of $\text{mol} \times s^{-1} \times L^{-1}$. The molar concentration of an enzyme solution may be found by combining Eqs. 2.44 and 2.45

$$[E]_0 = (1.67 \times 10^{-8}) \times (\text{I.U.}/V_T) \times (1/k_{cat}) \qquad (2.46)$$

While Eq. 2.46 is preferred for the calculation of $[E]_0$, it requires that the k_{cat} value be known for the enzyme of interest under the conditions required for its use, and this value is not always available in the literature.

Specific activity is defined as I.U. of enzyme activity per unit weight (under given conditions of T and pH), and is considered a measure of enzyme purity: the higher the specific activity of a given enzyme preparation, the greater is its purity. Specific activity is generally reported in I.U. per milligram of solid enzyme. If the molecular weight of an enzyme is known, and if a given preparation can be assumed to be 100% pure, then its specific activity can be used to estimate k_{cat} through Eq. 2.47:

$$k_{cat} = (1.67 \times 10^{-5}) \times (\text{Specific activity}) \times (\text{Molecular weight}) \qquad (2.47)$$

where the constant has been calculated for specific activity (in I.U./mg), molecular weight [in daltons (Da)], and k_{cat} (in s^{-1}).

Because of the nonstandard units associated with the I.U. system for defining enzyme concentrations, an equivalent International system of units (Systéme International) (S.I.) unit has been defined, and is called the *katal*. One katal (kat) of enzyme activity is that quantity that will consume 1-mol substrate/s; 1 μkat = 60 I.U.

Enzyme concentrations expressed as activity per unit volume must specify the temperature at which activity is measured. Even simple enzyme-catalyzed reactions consist of at least three stages, all of which are temperature dependent: the formation of the enzyme–substrate complex, the conversion of this complex to the enzyme–product complex, and the dissociation of the enzyme–product complex.

The overall effect of temperature on reaction rates is a combination of the effects produced at each stage. In general, a 10 °C increase in T will double the rate of an enzymatic reaction. Temperature control to within 0.1 °C is necessary to ensure the reproducible measurement of reaction rates. Temperatures <40 °C are generally employed, to avoid protein denaturation.

SUGGESTED REFERENCES

N. C. Price and L. Stevens, *Fundamentals of Enzymology*, 2nd ed., Oxford Science Publishers, London, 1989.

R. A. Copeland, *Enzymes*, 2nd ed., Wiley-VCH, New York, 2000.

International Union of Biochemistry, *Enzyme Nomenclature*, Academic Press, Orlando, FL, 1984.

A. Cornish-Bowden, *Fundamentals of Enzyme Kinetics*, Butterworths, London, 1979.

D. L. Purich, *Contemporary Enzyme Kinetics and Mechanism*, Academic Press, New York, 1983.

K. J. Laidler and P. S. Bunting, *The Chemical Kinetics of Enzyme Action*, Oxford University Press, London, 1973.

REFERENCES

1. M. Husain and V. Massey, *Methods Enzymol.* **53**, 1978, 429–437.

2. International Union of Biochemistry and Molecular Biology, *Enzyme Nomenclature 1992*, Academic Press, Orlando, FL, 1992.

3. M. Dixon, and E. C. Webb, *Enzymes*, 3rd ed., Academic Press, New York, 1979. p. 243.

4. N. C. Price and L. Stevens, *Fundamentals of Enzymology*, 2nd ed., Oxford Science Publishers, London, 1989, p. 4.

5. G. S. Eadie, *J. Biol. Chem.* **146**, 1942, 85–93.

6. B. H. J. Hofstee, *Science* **116**, 1952, 329–331.

7. C. S. Hanes, *Biochem. J.* **26**, 1932, 1406–1421.

8. H. Lineweaver and D. Burk, *J. Am. Chem. Soc.* **56**, 1934, 658–666.

9. A. Cornish-Bowden and R. Eisenthal, *Biochem. J.* **139**, 1974, 721–730.

10. G. L. Atkins and I. A. Nimmo, *Biochem. J.* **149**, 1975, 775–777.

11. T. M. Devlin, Ed., *Textbook of Biochemistry with Clinical Correlations*, J. Wiley & Sons, Inc., New York, 1982, p. 166.

PROBLEMS

1. The enzyme alanopine dehydrogenase reversibly catalyzes the reaction shown below. It is present in bivalve molluscs, such as the periwinkle, that live in intertidal regions. These species survive exposure during low tides through valve closure, and their metabolism switches from the aerobic lactate dehydrogenase pathway to the anaerobic alanopine dehydrogenase pathway. Exposed mollusks accumulate alanopine in their tissues, and alanopine levels have been related to exposure times.

$$NH[CH(CH_3)(COOH)]_2 + NAD^+ + H_2O \leftrightarrow H_2NCH(CH_3)COOH + CH_3COCOOH + NADH + H^+$$

  Alanopine   Alanine   Pyruvate

(a) In the forward direction as written above, should alanopine dehydrogenase be considered a one-, two-, or three-substrate enzyme? Why?

(b) Alanopine dehydrogenase from periwinkle has been shown to possess K_m values for alanopine and NAD^+ of 17 and 0.22 mM, respectively. What minimum NAD^+ concentration should be used to obtain a linear dependence of initial reaction rate on alanopine concentration? Over what alanopine concentration range is the linear dependence expected to hold?

(c) In the reverse direction, it has been found that the product NAD^+ acts as a reversible inhibitor of alanopine dehydrogenase. Under conditions of excess alanine and pyruvate, the initial rate of the reaction was measured as a function of NADH concentration, with NAD^+ concentrations of 0, 1, and 2 mM initially present. Lineweaver–Burk plots of $1/v$ versus $1/[NADH]$ were constructed from these data, and these plots showed that the y intercept was identical at all three initial NAD^+ concentrations. The x intercepts, however, were negative, and tended toward zero at higher initial NAD^+ concentrations. What kind of inhibitor is NAD^+? What effect does NAD^+ have on the apparent K_m and V_{max} for NADH?

2. An enzyme has a molecular weight of 47 kDa and a preparation that is 100% pure has a specific activity of 700 I.U./mg, at 25 °C in a pH 7.5 phosphate buffer. The enzyme is known to be monomeric (one active site per enzyme molecule).

(a) Calculate the mass of the pure enzyme preparation needed in 100 mL of buffer (saturated with substrates) to produce a solution that will initially consume substrate at the rate of $5 \times 10^{-7} M/min$.

(b) Calculate k_{cat} for the enzyme.

(c) A second preparation of this enzyme was tested for activity. In 100 mL of buffer, 1.6 mg of the solid enzyme was dissolved. This enzyme solution was mixed with a substrate solution in the ratio 500:1500 mL, and in the final solution, all substrates were present in great excess of their K_m values. Absorbance spectroscopy showed that 0.32-mM substrate was consumed per minute. Calculate the specific activity of the enzyme preparation.

3. The enzyme methylamine dehydrogenase catalyzes the oxidation of amino-methane to formaldehyde according to the following equation:

$$CH_3NH_2 + H_2O + 2(AC-Cu^{2+}) \leftrightarrow HCHO + NH_3 + 2H^+ + 2(AC-Cu^+).$$

The abbreviation AC represents amicyanin, a "blue" copper protein that acts as the physiological electron acceptor. Amicyamin can be replaced *in vitro* by the low molecular weight electron acceptor phenazine ethosulfate (PES). PES is oxidized in a single two-electron step, so that 1:1 stoichiometry prevails between aminomethane and PES. The reduced form of PES reacts rapidly with 2,6-dichloroindophenol, again in 1:1 stoichiometry, to produce an intensely colored product with an absorbance maximum at 600 nm and a molar absorptivity of $2.15 \times 10^4 \, M^{-1} \, cm^{-1}$. Calculate the initial change in absorbance with time that would be expected at 600 nm in a 1-cm cuvette, for a solution containing excess aminomethane, PES, and 2,6-dichloroindophenol, if the solution also contained 20 ng/mL of a methylamine dehydrogenase preparation known to have a specific activity of 16 I.U./mg.

4. An enzyme converts substrate S to product P, and obeys simple one-substrate Michaelis–Menten behavior. The following rate data were obtained with a $5 \times 10^{-8} \, M$ enzyme at 25 °C in a pH 7.2 phosphate buffer:

[S] (mM)	4.000	1.336	0.800	0.568	0.448	0.368
Initial rate (mmol/min)	0.648	0.488	0.418	0.353	0.310	0.275

Using the Eadie–Hofstee and the Cornish–Bowden–Eisenthal approaches, determine the values of K_m and V_{max} for this enzyme. Which is likely to yield more precise estimates? From these results, calculate k_{cat} for the enzyme.

5. A Hanes plot of $[S]/v$ against $[S]$, obtained at constant enzyme concentration, T, and pH, yields a positive slope of 100 min/mM and a y intercept of 0.0020 min.
(a) Calculate V_{max} and K_m for the enzyme.
(b) If the solution was known to contain 10 ng/mL of a solid (lyophilized) enzyme, calculate the specific activity of the enzyme preparation.

CHAPTER 3

Quantitation of Enzymes and Their Substrates

3.1. INTRODUCTION

The quantitation of enzymes and substrates has long been of critical importance in clinical chemistry, since metabolic levels of a variety of species are known to be associated with certain disease states. Enzymatic methods may be used in complex matrices, such as serum or urine, due to the high selectivity of enzymes for their natural substrates. Because of this selectivity, enzymatic assays are also used in chemical and biochemical research. This chapter considers quantitative experimental methods, the biochemical species that is being measured, how the measurement is made, and how experimental data relate to concentration. This chapter assumes familiarity with the principles of spectroscopic (absorbance, fluorescence, chemi- and bioluminescence, nephelometry, and turbidimetry), electrochemical (potentiometry and amperometry), calorimetry, and radiochemical methods. For an excellent coverage of these topics, the student is referred to Daniel C. Harris, *Quantitative Chemical Analysis (6th ed.)*. In addition, statistical terms and methods, such as detection limit, signal-to-noise ratio (S/N), sensitivity, relative standard deviation (RSD), and linear regression are assumed familiar; Chapter 16 in this volume discusses statistical parameters.

Biochemical literature often uses the terms *detection limit* and *sensitivity* interchangeably. We will use *detection limit* to describe the minimum detectable amount of analyte, which is that amount required to generate a signal that is two or three standard deviations in magnitude above the signal obtained for a reagent blank solution (i.e., a blank that contains everything except the analyte). The slope of the calibration curve of signal versus analyte concentration is defined as the *sensitivity* of a quantitative assay; this slope describes how sensitive the assay is to changes in analyte concentration. It follows from this definition that, for nonlinear calibration curves, sensitivity is a function of concentration. These definitions conform to conventions in analytical chemistry.

Bianalytical Chemistry, by Susan R. Mikkelsen and Eduardo Cortón
ISBN 0-471-54447-7 Copyright © 2004 John Wiley & Sons, Inc.

3.2. SUBSTRATE DEPLETION OR PRODUCT ACCUMULATION

When a substrate can be directly measured, the reason for using an enzymatic assay for its quantitation may not be immediately apparent. However, complex biological media matrices may contain a variety of species that interfere with the direct measurement of analyte concentration. For example, if the analyte absorbs in the visible (vis) or ultraviolet (UV) region, direct quantitation may result in erroneously high values if interfering species absorb at the measurement wavelength. An enzymatic method, on the other hand, can monitor the absorbance decrease that occurs as a result of the selective consumption of analyte by the enzyme, thus avoiding the spectral interference.

An assay for uric acid involves the enzyme urate oxidase, which catalyzes the following reaction:

$$\text{Uric acid} + O_2 + 2H_2O \longrightarrow \text{Allantoin} + H_2O_2 + CO_2 \tag{3.1}$$

While allantoin is difficult to measure spectrophotometrically, uric acid possesses a strong UV absorption with a maximum at 293 nm (molar absorptivity $1.22 \times 10^4\,M^{-1}\text{cm}^{-1}$).[1] Uric acid quantitation thus involves monitoring the *decrease* in A_{293}.

Coenzyme A (CoA) may be quantitated using the enzyme phosphotransacetylase, which catalyzes the acetylation of CoA by acetylphosphate:

$$\text{CoA--SH} + CH_3CO\text{--}OPO_3H \longrightarrow \text{CoA--S--COCH}_3 + H_3PO_4 \tag{3.2}$$

In this case, CoA–S–COCH$_3$ accumulation is measured, since it possesses an absorption maximum at 232 nm (molar absorptivity $4.5 \times 10^3\,M^{-1}\,\text{cm}^{-1}$).[2] The parameter A_{232} will therefore *increase* as the reaction proceeds.

With some enzymatic reactions, both substrate and product may be readily measured, so that reactions may be followed by measuring either the depletion of substrate or the accumulation of product. The optimum choice of species for measurement then depends on the measurable properties of the substrate and product and on the measurement technique employed, so that a comparison of sensitivities and detection limits is necessary.

For example, if substrate and product are both fluorescent with identical molar absorptivities and quantum yields but different excitation or emission wavelengths, product quantitation is preferred because a small increase in fluorescence intensity is readily measured when superimposed on a background signal near zero, where the noise magnitude is very low. A more complicated decision is required if absorbance spectroscopy is to be used, if substrate and product have equal molar absorptivities. Absorbance measurements are generated by instruments that actually measure the quantity of light transmitted by a sample, so that the largest signals (with poorest precision) are generated at low analyte concentrations. The best RSD values are obtained for absorbance values in the 0.1–2.0 range. Although the measurement of product accumulation is usually chosen, a rigorous comparison of calibration data for substrate and product measurements is necessary for an optimized direct absorbance-based assay. In practice, product accumulation is almost invariably measured in an indirect manner, since it is often possible to couple a second reaction in series with the first substrate-selective reaction in order to generate an easily monitored secondary product.

3.3. DIRECT AND COUPLED MEASUREMENTS

Some enzymatic reactions can be followed directly, either by substrate depletion or product accumulation measurements, with adequate precision for direct enzymatic assays. However, many enzymes catalyze reactions involving species that are not themselves readily measured. In these situations, products are converted to species that are measurable, in a *coupled*, or *indicator* reaction. The indicator reaction may be chemical or enzymatic, and quantitatively converts the product of the primary reaction into a readily measurable species. The main requirement for the indicator reaction, whether it is chemical or enzymatic in nature, is that the conversion of the primary product into the measured product must be rapid and quantitative.

$$\text{Substrate} \xrightarrow{\text{E}_{prim}} \text{Primary product} \xrightarrow{\text{E}_{ind}} \text{Measured product} \qquad (3.3)$$

An assay for adenosine involves the primary enzyme adenosine deaminase, and a chemical indicator reaction that consumes ammonia by reaction with the ninhydrin reagent (Eq. 3.4):

$$(3.4)$$

The product of this reaction absorbs visible light, with a maximum at 546 nm. The ninhydrin indicator reaction[3] may be employed with virtually any primary reaction that produces NH_3.

The most commonly used indicator enzymes are dehydrogenases and peroxidases. The reactions catalyzed are shown in Eq. 3.5.

$$\text{Reduced substrate} + \text{NAD(P)}^{+} \xrightarrow{\text{dehydrogenase}} \text{Oxidized substrate} + \text{NAD(P)H} + \text{H}^{+} \qquad (3.5)$$

$$\text{Reduced dye} + \text{H}_2\text{O}_2 \xrightarrow{\text{peroxidase}} \text{Oxidized dye} + \text{H}_2\text{O} \qquad (3.6)$$

Dehydrogenases are used in cases where the primary enzymatic reaction produces the reduced substrate for a particular dehydrogenase enzyme reaction. These species are then converted to their oxidized forms in the indicator reaction, where the

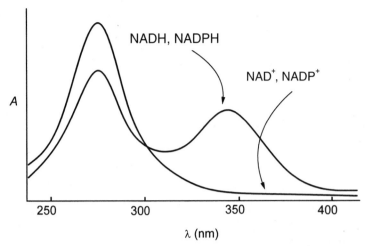

Figure 3.1. Absorption spectra of the nicotinamide coenzymes.[4] [Reprinted, with permission, from H. Netheler, in *Methods of Enzymatic Analysis*, Wiley-VCH, 1983, Vol. 1, Edited by Hans Ulrich Bergmeyer, 3rd English Edition, © Verlag Chemie GmbH, Weinheim, 1983.]

formation of reduced nicotinamide coenzyme (NADH or NADPH) allows absorbance measurements at 340 nm, where the molar absorptivities[4] of the reduced coenzymes are $6.2 \times 10^3 \, M^{-1} \, cm^{-1}$ (Fig. 3.1). Dehydrogenase indicator reactions are common because few interferences exist for absorbance measurements at 340 nm.

Peroxidase indicator reactions may be used to follow any primary reaction that produces hydrogen peroxide. Peroxidases are very specific for H_2O_2, but will react with a variety of organic dye species that are colorless in the reduced form, but highly absorbing in the oxidized form. Examples include 2,4-dichlorophenol, *O*-dianisidine, malachite green, and benzidine.

Two commercial enzymatic assays exist for the determination of glucose in serum. One involves the primary enzyme glucose oxidase and the indicator enzyme peroxidase:

$$\text{Glucose} + O_2 \longrightarrow \delta\text{-D-Gluconolactone} + H_2O_2 \qquad (3.7)$$

$$O\text{-Dianisidine} + H_2O_2 \longrightarrow 2, 2'\text{-Dimethoxydiphenylquinonediimine} + 2H_2O$$

$$(3.8)$$

The oxidized form of *O*-dianisidine is red, and shows an absorbance maximum at 450 nm with a molar absorptivity[5] of $8.6 \times 10^3 \, M^{-1} \, cm^{-1}$.

The second glucose assay involves hexokinase as a primary enzyme, and uses glucose-6-phosphate (G6P) dehydrogenase as an indicator enzyme:

$$\text{Glucose} + \text{ATP} \longrightarrow \text{G6P} + \text{ADP} \qquad (3.9)$$

$$\text{G6P} + \text{NADP}^+ \longrightarrow 6\text{-Phosphoglyceric acid} + \text{NADPH} + H^+ \qquad (3.10)$$

Based on the molar absorptivities of the products, the glucose oxidase–peroxidase assay may be expected to yield higher sensitivity and a lower detection limit.

Ideally, the indicator enzyme converts primary product into measured product in a "linear" manner, meaning that every molecule of primary product is instantaneously converted, regardless of substrate concentration. To accomplish this, primary product concentrations are kept low, so that they fall into the linear region of the saturation kinetics curve. For linear conversion at all analyte (primary substrate) concentrations, the effective rate of the indicator reaction, $(V_{eff})_{ind}$, must equal V_{max} for the primary reaction:

$$(V_{max})_{prim} = (V_{eff})_{ind} = (V_{max})_{ind} / \{1 + K_{m,P1}/[P1] + K_{m,S2}/[S2]\} \qquad (3.11)$$

Since $K_{m,P1}$ is a characteristic of the indicator enzyme, and [P1] is dictated by the analyte concentration, the only variables that can be controlled experimentally are $(V_{max})_{ind}$, which is equal to $k_{cat}[E_{ind}]$ (cf. Chapter 2) and the cosubstrate concentration [S2]. For this reason, a large excess (100-fold or more) of indicator enzyme is employed, in addition to saturating levels of cosubstrate (*O*-dianisidine in Eq. 3.8, and $NADP^+$ in Eq. 3.10).

When making initial rate measurements with coupled enzyme systems, there is often a significant lag time during which the linkage products build up to steady-state concentrations.[6] Remember that some dehydrogenase reactions possess unfavorable equilibria; lactate dehydrogenase, for example, which catalyzes the reaction shown in Eq. 3.12:

$$\text{Lactate} + NAD^+ \rightleftharpoons \text{Pyruvate} + NADH + H^+ \qquad (3.12)$$

prefers to convert pyruvate to lactate $(K = 5 \times 10^{-5})$ at pH 7. For this reason, coupling lactate dehydrogenase to a lactate-producing enzyme will not be effective, unless a trapping reagent is used to remove pyruvate and force the reaction towards products. For example, at pH 9.5, phenylhydrazine will trap pyruvate to produce a phenylhydrazone, shifting the lactate dehydrogenase equilibrium to $K = 2 \times 10^{-2}$.[7] Alternatively, NADH may be trapped with an electron-transfer reagent such as phenazine methosulfate (PMS), which is colorless in the oxidized form but absorbs at 388 nm in the reduced form.

3.4. CLASSIFICATION OF METHODS

Enzymatic assay methods are classified as fixed-time assays, fixed-change assays, or kinetic (initial rate) assays. Kinetic assays continuously monitor concentration as a function of time; pseudo-first-order conditions generally apply up to $\sim 10\%$ completion of the reaction to allow the initial reaction rate to be determined. If the initial substrate concentration is $>10K_m$, then the initial rate is directly proportional to enzyme concentration. At low initial substrate concentrations ($< 0.1\,K_m$), the initial rate will be directly proportional to initial substrate concentration (cf. Chapter 2). For enzyme quantitation, a plot of initial rate against [E] provides a linear calibration curve, while a linear plot of initial rate against [S] is used for substrate quantitation.

A commercial serum creatine kinase assay[8] employs the kinetic method for enzyme quantitation. This three-enzyme, coupled assay involves the following sequence of reactions:

$$\text{Creatine phosphate} + \text{ADP} \xrightarrow{\text{creatine kinase}} \text{Creatine} + \text{ATP} \tag{3.13}$$

$$\text{ATP} + \text{Glucose} \xrightarrow{\text{hexokinase}} \text{ADP} + \text{G6P} \tag{3.14}$$

$$\text{G6P} + \text{NADP}^+ \xrightarrow{\text{G6P dehydrogenase}} \text{6-Phosphogluconate} + \text{NADPH} + \text{H}^+ \tag{3.15}$$

In this assay, diluted serum is preincubated with glucose, hexokinase, NADP^+, and G6P dehydrogenase to allow any creatine phosphate and ADP present in the serum sample to be consumed. When a constant A_{340} value is achieved, a reagent solution consisting of concentrated creatine phosphate and ADP is added, and the increase in A_{340} is monitored as a function of time. A typical trace is shown in Figure 3.2. The slope of the initial linear section of this curve is directly proportional to enzyme concentration.

Fixed-time assays include the so-called "endpoint" assays, and measure the change in [S] or [P] that occurs over a fixed, relatively long, period of time. They rely on quantitative (or near-quantitative) conversion of substrate to product, and are used exclusively for substrate quantitation. A linear calibration curve of signal change against initial substrate concentration is used to quantitate substrate. Fixed-time assays do not require $[S]_0 < 0.1\,K_m$, since essentially complete conversion

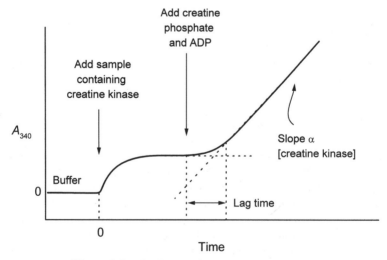

Figure 3.2. Kinetic trace for creatine kinase assay.

occurs. This type of assay cannot be used for enzyme quantitation since saturating substrate conditions are not maintained over the duration of the reaction.

A commercial endpoint assay for total serum cholesterol[9] employs three enzymes, cholesterol esterase, cholesterol oxidase, and peroxidase:

$$\text{Cholesterol ester} + H_2O \xrightarrow{\text{cholesterol esterase}} \text{Cholesterol} + \text{Fatty acid} \tag{3.16}$$

$$\text{Cholesterol} + O_2 \xrightarrow{\text{cholesterol oxidase}} \text{Cholest-4-en-3-one} + H_2O_2 \tag{3.17}$$

$$H_2O_2 + \text{4-Aminoantipyrene} + p\text{-Hydroxybenzenesulfonate} \xrightarrow{\text{peroxidase}}$$

$$\text{quinoneimine dye} + 2H_2O \tag{3.18}$$

Serum cholesterol exists as a mixture of fatty acid esters and free cholesterol. Quantitation of total cholesterol involves the initial conversion of the esters to free cholesterol, followed by the total conversion of free cholesterol to its oxidation product. This reaction is coupled to the familiar dye-peroxidase indicator reaction. The parameter A_{500} measurements using stock cholesterol solutions provide a calibration curve. A reagent blank solution is prepared using all components except cholesterol, and this value is subtracted from all measured A_{500} values, correcting for any background oxidation of the dye.

Fixed-change assays are relatively uncommon, and are used for enzyme quantitation. These assays monitor the time required for the generation of a given concentration of product. The enzyme concentration is inversely related to the time required for this extent of reaction to occur, so that a linear plot of $1/t$ against $[E]$ is used as a calibration curve.

3.5. INSTRUMENTAL METHODS

A wide variety of instrumental methods have been used to quantitate enzymes and their substrates. The choice of method depends primarily on the physical properties of the species being measured, and this is generally the product of the enzymatic or indicator reaction. In this section, instrumental detection methods are broadly classified as optical, electrochemical or "other", where other techniques include radiochemical and manometric methods.

3.5.1. Optical Detection

3.5.1.1. Absorbance. The detection of substrate depletion or product accumulation through the measurement of visible or UV light absorbance is based on the Beer–Lambert law, which directly relates absorbance at a given wavelength to concentration:

$$A = \varepsilon b C \tag{3.19}$$

where b is the path length of the sample through which light travels, and ε is a constant for a particular molecular species at a given wavelength under fixed conditions of temperature, solvent, and pH. This law is valid for virtually all light-absorbing species at low (< 0.1 mM) concentrations. Apparent deviations from the Beer–Lambert law occur if analytes are involved in reversible equilibria (such as pH dependent protonations or association reactions) or if they decompose in solution.

The sensitivity of an absorbance-based assay depends directly on ε, the molar absorptivity of the species being measured. For example, we will consider the two glucose assays in Eqs. 3.7 and 3.8 and 3.9 and 3.10. In the former assay, peroxidase produces the dye 2,2'-dimethoxy-p-phenylenediimine, in a 1:1 stoichiometric ratio with glucose consumed in the glucose oxidase reaction. In the latter reaction scheme, NADPH is the species being measured, and it too is produced in a 1:1 stoichiometric ratio to glucose. Sensitivity is the slope of the plot of signal against concentration, so that a comparison of the two assays requires a comparison of these slopes. If the indicator reactions have been configured for linear conversion of primary substrate into measured product, then over a given time period, identical quantities of glucose will yield identical quantities of measured product for the two reaction schemes. Since $\Delta A = \varepsilon b \Delta C$, and ΔC is identical for the two assays, the magnitude of the signal generated will depend on ε. For the peroxidase indicator reaction, the oxidized dye has $\varepsilon = 8.6 \times 10^3 \, M^{-1}\text{cm}^{-1}$ at 450 nm, while the dehydrogenase reaction produces NADPH, with $\varepsilon = 6.2 \times 10^3 \, M^{-1}\text{cm}^{-1}$ at 340 nm. Clearly, the slope of the calibration curve of ΔA (for fixed-time assays) or $\Delta A / \Delta t$ (for kinetic assays) against [glucose] (i.e., the sensitivity of the assay) will be greater for the peroxidase indicator reaction.

The detection limits of absorbance-based assays are also dependent on molar absorptivity. We have defined the detection limit of an assay as being the minimum analyte concentration required to generate a signal two or three times as large as the standard deviation in the blank measurement ($A_{\text{meas}}/\sigma_A = 2$ or 3). If the magnitude of the blank noise is comparable for measurements made at 340 and 450 nm, then comparable absorbance values will be measured at the detection limit. Under these conditions, since $A = \varepsilon b C$, higher molar absorptivities yield lower detectable concentrations of analyte. The detection limit of the dye–peroxidase based assay is therefore expected to be lower than that of the dehydrogenase assay by a factor of 8600/6200 ($\sim$1.4). The detection limits of enzymatic assays for substrates are generally in the micromolar concentration range.

Serum alkaline phosphatase, which is an important metabolic indicator, is generally quantitated by absorbance methods. Alkaline phosphatase catalyzes the dephosphorylation of NADP$^+$ and a variety of other substrates *in vivo*, but *in vitro*, the synthetic substrate p-nitrophenylphosphate (Eq. 3.20) can be used.

$$O_2N\!-\!\!\left\langle\ \right\rangle\!\!-\!O\!-\!\overset{\overset{\displaystyle O}{\|}}{\underset{\underset{\displaystyle O^-}{|}}{P}}\!-\!O^- \ + \ H_2O \ \xrightarrow[\text{phosphatase}]{\text{alkaline}} \ O_2N\!-\!\!\left\langle\ \right\rangle\!\!-\!OH \ + \ HPO_4{}^{2-} \quad (3.20)$$

The product, nitrophenol, absorbs at 405 nm[10] with $\varepsilon = 1.85 \times 10^4 \, M^{-1}\text{cm}^{-1}$. When a diluted serum sample is subjected to saturating concentrations of p-nitrophenylphosphate, and incubated at room temperature for 15 min, and absorbance measured at 405 nm, the detection limit of the assay is $4 \times 10^{-12} M$ alkaline phosphatase in the original serum sample. While this may seem like a very low concentration, diagnostic tests often require an improved detection limit.

A cascade amplification method has been proposed for improving the detection limit of the alkaline phosphatase assay, and involves a different synthetic substrate (Eq. 3.21), (oxotrifluorobutyl)phenylphosphate:[11]

$$(3.21)$$

When this substrate is dephosphorylated, its product acts as an *inhibitor* of a second enzyme, rabbit liver esterase. Thus, as alkaline phosphatase activity increases, there is a corresponding increase in the concentration of inhibitor produced, and consequently a decrease in the activity of the esterase. A second synthetic substrate was prepared for the (relatively nonselective) esterase, so that the indicator reaction is as shown in Eq. 3.22 below:

$$(3.22)$$

The substrate for the indicator reaction, dichloroindophenylbutyrate, is pale yellow. The product of the reaction is deep blue, with an absorbance maximum at 620 nm. The ingenuity of this method lies in the production of an enzyme inhibitor, (oxotrifluorobutyl)phenol, rather than a species that is directly quantitated. The inhibitor affects the *activity* of the esterase, resulting in catalytic amplification of the initial alkaline phosphatase activity. The detection limit of the amplified assay has been reported as $3.2 \times 10^{-14} M$, a 100-fold improvement over the detection limit of the standard alkaline phosphatase assay.

3.5.1.2. *Fluorescence.*

The use of molecular fluorescence spectroscopy for the quantitation of enzyme reaction products has resulted in detection limits that are several orders of magnitude lower than those achieved by standard absorbance methods. At low analyte concentrations, fluorescence emission intensity is directly proportional to concentration, and its value depends on both the molar absorptivity of the analyte at the excitation wavelength, and the fluorescence quantum yield of the analyte, under the assay conditions.

Fluorescence methods are inherently capable of detecting much lower concentrations of analytes, because of the instrumental principles involved. Consider, for example, a comparison of the light intensities measured in blank solutions by absorption and fluorescence spectrophotometers. Incident radiation passes through the sample cuvette essentially unabsorbed in both cases. The absorption detector, a photodiode or a photomultiplier tube, measures a high transmitted light intensity. Detection of light involves the random process of the arrival of photons at the detection element; such a random process exhibits a Gaussian distribution of events per unit time, with noise increasing as $(events)^{1/2}$. This result means that the high intensity transmitted by the blank solution possesses an inherently high-noise level. A small increase in sample concentration results in a slight decrease in transmitted light intensity, superimposed on this inherently high level of noise. Fluorescence emission is measured at right angles to the incident light beam, so that only emitted light arrives at the detection element. Because of this, a blank solution will generate a very low detector signal, with an inherently low noise level. Small concentration increases are thus much easier to quantitate precisely.

Fluorescence detection is also inherently more selective than absorbance detection, since both the excitation and emission wavelengths may be chosen to suit a particular reaction product. For example, assays employing dehydrogenase enzymes may monitor NAD^+ or nicotinamide adenine dinucleotide phosphate $(NADP^+)$ absorbance at 340 nm with reasonable sensitivity and selectivity. However, if excited at 340 nm, the nicotinamide coenzymes fluoresce at 460 nm. Not only do the fluorescence measurements inherently have lower detection limits, but they also provide selectivity against potential interferents that may also absorb at 340 nm but do not emit at 460 nm.

Fluorescence measurements are subject to certain difficulties, the most commonly encountered being quenching and inner-filter effects. Any compounds that absorb near the emission wavelength will quench emitted light and reduce the apparent emission intensity. Polypeptides and oligonucleotides possess absorption maxima <300 nm and so do not usually act as quenchers; however, prosthetic groups such as flavins, hemes, and coordinated metal atoms will effectively quench fluorescence emission in certain regions of the visible spectrum. Inner-filter effects must be considered if the selected excitation wavelength is absorbed by species other than the analyte, again resulting in reduced apparent emission intensity. The tyrosine and tryptophan residues of proteins contribute to inner-filter effects for excitation wavelengths near 300 nm.

The principles of fluorescence quenching have been successfully applied to assays for protease activity. Proteases are digestive enzymes that degrade polypeptides into smaller oligopeptides or constituent amino acids. A general assay[12] for protease activity employs a substrate prepared by covalently derivatizing a protein, transferrin, with a number of fluorescein isothiocyanate (FITC) labels. The FITC labeled proteins (Fig. 3.3) exhibit absorbance maxima at 495 nm and emission maxima at 525 nm.

When transferrin is labeled with FITC, the coordinated iron atom at the active site acts as an effective quencher of emitted light, so that this prepared substrate exhibits low emission intensity at 525 nm. Upon exposure to protease, cleavage of the polypeptide occurs, resulting in the release of FITC labeled amino acids and oligopeptides. Because the fluorophore and the quencher are no longer fixed in

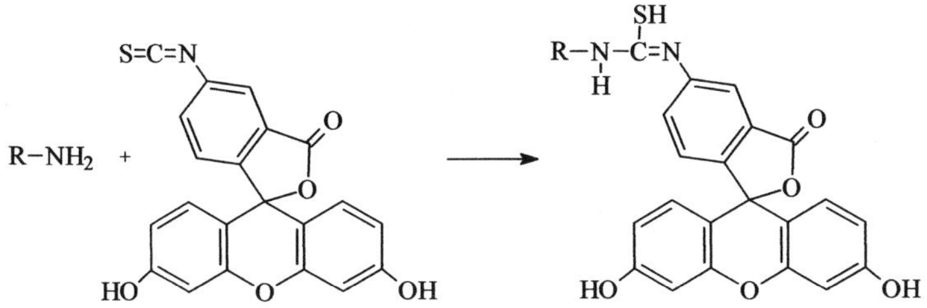

Figure 3.3. Labeling of protein Lys residues (R–NH$_2$) with FITC.

close proximity, the intensity of emitted light increases. The assay therefore involves monitoring the increase over time of emission intensity at 525 nm.

A second protease assay based on fluorescence quenching was designed for *Astacus* protease,[13] a specific endopeptidase found in freshwater crayfish that cleaves between arginine and alanine residues. A synthetic substrate (Fig. 3.4) was prepared by linking an oligopeptide containing one fluorescent tryptophan residue with a terminal dansyl group that is an effective quencher of tryptophan fluorescence. Note that the cleavage site occurs *between* tryptophan and dansyl groups.

Tryptophan is excited at 285 nm, and emits at 360 nm. In the synthetic substrate, tryptophan fluorescence is quenched by the dansyl group by radiationless energy transfer; the fluorophore and the quencher are fixed in close proximity, allowing this process to occur. Upon cleavage of the Arg-Ala bond, the fluorescent product, Ala-Pro-Trp-Val, shows much higher emission intensities at 360 nm, since the dansyl quenching group is no longer held in close proximity.

3.5.1.3. Luminescence. Bioluminescence methods rely on the production of light by an enzyme-catalyzed reaction. The enzymatic reaction yields an excited-state product, which returns to the ground state upon emission of light, as shown in Eq. 3.23.

$$\text{Substrate(s)} \xrightarrow{\text{enzyme}} \text{Excited product(s)} \longrightarrow \text{Product(s)} + h\nu \qquad (3.23)$$

Bioluminescence methods are distinct from absorbance and fluorescence methods because the light measured is now transient rather than steady state. Because of this, initial rate measurements yield a *constant* level of detected light, rather than the

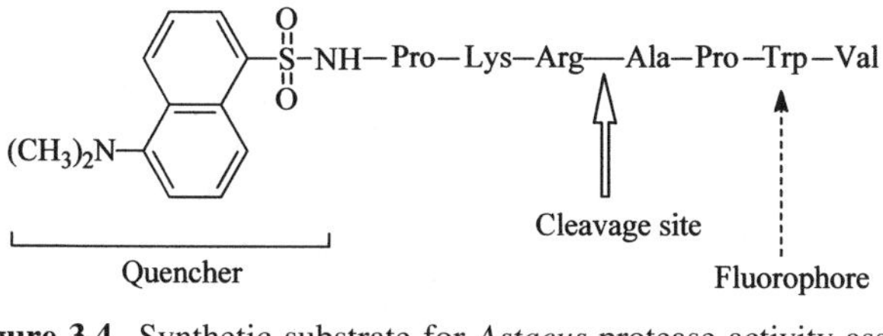

Figure 3.4. Synthetic substrate for *Astacus* protease activity assay.

linear increases observed with absorbance and fluorescence measurements. Since photons are a product of the reaction, they are produced in stoichiometric proportion to the amount of substrate consumed. Absorbance and fluorescence measurements, on the other hand, produce a steady stream of detectable photons. For this reason, light intensities are much lower with luminescence reactions, and signals are generally integrated over several minutes to yield acceptable S/N levels. Instrumental requirements for luminescence measurements are fairly simple: A standard luminescence photometer consists of a light detector (photon counter) surrounding a sample holder inside an enclosure that eliminates ambient light; neither light source nor monochromator are required.

Primary reactions that produce hydrogen peroxide have been tested[14] with a luminol–peroxidase indicator reaction, Eq. 3.24:

$$\text{(luminol)} + H_2O_2 \longrightarrow \text{(oxidized luminol)} + 2H_2O + h\nu \tag{3.24}$$

The oxidized form of luminol is produced in the excited state, and the emitted light is integrated over 4 min. When this reaction was used as an indicator reaction for a glucose assay using glucose oxidase, linearity was observed between 0.1- and 2.0-mM glucose, while a cholesterol assay using cholesterol oxidase yielded a linear calibration curve between 50 and 100 mg/mL. The sensitivity of the indicator reaction, and thus of both of the assays, depended strongly on the particular peroxidase used. *Arthromyces* (bacterial) peroxidase was found to yield calibration curve slopes that were 28 times greater for the glucose assay and 134 times greater for the cholesterol assay than horseradish peroxidase.

Firefly luciferase[15] may be employed as an indicator reaction for any primary enzymatic reaction that produces ATP. Luciferase catalyzes the oxidative decarboxylation of luciferin according to Eq. 3.25:

$$\text{(luciferin)} + O_2 + ATP \longrightarrow \text{(product)} + AMP$$
$$+ PPi + CO_2 + h\nu \tag{3.25}$$

This reaction results in light emission between 540 and 600 nm. Luciferase is activated by divalent magnesium, so experiments are carried out in the presence of excess Mg^{2+} (1 mM). This reaction does not yield a constant light output under pseudo-first-order conditions, as shown in Figure 3.5, which is thought to be due to product inhibition of luciferase. Figure 3.5 also shows the effect of hydrophobic and amphipathic species on light output; these species may be quantitated by their inhibitory effect on the luciferase reaction.

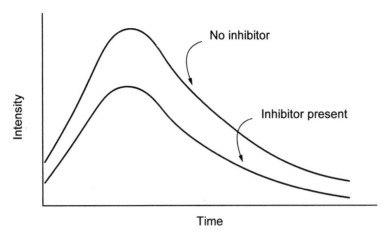

Figure 3.5. Light intensity versus time for luciferin/luciferase reaction.

The light output is integrated over 1 s to yield a total emission value that is used for quantitation. This reaction has been used to study the partitioning of anesthetics into lipid bilayers, where the particular anesthetic species studied acted as competitive inhibitors of luciferase.

3.5.1.4. Nephelometry. Nephelometry, or light scattering, may be employed to monitor enzymatic reactions when the turbidity of the reaction medium changes as the reaction proceeds. At low turbidities, scattered light intensity is proportional to concentration. Standard spectrofluorometric instrumentation is used to measure light scattered perpendicular to the incident light beam. This method has been used to quantitate lipids, using the enzyme lipase.[16] This enzyme catalyzes the hydrolysis of lipids into their constituent fatty acids, which are more soluble than lipids in aqueous media. The turbidity of the sample therefore decreases as the reaction proceeds.

3.5.2. Electrochemical Detection

3.5.2.1. Amperometry. Amperometric methods measure the current produced at a working electrode in response to an applied potential. Amperometric enzyme assays rely on the production of an oxidizable or reducible species from an enzyme-catalyzed reaction. The applied potential is extreme enough to completely oxidize (at positive potentials) or reduce (at negative potentials) any analyte that contacts the working electrode. In stirred or unstirred solutions, the current produced under such mass-transport-controlled conditions is directly proportional to analyte concentration.

Amperometry is most commonly employed in enzymatic assays in which hydrogen peroxide is one of the products. Hydrogen peroxide is oxidized at a constant potential of $+700$ mV versus saturated calomel electrode (SCE) to produce molecular oxygen:

$$H_2O_2 = 2\,H^+ + O_2 + 2\,e^- \tag{3.26}$$

The electrons produced in this oxidation reaction result in a measured current that is directly proportional to H_2O_2 concentration.

Amperometry has been used to quantitate inorganic phosphate $(P_i)^{17}$ in a dual-enzyme assay, shown in Eqs. 3.27 and 3.28:

$$\text{Phosphate} + \text{Inosine} \xrightarrow{\text{nucleoside phosphorylase}} \text{Ribose-1-phosphate} + \text{Hypoxanthine} \tag{3.27}$$

$$\text{Hypoxanthine} + 2\,H_2O + O_2 \xrightarrow{\text{xanthine oxidase}} \text{Uric acid} + H_2O_2 \tag{3.28}$$

In addition to peroxide-producing reactions, amperometry may be used in conjunction with a variety of oxidase and dehydrogenase enzymes that employ low molecular weight *mediators* (Med) as electron acceptors in place of molecular oxygen.

$$S_{red} + n(Med)_{ox} \xrightarrow[\text{(cofactor)}]{\text{enzyme}} P_{ox} + n(Med)_{red} \tag{3.29}$$

$$(Med)_{red} = (Med)_{ox} + e^- \tag{3.30}$$

Ferricinium derivatives, ferricyanide, a variety of quinone derivatives, and other organic oxidants have been used as mediators of oxidase and dehydrogenase reactions. Following their reduction in solution, these mediators are reoxidized at a working electrode, yielding measurable currents. Electron-transfer mediators that posses fast rates of heterogeneous (electrochemical) electron transfer have been shown to yield better S/N than oxygen/hydrogen peroxide mediation.

3.5.2.2. Potentiometry.
Potentiometric methods rely on the logarithmic relationship between measured potential and analyte concentration. The most common involves an instrument called a "pH-Stat", in which a glass (pH) electrode follows reactions that either consume or produce protons. Since pH changes cause changes in enzyme activity, the pH is maintained at a constant value by the addition of acid or base. The rate of titrant addition is then proportional to the rate of the enzymatic reaction. Precise measurements using the pH-Stat require low buffer concentrations in the enzymatic assay mixture.

Other potentiometric methods employ gas-sensing electrodes for NH_3 (for deaminase reactions) and CO_2 (for decarboxylase reactions). Ion-selective electrodes have also been used to quantitate penicillin, since the penicillinase reaction may be mediated with I^- or CN^-.

3.5.2.3. Conductimetry.
These methods involve the application of an alternating voltage across two electrodes in solution, and the measurement of the

magnitude of the alternating current response. This current is directly proportional to the conductivity of the solution, which, in turn, is dictated by ionic strength.

The urease reaction, in particular, is ideally suited to conductimetric quantitation.[18] Urea is hydrolyzed by urease according to Eq. 3.31:

$$H_2NCONH_2 + 3\,H_2O \xrightarrow{\text{urease}} 2\,NH_4^+ + HCO_3^- + OH^- \qquad (3.31)$$

In this reaction, four ions are produced from a single uncharged molecule of substrate, resulting in a significant increase in ionic strength. Good S/N may be achieved if the initial ionic strength (buffer concentration) of the assay medium is low, in the 5–10-mM range.

3.5.3. Other Detection Methods

3.5.3.1. Radiochemical. Radiochemical methods are increasingly becoming supplanted by methods that do not require specialized facilities, although in some cases, the method of choice still involves the use and quantitation of radioactive species. Examples include the endpoint assays for guanosine-5′-triphosphate (GTP) and guanosine-5′-diphosphate (GDP).[19] These species are involved in hormone regulation, protein synthesis and secretion, and gluconeogenesis. The GTP assay involves Reactions 3.32 and 3.33:

$$\text{Aspartate}(^{14}C) + \alpha\text{-Ketoglutarate} \xrightarrow{\text{aspartate aminotransferase}} \text{Oxaloacetate}(^{14}C) + \text{Glutamate}$$
$$(3.32)$$

$$\text{Oxaloacetate}(^{14}C) + GTP \xrightarrow{\text{PEP carboxykinase}} PEP(^{14}C) + GDP + CO_2 \qquad (3.33)$$

where Phosphoenolpyruvate $=$ PEP.

All reagents are added in excess, and the mixture is equilibrated overnight with the GTP sample, so that the quantity of PEP(^{14}C) produced is limited by the quantity of GTP present in the sample and the equilibrium constants of the enzymatic reactions. The labeled PEP is then separated by anion-exchange chromatography and quantitated in a scintillation counter. The assay for GDP is conducted using the same reactions in reverse (both enzymes are reversible, and yield an equilibrium mixture of reactants and products), and labeled PEP as a reagent. This reaction is followed by the separation and quantitation of labeled aspartate. Note that all radiochemical enzymatic assays require a separation step prior to quantitation.

3.5.3.2. Manometry. Simple manometric methods may be used in endpoint assays, to measure the total quantity of a gas (such as NH_3 or CO_2) produced in an enzymatic reaction. Automated manometers may be used in a kinetic mode, to quantitate gas evolution as a function of time. These methods are based on the ideal gas law, which states that the volume occupied by a gas is directly proportional to the number of moles of the gas at constant pressure and temperature. The gas volume is thus measured as a function of time or following completion of the reaction. Manometric methods have been largely supplanted by more selective instrumental techniques.

3.5.3.3. Calorimetry. Isothermal titration calorimetry (ITC) has recently been proposed as a general method for the determination of enzymatic reaction rates.[20] This method is based on the relationship between the power needed to maintain constant temperature and the number of moles of substrate converted. Power, the measured quantity, is related to heat (Q) by differentiation:

$$Power = dQ/dt \tag{3.34}$$

Heat is then directly related to the number of moles of substrate converted, with a proportionality constant of ΔH_{app}, an experimentally determined apparent molar enthalpy for the reaction:

$$Q = n\,\Delta H_{app} \tag{3.35}$$

Thus, the measured thermal power for an enzyme reaction may be directly related to the enzymatic reaction rate:

$$Rate = (d[P]/dt) = (1/(V\,\Delta H_{app}))(dQ/dt) \tag{3.36}$$

where V is the solution volume and $[P]$ is the concentration of reaction product.

The sensitivity of this method is directly related to the apparent molar enthalpy of reaction, so that very endo- or exothermic reactions will be most readily followed. Examples of the application of this method to the determination of enzyme kinetic parameters include dihydrofolate reductase, creatine phosphokinase, hexokinase, urease, trypsin, HIV-1 protease, heparinase, and pyruvate carboxylase.

3.6. ULTRA-HIGH-THROUGHPUT ASSAYS (HTA)[21–23]

Recent developments in the design of miniaturized well plates have demonstrated that many reactions that can be followed using conventional 96-well microtiter plates can be adapted to newer 384- and 1536-well microplates. Motivation for this increased assay capacity has come largely from the pharmaceutical industry, where screening of a large number of drug candidates produced by combinatorial synthesis allows higher productivity and more rapid identification of lead compounds. Challenges in high-throughput assay development involve the reduction in assay volume from 200 (96-well) to <1 μL (1536-well) with the concomitant problems of solvent evaporation and the development of instrumentation to reliably determine assay signals from such small sample volumes. Absorbance-based assays are currently the most readily adapted, since path length decreases by only a factor of 2 for the 1536-well plates as compared to the conventional 96-well plates. Such assays have been demonstrated for phosphatase, ATPase, and beta-lactamase, as well as their inhibitors, under conditions where the 1536-well plates had completely filled wells (10-μL total volume).[21,22] Further developments are underway to produce 9600-well plates for ultra-HTS methods. An excellent summary of HTS and ultra-HTS robotics and instrumentation has recently been published.[23]

3.7. PRACTICAL CONSIDERATIONS FOR ENZYMATIC ASSAYS

With any assay involving enzymes, care must be taken to avoid denaturation and loss of activity due to improper handling and storage conditions. In particular, the following conditions should be avoided for most enzymes: temperatures in excess of 40 °C, pH values >9 or <5, and the presence of organic solvents, surfactants, and trace metals. When adjusting the pH of an enzyme solution, acids or bases should be added dropwise along the sides of the vessel, with stirring, to avoid local pH extremes.

Enzyme activity is preserved by cold storage. Lyophilized (freeze-dried) proteins are stored in a freezer or refrigerator, while dilute solutions are stored at 2–5 °C. Concentrated suspensions of enzymes in ammonium sulfate are usually stable for long periods at 2–5 °C.

Some enzymes strongly adsorb to glassware. If this is the case, solutions are commonly prepared in the presence of a large (100-fold) excess of an inert protein, such as albumin, to avoid activity losses that would lead to miscalibrations.

The reduced forms of the pyridine coenzymes, NADH and NADPH, are sensitive to light (20% loss over 7 weeks), moisture (50% loss in 24 h), and high temperatures (10% loss over 3 weeks at 33 °C). The oxidized forms of these coenzymes are more stable, but should also be stored at low temperature. Flavin cofactors are light sensitive, and should be stored in the dark.

Buffered solutions should be prepared under sterile conditions, and stored after ultrafiltration. Many biochemical buffer systems are excellent growth media for bacteria.

SUGGESTED REFERENCES

H. Bergmeyer, Ed. *Methods of Enzymatic Analysis*, 3rd ed., Vols. I–XII, VCH Publishers, Weinheim, Germany, 1986.

R. Eisenthal and M. J. Danson, Eds., *Enzyme Assays: A Practical Approach*, Oxford University Press, New York, 1992.

D. C. Harris, *Quantitative Chemical Analysis* (*6th ed.*), W. H. Freeman, New York, 2003.

G. G. Guilbault, *Enzymatic Methods of Analysis*, Pergamon Press, New York, 1970.

REFERENCES

1. P. Plesner and H. M. Kalckar, in *Methods of Biochemical Analysis*, Vol. 3, D. Glick, Ed., Wiley, New York, 1956, p. 99.

2. G. D. Novelli, in *Methods of Biochemical Analysis*, Vol. 2, D. Glick, Ed., Wiley, New York, 1955, p. 208.

3. W. Appel, in *Methods of Enzymatic Analysis*, Vol. 2, H. U. Bergmeyer, Ed., Academic Press, New York, 1974, pp. 989–992.

4. H. Netheler, in *Methods of Enzymatic Analysis*, Vol. 1, H. U. Bergmeyer, Ed., Wiley-VCH, Deerfield Beach, 1983, p. 165.

5. A. Claiborne and I. Fridovich, *Biochemistry* **18**, 1979, 2324–2329.

6. S. P. J. Brooks and C. H. Suelter, *Anal. Biochem.* **176**, 1989, 1–14.

7. J. King, *Practical Clinical Enzymology*, Van Nostrand, Princeton, NJ, 1965.

8. Sigma Chemical Company, St. Louis, MO, 1994 Catalogue, p. 2197.

9. Sigma Chemical Company, St. Louis, MO, 1994 Catalogue, p. 2194.

10. H. U. Bergmeyer, K. Gawehn, and M. Grassl, in *Methods of Enzymatic Analysis*, Vol. 1, H. U. Bergmeyer, Ed., Academic Press, New York, 1974, p. 496.

11. P. D. Mize, R. A. Hoke, C. P. Linn, J. E. Reardon, and T. H. Schulte, *Anal. Biochem.* **179**, 1989, 229–235.

12. K. A. Homer and D. Beighton, *Anal. Biochem.* **191**, 1990, 193–197.

13. W. Stocker, M. Ng, and D. S. Auld, *Biochemistry* **29**, 1990, 10418–10425.

14. B. B. Kim, V. V. Pisarev, and A. M. Egorov, *Anal. Biochem.* **199**, 1991, 1–6.

15. S. Naderi and D. L. Melchior, *Anal. Biochem.* **190**, 1990, 304–308.

16. L. Zinterhofer, S. Wardlow, P. Jatlow, and D. Seligson, *Clin. Chim. Acta* **44**, 1973, 173–178.

17. S. D. Haemmerli, A. A. Suleiman, and G. G. Guilbault, *Anal. Biochem.* **191**, 1990, 106–109.

18. P. Duffy, C. Mealet, J. M. Wallach, and J. J. Fombon, *Anal. Chim. Acta* **211**, 1988, 205–211.

19. P. F. Cerpovicz and R. S. Ochs, *Anal. Biochem.* **192**, 1991, 197–202.

20. M. J. Todd and J. Gomez, *Anal. Biochem.* **296**, 2001, 179–187.

21. J. J. Burbaum, *J. Biomol. Screening* **5**, 2000, 5–8.

22. P. Lavery, M. J. B. Brown, and A. J. Pope, *J. Biomol. Screening* **6**, 2001, 3–9.

23. D. A. Wells, *High Throughput Bioanalytical Sample Preparation*, Elsevier, New York, 2003.

PROBLEMS

1. For which type of substrate assay, the kinetic or the endpoint, is it necessary to maintain the substrate–analyte concentration at levels well below the K_m of the enzyme used?

2. In a direct absorbance-based kinetic assay in which a nonabsorbing substrate is enzymatically converted to an absorbing product, over what substrate concentration range can the assay be used to quantitate substrate? Over what substrate concentration range can the assay be used to quantitate enzyme activity?

3. Indicator reactions are employed in series with an enzymatic reaction in order to generate an easily measured product. Would you expect detection limits to be better (lower) if the indicator reaction generates a fluorophore directly, or if the indicator reaction produces a species that inhibits an additional enzymatic reaction that generates a fluorophore?

4. A new assay has recently been reported for the enzyme alkaline phosphatase. The reactions upon which the assay is based are shown below:

$$FADP + H_2O \xrightarrow{\text{alkaline phosphatase}} FAD + P_i$$

$$FAD + \text{Apo-D-Amino acid oxidase} \longrightarrow \text{D-Amino acid oxidase}$$
$$\text{(inactive)} \qquad\qquad\qquad \text{(active)}$$

$$\text{D-Proline} + O_2 \xrightarrow{\text{D-amino acid oxidase}} \text{2-Oxo-5-Aminopentanoic acid} + H_2O_2$$

$$\text{4-Aminoantipyrene} + 3,5\text{-Dichloro-2-hydroxybenzenesulfonate} \xrightarrow{\text{peroxidase}}$$
$$N\text{-(4-Antipyryl)-3-chloro-5-sulfonate-}p\text{-benzoquinoneimine}$$

FADP = flavin adeninedinucleotide phosphate

The product of the fourth reaction absorbs strongly at 520 nm, with a molar absorptivity of $2.3 \times 10^4\ M^{-1}\text{cm}^{-1}$. Following a 10-min incubation, 100 amol (1 amol = 10^{-18} mol) of alkaline phosphatase in 0.100-mL total volume hydrolyzes 0.1 mM of the FADP (0.5% of the initial saturating substrate concentration of 20 mM), and converts 12% of the originally 100 nm apo-D-amino acid oxidase to the active holoenzyme.

(a) Which of the four reactions shown above is/are critical to the assay for alkaline phosphatase? Which are indicator reactions that may be replaced by others?

(b) Calculate the concentration of alkaline phosphatase in I.U./mL.

(c) Given that the k_{cat} for D-amino acid oxidase is 3.6 s^{-1}, and assuming that all substrates are present in great excess, calculate the absorbance change that would occur for 12 nM holo-D-amino acid oxidase over 10 min.

(d) Given that the measured absorbance change in the assay is one-half of the value calculated in (c) (since the holoenzyme is generated in a linear fashion with time over the 10-min incubation period), calculate the working concentration range over which alkaline phosphatase may be quantitated. Assume that final absorbance values are to be between 0.1 and 2.0 absorbance units, and report the alkaline phosphatase concentration range in I.U./mL.

5. 1,5-Anhydro-D-glucitol (AG, 164 g/mol) is one of the main sugar alcohols in human cerebrospinal fluid and blood. In plasma, the normal AG level is 24.6 ± 7.2 mg/L (539 patients tested), while patients with diabetes mellitus show reduced AG levels of 7.3 ± 7.1 mg/L (808 patients tested). Because reduced AG levels are specific indicators of this type of diabetes, a diagnostic enzyme assay for AG has been developed. The assay employs the enzymes pyranose oxidase (PROD) and horseradish peroxidase (HRP) in the following reaction scheme:

$$AG + O_2 \xrightarrow{\text{PROD}} AG\text{-Lactone} + H_2O_2$$
$$ABTS_{red} + H_2O_2 \xrightarrow{\text{HRP}} ABTS_{ox} + 2\,H_2O$$

The oxidized form of ABTS exhibits an absorbance maximum at 420 nm, with a molar absorptivity of $3.48 \times 10^4\,M^{-1}cm^{-1}$. The AG assay was developed as an endpoint assay, relying on the complete conversion of substrate to product over a 1-h period at 37 °C and pH 5.9, the optimum pH for the complete system. The specific activity of PROD at this pH is 2.11 I.U./mg.

A plasma sample from a patient suspected of having diabetes mellitus was diluted 10-fold during sample preparation and reagent addition, and was incubated 1 h at 37 °C. During this time, O_2, PROD, $ABTS_{red}$, and HRP were present in large excess to ensure complete conversion of normal levels of AG. Following incubation, the absorbance of the resulting solution at 420 nm was measured in a 1-cm cuvette, yielding $A = 0.174$. Is the patient diabetic?

CHAPTER 4

Immobilized Enzymes

4.1. INTRODUCTION

Soluble enzymes are employed in a wide variety of substrate and enzyme activity assays, and specialized instrumentation has been developed to automate reagent addition and quantitation. However, several disadvantages exist in the analytical use of soluble enzymes for substrate assays. Soluble enzymes are not reused or recycled, unless the cost of the enzyme justifies the lengthy repurification procedure. Furthermore, the activities of soluble enzymes decreases significantly with time, so that fresh assay solutions are frequently required. For these reasons, many assays that employ soluble enzymes have been adapted for use with immobilized enzymes. These immobilized enzymes are usually incorporated onto or into a stationary phase in a flow system; substrate is introduced via a mobile, or buffer phase, and conversion into products occurs as the mobile phase flows through a column containing immobilized enzyme. A postcolumn detector allows product quantitation.

Enzyme reactors, the columns containing a stationary phase with immobilized enzyme, may be reused many times, often for several months. Assays employing enzyme reactors are readily automated using robotics and software originally designed for high-performance liquid chromatography (HPLC) automation. Furthermore, the stability of immobilized enzymes with respect to time, temperature, and pH is almost always greater than that of soluble enzymes, due to the effects of the local environment in which the immobilized enzyme exists. For these reasons, substrate assays that employ immobilized enzyme reactors are preferred over soluble enzyme assays.

4.2. IMMOBILIZATION METHODS

Enzyme immobilization methods are classified as chemical or physical. Chemical methods involve the formation of covalent bonds between functional groups on the

Bianalytical Chemistry, by Susan R. Mikkelsen and Eduardo Cortón
ISBN 0-471-54447-7 Copyright © 2004 John Wiley & Sons, Inc.

support material (also called the matrix or the carrier) and functional groups on the enzyme. Chemical methods are subclassified as either nonpolymerizing or cross-linking methods. Nonpolymerizing methods involve the formation of covalent bonds only between enzyme and support, but not between individual enzyme molecules, while cross-linking methods allow the formation of both enzyme-support bonds as well as enzyme–enzyme cross-links. Chemical immobilization methods may be represented by the illustrations shown in Figure 4.1 (*a* and *b*).

Physical immobilization methods do not involve covalent bond formation with the enzyme, so that the native composition of the enzyme remains unaltered. Physical immobilization methods are subclassified as adsorption, entrapment, and encapsulation methods. Adsorption of proteins to the surface of a carrier is, in principle, reversible, but careful selection of the carrier material and the immobilization conditions can render desorption negligible. Entrapment of enzymes in a cross-linked polymer is accomplished by carrying out the polymerization reaction in the presence of enzyme; the enzyme becomes trapped in interstitial spaces in the polymer matrix. Encapsulation of enzymes results in regions of high enzyme concentration being separated from the bulk solvent system by a semipermeable membrane, through which substrate, but not enzyme, may diffuse. Physical immobilization methods are represented in Figure 4.1 (*c–e*).

The following sections consider each immobilization method in detail.

4.2.1. Nonpolymerizing Covalent Immobilization

Covalent immobilization methods rely on functional groups on both the enzyme and the support material for the formation of stable covalent bonds. For this reason, the choice of a support is crucial in that it determines the immobilization chemistry

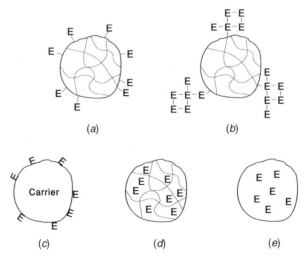

Figure 4.1. Enzyme immobilization methods. (*a*) Nonpolymerizing, (*b*) cross-linking, (*c*) adsorption, (*d*) entrapment, and (*e*) encapsulation.

TABLE 4.1. Representative Support Materials for Enzyme Immobilizations

Functional Group	Support Materials Used
$-Si(OH)_3$	Controlled-pore glass
$-CONH_2$	Polyacrylamide
$-COOH$	Polyacrylic acid
	Polyaspartic acid
	Polyglutamic acid
	Carboxymethylcellulose
$R-COO-CO-R'$	Maleic anhydride–ethylene copolymer
$-NH_2$	Polystyrene (nitrated and reduced)
	Nylon (after cleavage in HCl)
$-OH$	Cellulose
	Sephadex
	Agarose
	Sepharose
	Carboxymethylcellulose

and the stability of the enzyme-support bonds. Carriers may be activated under harsh conditions, to convert relatively unreactive functional groups into activated species that will then readily react with proteins under mild conditions of T, ionic strength, and pH. Table 4.1 gives a representative listing of support materials onto which enzymes have been immobilized, and shows the reactive functional groups present on the support that may be used for coupling.

The amino acid residues constituting the polypeptide component of enzymes provide sites at which covalent attachment to the support material may occur. Functional groups present on the side chains include primary amines (lysine), phenol hydroxyls (tyrosine), carboxylic acids (aspartate, glutamate), thiols (cysteine), hydroxyls (serine, threonine), and imidazole nitrogen (histidine). Of these, primary amines, carboxylic acids, and hydroxyl groups are most commonly used for protein immobilizations, because their hydrophilic character suggests that they will occur in an accessible area of the protein, at the aqueous interface, and will therefore be readily available for chemical reactions. Mild coupling conditions are essential in order to avoid the chemical modification of amino acid residues near the active site, as well as any change in tertiary structure that may affect the activity. It is sometimes possible to include a substrate analogue, or competitive inhibitor, in the enzyme immobilization reaction to protect the enzyme's active site from coupling reagents.

4.2.1.1. Controlled-Pore Glass.

Silanol groups on the glass surface provide sites at which silanization may be performed. Silanization may be effectively performed with a 5% aqueous solution of aminopropyltri(ethoxy)silane (APTES),[1] which results in the provision of more reactive primary amino groups for further derivatization:

$$-Si(OH)_3 + (CH_3CH_2O)_3Si(CH_2)_3NH_2 \longrightarrow -Si(OH)_2OSi(OH)_2(CH_2)_3NH_2$$
$$+ 3\,CH_3CH_2OH \qquad (4.1)$$

Since primary amine groups are reasonably reactive as nucleophiles, a number of reactions may follow this initial activation step. Dehydrating reagents, such as water-soluble carbodiimides, may be used to activate protein carboxylic acids to O-acylisourea intermediates under mild (pH 7) conditions, to allow amide bond formation with the APTES-derivatized glass.[2] If carried out in the presence of sulfonated N-hydroxysuccinimide, this reaction is highly selective for primary amine groups, but may still result in significant protein cross-linking. For more selective nonpolymerizing immobilization, however, further derivatization of the surface is necessary. Exposure to p-nitrobenzoyl chloride results in the addition of a nitrophenyl group, which is then reduced to the aminophenyl group (Eq. 4.2):

$$-RNH_2 + ClCO(C_6H_4)NO_2 \longrightarrow -RNHCO(C_6H_4)NO_2 \longrightarrow -RNHCO(C_6H_4)NH_2$$
$$(4.2)$$

One final activation step is required, and is the conversion of the aminophenyl group to the diazonium salt:

$$-RNHCO(C_6H_4)NH_2 \xrightarrow{\text{HONO}} -RNHCO(C_6H_4)N_2^+ \qquad (4.3)$$

The diazonium salt is readily prepared from the aminophenyl group, but does not result from treatment of an aliphatic amine (such as the APTES-derivatized surface) with nitrous acid. The activated surface is now ready for enzyme coupling, since diazonium salts are very reactive toward protein tyrosine residues (Eq. 4.4):

$$(4.4)$$

Although four activation steps are required prior to enzyme immobilization, this method possesses the advantages of not only providing selective covalent immobilization through tyrosines, but also of avoiding enzyme–enzyme crosslinks that can result in activity losses.[4]

4.2.1.2. Polysaccharides.
Polysaccharides possess readily accessible hydroxyl groups that may be derivatized for selective protein immobilization. We will consider the two most commonly used immobilization reactions, the triazine and the cyanogen bromide reactions.

The triazine method[5] involves the direct reaction of polysaccharide hydroxyl groups with a substituted triazine, to form a derivative that is reactive toward the lysine primary amine groups of the enzyme (Eq. 4.5). The R group on the triazine may be —Cl (cyanuric chloride), OCH_2COOH, $NHCH_2COOH$, or NH_2. The activation step is usually carried out using a nonaqueous solvent, while the enzyme coupling step is performed under mild aqueous conditions.

$$R'OH + Cl-C\underset{N-C}{\overset{N=C}{<}}\underset{R}{N} \longrightarrow R'-O-C\underset{N-C}{\overset{N=C}{<}}\underset{R}{N} \longrightarrow R'-O-C\underset{N-C}{\overset{N=C}{<}}\underset{R}{N} \quad (4.5)$$

The cyanogen bromide method[6] is not as efficient as the triazine immobilization method, since side reactions deactivate some of the activated functional groups. However, it is a method used to prepare commercially available derivatives of polysaccharides in large quantities, so that users may reproducibly immobilize proteins without performing the activation reactions. This method also employs hydroxyl groups on the polysaccharide, and reaction with cyanogen bromide produces a cyanate derivative, which rearranges to a reactive imidocarbonate group that is selective for lysine primary amino groups (Eqs. 4.6 and 4.7):

$$R\overset{OH}{\underset{OH}{<}} + CNBr \longrightarrow R\overset{OC\equiv N}{\underset{OH}{<}} \longrightarrow R\overset{O}{\underset{O}{<}}C=NH \quad (4.6)$$

$$R\overset{O}{\underset{O}{<}}C=NH \overset{E-NH_2}{\longrightarrow} R\overset{O-\overset{O}{\overset{\|}{C}}-NH-E}{\underset{OH}{<}} + R\overset{O}{\underset{O}{<}}C=N-E + R\overset{OH}{\underset{\underset{NH}{O-\overset{\|}{C}-NH-E}}{<}} \quad (4.7)$$

The side reaction that deactivates the surface involves hydrolysis of the cyanate group to form an unreactive carbamate on the polysaccharide surface (Eq. 4.8):

$$=R(OH)(OCN) + H_2O \longrightarrow =R(OH)(OCONH_2) \quad (4.8)$$

Following the initial reaction of the polysaccharide with cyanogen bromide, the support material is stored under dry conditions as the reactive imidocarbonate. Cyanogen bromide derivatives of cellulose and sephadex are commercially available from a number of suppliers. Exposure of these materials to enzyme solutions under mild conditions results in the covalent coupling of the enzyme to the support with no cross-linking.

4.2.1.3. Polyacrylamide.

Polyacrylamide possesses free amide groups that will undergo substitution with primary amines under conditions where a large excess of the primary amine is available. If reacted with neat ethylenediamine, for example, free primary amine groups are generated:[7]

$$R-CONH_2 + NH_2CH_2CH_2NH_2 \longrightarrow R-CONHCH_2CH_2NH_2 \quad (4.9)$$

Under extreme conditions, the free amine group can be activated by diazotization and then coupled to protein tyrosine residues in a mild subsequent immobilization step.

4.2.1.4. Acidic Supports.

Three methods will be considered for enzyme immobilizations onto carboxylic acid-containing supports. All result in the formation of amide bonds to the primary amines of lysine residues on the protein, but the selectivity for lysine varies with the method used. In the acyl azide method,[8] the carboxylic acid group is sequentially converted to its methyl ester, acyl hydrazide and acyl azide (Eq. 4.10):

$$R-\overset{O}{\underset{\Vert}{C}}-OH \xrightarrow[CH_3OH]{H^+} R-\overset{O}{\underset{\Vert}{C}}-OCH_3 \xrightarrow{H_2NNH_2} R-\overset{O}{\underset{\Vert}{C}}-NH-NH_2 \xrightarrow{HONO} R-\overset{O}{\underset{\Vert}{C}}-N_3 \quad (4.10)$$

The acyl azide is subject to nucleophilic attack, and is most reactive toward primary amino groups (lysine), but will also react with tyrosine hydroxyl, cysteine sulfhydryl, and serine hydroxyl groups (Eq. 4.11):

$$R-\overset{O}{\underset{\Vert}{C}}-N_3 + H_2N-Enzyme \longrightarrow R-\overset{O}{\underset{\Vert}{C}}-NH-Enzyme \quad (4.11)$$

Woodward's reagent K[9] is a water-soluble dehydrating reagent that activates carboxylic acid groups toward nucleophilic attack. It may be used under mild, aqueous conditions, and sequential activation and coupling steps prevent enzyme cross-linking (Eqs. 4.12 and 4.13):

$$(4.12)$$

$$(4.13)$$

If cross-linking is not expected to occur, then Woodward's reagent K may be included in an enzyme solution, and a single-step immobilization reaction may be performed.

Carbodiimide dehydrating reagents[10] activate carboxylic acid groups toward nucleophilic attack by the formation of an O-acylisourea intermediate (Eq. 4.14):

$$\begin{array}{ccc} \underset{\underset{R-C-OH}{\|}}{O} & + & \underset{\underset{C}{\|}}{\overset{NR'}{C}} \\ & & \underset{NR''}{} \end{array} \longrightarrow \underset{\underset{R-C-O-C}{\|}}{\overset{O}{}\overset{NHR'}{}} \qquad (4.14)$$

The intermediate is reactive toward lysine, tyrosine, cysteine, serine, and methionine, to result in amide, ester, and thioester linkages. The selectivity of this reaction is significantly improved by adding an N-hydroxysuccinimide derivative during the activation step, resulting in the formation of an N-hydroxysuccinimide (NHS) ester (Eq. 4.15):

$$\qquad (4.15)$$

The NHS ester requires a very strong nucleophile, such as a primary amine group, for displacement, and is selective for the formation of amide bonds to lysine residues.

4.2.1.5. Anhydride Groups.
Anhydride groups present on commercially available ethylene–maleic anhydride copolymers are readily reactive with the nucleophile-containing amino acid residues of proteins.[11] These groups react directly with lysine residues, for example, to form amide bonds (4.16):

$$\qquad (4.16)$$

4.2.1.6. Thiol Groups.
Thiol groups generated on the surface of a polymeric support material[12] allow the chemically reversible immobilization of enzymes, through a redox process involving the formation of disulfide bonds with cysteine residues:

$$-RCH_2SH + Enzyme-SH + 2[Fe(CN)_6]^{3-} \longrightarrow -RCH_2S-S-Enzyme$$
$$+ 2[Fe(CN)_6]^{4-} + 2H^+ \qquad (4.17)$$

With this scheme, the subsequent addition of a disulfide reducing agent, such as mercaptoethanol or dithiothreitol, allows the recovery of the immobilized enzyme.

Table 4.2 shows a summary of the selectivities of the immobilization reactions discussed in Section 4.2.1.

TABLE 4.2. Selectivity of Immobilization Methods

Method	Amino Acid Residues Modified
Diazotization	Lys, His, Tyr, Arg, Cys
Triazine	Lys
Cyanogen bromide	Lys
Acyl azide	Lys, Tyr, Cys, Ser
Carbodiimide	Lys, Tyr, Cys, Ser, Met
Carbodiimide + NHS	Lys
Anhydride	Lys
Disulfide	Cys

4.2.2. Cross-Linking with Bifunctional Reagents

Cross-linking reagents are used for the covalent immobilization in cases where the enzyme is available in large quantities, and retention of a high percentage of the original activity is not crucial. Cross-linking usually results in a high quantity of immobilized protein, since bonds between enzyme molecules form in addition to enzyme-support bonds. However, loss of activity on immobilization is generally higher if cross-linking occurs, since the polymeric network of enzyme that forms may obscure the active sites, slow the diffusion of substrates and products, and distort the tertiary structure of the enzyme. Cross-linking immobilizations are popular because of their speed and simplicity.

Cross-linking reagents consist of two reactive functional groups, separated by a spacer group. The reactive groups may be identical, such as in glutaraldehyde [OHC—$(CH_2)_3$ —CHO], in which case it is called a homobifunctional reagent. If the two functional groups are different, the reagent is a heterobifunctional cross-linker.

Common homobifuctional cross-linking reagents are shown in Figure 4.2. They include glutaraldehyde,[13] which is selective for primary amines, forming imide linkages that may be subsequently reduced to secondary amines by borohydride reduction; diazobenzidine[14] is reactive toward tyrosine residues, but also reacts with histidine nitrogen and lysine; hexamethylene-bis(iodoacetamide),[15] which is

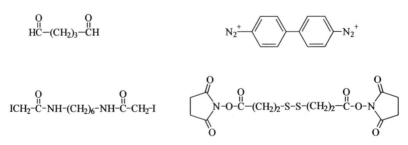

Figure 4.2. Homobifunctional cross-linking reagents.

Figure 4.3. Heterobifunctional cross-linking reagents.

selective for lysine amino groups by nucleophilic substitution of iodide; and bis(*N*-hydroxysuccinimidyl)dithiodipropionate,[16] which is not only selective for lysine amino groups, but is also a reversible cross-linking reagent, since the disulfide bond may be broken by common reducing agents.

Heterobifunctional cross-linking reagents are less commonly used, but provide an additional level of selectivity due to the different reactivities of the terminal groups. Some commercially available heterobifunctional reagents are shown in Figure 4.3. Reagents with isocyanate and isothiocyanate groups are selective toward primary amines, with the isocyanate group being reactive at room temperature, and the isothiocyanate reacting at elevated temperatures to form substituted ureas and thioureas, respectively.[17] Reagents possessing maleimido groups are selective toward thiols, while the NHS ester functionalities react with primary amines.[18]

4.2.3. Adsorption

Adsorption is the simplest of methods for enzyme immobilization. An enzyme solution is incubated with an adsorbent for several hours; the adsorbent is then removed and rinsed with a buffer. Common adsorbents[19] include alumina, carbon or charcoal, ion-exchange resins, clays, collagen, metals, and glass. Adsorptive interactions may be ionic, polar or hydrogen bonding, or they may involve hydrophobic or aromatic stacking interactions. These interactions are all noncovalent, and are, in principle, reversible. With adsorptive immobilizations, the activity yield of the immobilized enzyme is not directly related to the amount of protein adsorbed, since drastic conformational changes can occur upon adsorption that cause significant losses in activity.

For practical applications, essentially irreversible adsorption is required to prevent leaching, or desorption, which leads to activity loss. Optimum conditions are determined empirically, by varying pH, temperature, ionic strength, and quantities of protein and adsorbent. Desorption of enzymes may also be substrate induced, where high substrate concentrations cause leaching due to conformational changes.

4.2.4. Entrapment

Entrapment, also called inclusion, occlusion, and lattice entrapment, involves the formation of a highly cross-linked polymer network in the presence of an enzyme, so that the enzyme is trapped in interstitial spaces. Smaller species, such as substrates and products, freely diffuse through the polymer network, while the large size of enzyme molecules prevent their leaching through diffusion. The most commonly used entrapment network is the polyacrylamide gel,[20] which is formed by cross-linking acrylamide with N,N'-hexamethylenebis(acrylamide) in the presence of potassium persulfate in the absence of oxygen, shown below (Eq. 4.18):

$$\underset{\substack{\text{H}_2\text{C=CH}\\|\\ \text{C=O}\\|\\ \text{NH}_2}}{} + \underset{\substack{\text{H}_2\text{C=CH}\\|\\ \text{C=O}\\|\\ \text{NH}\\|\\ \text{CH}_2\\|\\ \text{NH}\\|\\ \text{C=O}\\|\\ \text{H}_2\text{C=CH}}}{} \xrightarrow[\text{(no O}_2)]{\text{K}_2\text{S}_2\text{O}_8} \underset{\substack{\text{--H}_2\text{C-CH-H}_2\text{C-CH--}\\|\qquad\qquad|\\ \text{C=O}\qquad\text{C=O}\\|\qquad\qquad|\\ \text{NH}_2\qquad\text{NH}\\|\\ \text{CH}_2\\|\\ \text{NH}\\|\\ \text{C=O}\\|\\ \text{--H}_2\text{C-CH--}}}{} \qquad (4.18)$$

The resulting cross-linked polyacrylamide possesses characteristics that are determined by the quantities of monomer and cross-linker used. The [monomer]/[cross-linker] ratio determines the pore size in which enzyme is entrapped. The total [monomer] + [cross-linker] quantities used will determine the so-called mechanical properties of the gel: its stability and rigidity.

For example, cholinesterase may be entrapped in a gel prepared from 5% cross-linker and 15% monomer in aqueous solution.[21] Under these conditions, 56% of the total enzyme activity was retained. At higher total [monomer] + [cross-linker], the enzyme denatures, while higher [cross-linker] values yielded less entrapped enzyme.

4.2.5. Microencapsulation

Enzymes may be immobilized by encapsulation in nonpermanent (e.g., liposomes) or permanent (e.g., nylon) microcapsules. The enzyme is trapped inside by a semipermeable membrane, where substrates and products are small enough to freely diffuse across the boundary. While nonpermanent microcapsules are useful in biochemical research, only permanent microencapsulations yield analytically useful systems, because of their mechanical stability.

Nylon microcapsules are formed by the interfacial polymerization of hexamethylenediamine with sebacoyl chloride.[22] The hexamethylenediamine is initially present in an aqueous enzyme solution, while the sebacoyl chloride is present in an organic phase such as 4:1 hexane/chloroform. The two solutions are mixed to form an emulsion. The nylon-6,10 membrane forms around the emulsified microdroplets, at the organic-aqueous interface:

$$\text{H}_2\text{N(CH}_2)_6\text{NH}_2 + \text{ClCO(CH}_2)_8\text{COCl} \xrightarrow{-\text{HCl}} -\text{NH(CH}_2)_6\text{NHCO(CH}_2)_8\text{CO}-$$
$$(4.19)$$

The microcapsules are then rinsed and transferred to an aqueous solution containing a water-soluble surfactant, to prevent aggregation. This immobilization method requires a high-protein concentration in an aqueous phase (~ 10 mg/mL), resulting in a high osmotic pressure that prevents the collapse of the microcapsules, and also acts as an emulsifier. Particles are spherical, with diameters ranging from 2 μm to several millimeters (mm), while pore sizes are of the order of nanometers (nm).

4.3. PROPERTIES OF IMMOBILIZED ENZYMES

The behavior of immobilized enzymes differs from that of dissolved enzymes because of the effects of the support material, or matrix, as well as conformational changes in the enzyme that result from interactions with the support and covalent modification of amino acid residues. Properties observed to change significantly upon immobilization include specific activity, pH optimum, K_m, selectivity, and stability.[23] Physical immobilization methods, especially entrapment and encapsulation, yield less dramatic changes in an enzyme's catalytic behavior than chemical immobilization methods or adsorption. The reason is that entrapment and encapsulation result in the enzyme remaining essentially in its native conformation, in a hydrophilic environment, with no covalent modification.

When an enzyme is immobilized onto a support material, a diffusion layer is created around the particle so that, even with vigorous stirring, substrates must be transported from the bulk of the solution across this stagnant layer to reach the enzyme. Products must also diffuse from the particle surface to the bulk solution. Steric repulsion of substrates and products may also occur, since the enzyme-support system is a crowded molecular environment. Steric factors are particularly significant for enzymes that have high molecular weight substrates, since accessibility of the active site is less easily achieved following immobilization. Particle size is also an important consideration: smaller particles have been observed to yield catalytic properties more closely approximating those obtained with soluble enzymes. The flexibility of polymeric support materials also plays a role in determining the transport properties of substrates and products. Hydrophilic supports have been shown to have less drastic effects on enzyme properties than hydrophobic support materials. Ionic groups on polymeric supports have been shown to interact with enzymes, and also to affect local pH in the microenvironment of the polymer network.

Chemical immobilization methods may alter the local and net charges of enzymes, through covalent modification of charged residues such as lysine (NH_4^+), aspartate, and glutamate (COO^-). Conformational changes in secondary and tertiary protein structure may occur as a result of this covalent modification, or as a result of electrostatic, hydrogen-bonding or hydrophobic interactions with the support material. Finally, activity losses may occur as a result of the chemical transformation of catalytically essential amino acid residues.

The specific activity of an enzyme almost always decreases on immobilization. The active sites are less accessible to substrate, and the diffusion of substrates and products across the stagnant layer of solution at the particle surface, and within polymer networks, lowers apparent values of V_{max} and raises apparent K_m values. The activity of an immobilized enzyme should be expressed as specific activity

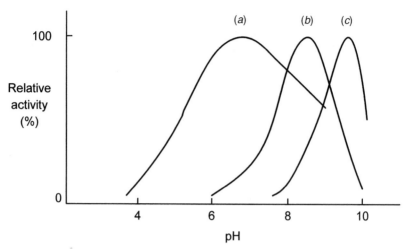

Figure 4.4. Relative activity versus pH for (*b*) native chymotrypsin (optimum pH 8.6) and for chymotrypsin bound to (*a*) polyornithine (optimum pH 7.0), (*c*) ethylene–maleic anhydride copolymer (optimum pH 9.4). [Reprinted, with permission, from L. Goldstein and E. Katchalski, *Fresnius's Z. Anal. Chem.* **243**, 1968, 375–396. "Use of Water-Insoluble Enzyme Derivatives in Biochemical Analysis and Separation".© 1968 by Springer.]

(i.e., units per mg protein) rather than activity per unit weight of the support plus protein. The reason is that the quantity of protein immobilized onto a support varies from a few micrograms to hundreds of milligrams per gram of support material, depending on the immobilization method and the support material chosen. The highest specific activities for immobilized enzymes have been achieved with hydrophilic support materials.

The optimum pH for an enzymatic reaction may shift by as much as 3 pH units upon immobilization.[24] This shift is a result of both the charge of the support material and the chemical modification of the enzyme. Figure 4.4 illustrates the dramatic shifts that occur in the pH optimum of chymotrypsin, a proteolytic enzyme, following covalent immobilizations onto a polyornithine (positive) carrier and an ethylene–maleic anhydride copolymer (negative).

This shift in the pH optimum can be explained by an uneven distribution of H^+ and OH^- between the bulk of the external solution and the polyelectrolyte (carrier) phase. Polyornithine, shown in Figure 4.5, possesses side chains with primary amino groups that are used for the covalent attachment of enzyme. Not all of these groups will react, and those that remain underivatized do so either for steric reasons or due to incomplete coupling reaction steps.

The primary amine groups that remain underivatized exist in the protonated form, and these positively charged $-NH_3^+$ groups attract OH^- from the bulk solution. This increases the local $[OH^-]$ relative to that in the bulk, so that

$$[OH^-]_{surface} > [OH^-]_{bulk} \quad \text{and} \quad pH_{surface} > pH_{bulk}$$

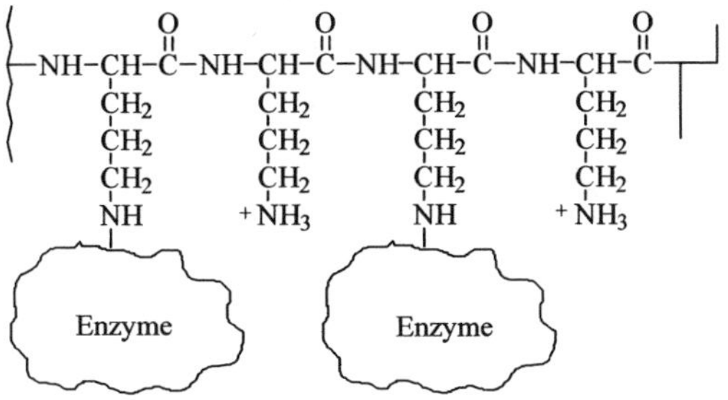

Figure 4.5. Enzyme immobilized onto a polyornithine carrier.

The *apparent* pH optimum therefore occurs at lower measured pH values, and this is observed in Figure 4.4 for chymotrypsin: The apparent pH optimum occurs at pH 7.0 instead of the native value of 8.6 following immobilization onto polyornithine.

When chymotrypsin is immobilized onto the ethylene–maleic anhydride copolymer, one acidic group is created on the surface for each enzyme molecule immobilized, as shown in Figure 4.6. This and other negatively charged supports attract H^+ from the bulk of the solution to the surface of the carrier, so that the local pH at the surface is lower

$$[H^+]_{surface} > [H^+]_{bulk} \quad \text{and} \quad pH_{surface} < pH_{bulk}$$

Because the local pH at the surface of the support is lower than the bulk, or measured pH, the *apparent* pH optimum shifts to higher pH values with this and other negatively charged support materials. Figure 4.4 shows that chymotrypsin immobilized onto an ethylene–maleic anhydride support exhibits a pH optimum of 9.4, almost one full pH unit higher than that observed for the native enzyme.

The increase in apparent K_m values observed following the immobilization of enzymes is also readily explained by considering local effects at the carrier surface. Recalling the Michaelis–Menten equation ($v = V_{max}[S]/\{K_m + [S]\}$), and its derivation (Chapter 2), we know that for soluble enzymes, K_m is independent of enzyme concentration and is a constant under a given set of conditions. Immobilized enzymes suspended in an aqueous medium have an unstirred solvent layer surrounding them, called the Nernst or diffusion layer. Substrates and products must diffuse across this layer, and, as a result, a *concentration gradient* is established for both substrates and products, as shown in Figure 4.7.

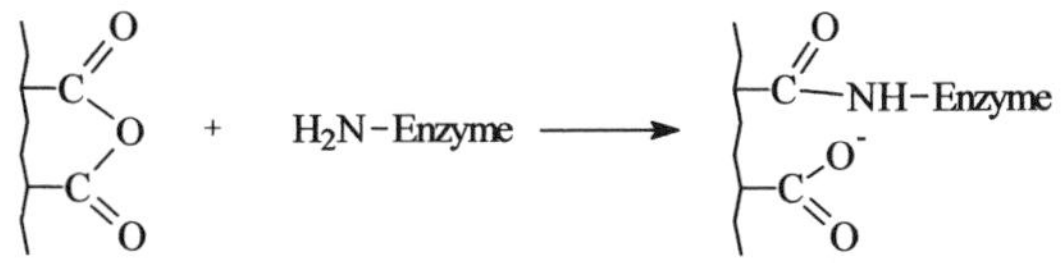

Figure 4.6. Enzyme bound to an ethylene–maleic anhydride carrier.

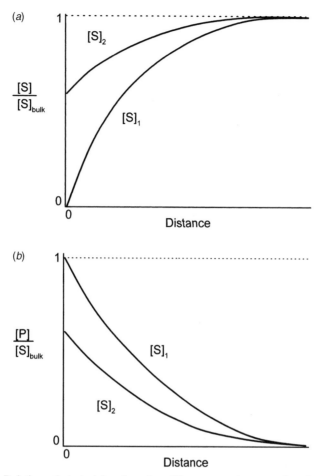

Figure 4.7. Relative substrate (*a*) and product (*b*) concentrations as a function of distance from the surface of a support particle. $[S]_{2,\text{bulk}} > [S]_{1,\text{bulk}}$ and $[P]_{\text{bulk}} = 0$.

The effects of this concentration gradient are most significant at low bulk concentrations of the substrate, since substrate is converted to product as soon as it reaches the surface of the particle, so that the surface concentration of substrate is zero. At very high bulk substrate concentrations, the enzymatic reaction rate is limited by enzyme kinetics rather than mass transport, so that surface concentrations do not differ significantly from those in the bulk. Because of the concentration gradient, however, enzyme saturation with substrate occurs at much higher bulk substrate concentrations than required to saturate the soluble enzyme. Apparent K_m values (K'_m) for immobilized enzymes are larger than K_m values obtained for the native soluble enzymes.

Diffusion layer thickness may be reduced by using smaller particles or by increasing the rate of stirring of the solution; K'_m values will then approach the

K_m values observed for the soluble enzyme. Remember that electrostatic and steric factors may also affect K'_m values if they lead to local concentrations that differ from bulk concentrations.

The selectivities of enzymes that catalyze reactions involving high molecular weight substrates have been found to change when these enzymes are immobilized, because the diffusion of macromolecular substrates is slower, and because steric factors lower the activity of the enzyme by preventing free access of substrate to the enzyme's active site. For example, the enzyme ribonuclease (RNase) catalyzes the hydrolytic cleavage of phosphodiester bonds linking the nucleotides of polymeric RNA (Eq. 4.20):

$$(4.20)$$

RNase will also catalyze phosphodiester bond cleavage in low molecular weight substrates, such as cyclic cytidine monophosphate (cCMP) to produce 5'-CMP (Eq. 4.21):

$$(4.21)$$

Following covalent immobilization onto the hydroxyl groups of agarose, RNase showed decreased activity toward RNA cleavage, when compared to the soluble enzyme.[25] In order to make the comparison, it was assumed that the rate of cCMP cleavage would not be affected significantly by the immobilization. Normalized rates could then be calculated as rates of RNA cleavage divided by the rate of cCMP cleavage. These rates were measured for RNA substrates over a wide range of molecular weights (MW), and an empirical correlation was observed between lower reaction rates and higher molecular weights:

$$\frac{\text{normalized rate with soluble enzyme}}{\text{normalized rate with immobilized enzyme}} = A + B \times \log(\text{MW}) \qquad (4.22)$$

This study definitively showed the effects of steric exclusion on reaction rates observed with immobilized enzymes.

Enzyme stability with respect to both storage time and denaturation temperature has generally been found to improve upon immobilization. Immobilized enzyme

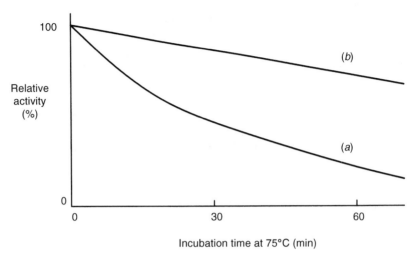

Figure 4.8. Thermal stability of soluble (*a*) and nylon-immobilized (*b*) urease at pH 7.0. [Reprinted, with permission, from P. V. Sundaram, and W. E. Hornby, *FEBS Letters* **10**, October 1970, 325–327. "Preparation and Properties of Urease Chemically Attached to Nylon Tube". © 1970 by Elsevier.]

reactors may be used for several months with little change in conversion efficiencies. Urease immobilized onto a nylon membrane has been studied with respect to its thermal denaturation properties. Figure 4.8 illustrates the comparative thermal stabilities of immobilized and soluble urease, determined by measuring relative activity as a function of incubation time at elevated temperatures.[26] Clearly, the immobilized urease exhibits improved thermal stability that results from the protective microenvironment at the surface of the support.

A spectacular example of stability enhancement through immobilization has been reported for the enzyme catechol-2,3-dioxygenase.[27] This enzyme, isolated from the thermophilic bacterium *Bacillus stearothermophilus*, catalyzes the conversion of catechol to 2-hydroxymuconic semialdehyde (which can be monitored by absorbance at 375 nm). The soluble enzyme exhibits maximal activity at 50 °C, but following immobilization on glyoxyl agarose beads with a borohydride reduction step, the optimum reaction temperature shifted to 70 °C. At a total protein concentration of 0.010 mg/mL and a temperature of 55 °C, the half-life of the soluble enzyme was 0.08 h, while the enzyme-modified beads had a half-life of 68 h. This represents a 750-fold enhancement of stability that has been attributed to the prevention of subunit dissociation upon immobilization.

4.4. IMMOBILIZED ENZYME REACTORS

The immobilization of enzymes onto particulate carriers that may be packed into a column (the "packed-bed" reactor), such as a typical HPLC column, facilitates

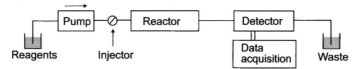

Figure 4.9. Diagram of an open enzyme reactor system.

repetitive use of the enzyme and also allows the automation of enzymatic assays.[28] Open-tubular reactors have also been constructed by covalently immobilizing an enzyme onto the inner wall of a nylon or polyethylene tube.[29] Immobilized enzyme reactors are used in conjunction with a pump, to force a buffer, or mobile phase, through the reactor at a steady rate, an injector located between the pump and the reactor to allow the introduction of substrate solutions, and a detector located close to the column exit. The optional inclusion of automatic sampling devices and data acquisition and processing systems provides a completely automated instrument for enzymatic substrate assays. The mobile phase contains all required cosubstrates and activators required for the enzymatic reaction, but does not contain the analyte substrate. A typical packed-bed system may use a 25-cm long reactor with a 5-mm inner-diameter, packed with the carrier-enzyme solid phase at high pressures. Flow rates of 0.5–2 mL/min and sample injection volumes of 10–100 μL are common. Detection involves the same principles used in homogeneous enzymatic assays, and flow-through optical absorbance and fluorescence detectors, and amperometric and potentiometric electrochemical detectors may be employed, with detector volumes of the order of tens of microliters being standard.

Enzyme reactor systems may be of the continuous flow or the stopped-flow variety. Continuous flow systems are further categorized as *open* or *closed* systems. The open system, shown in Figure 4.9, continuously pumps fresh buffer through the injector, reactor and detector, ultimately into a waste reservoir for discarding. This arrangement is preferred for the testing of enzyme reactors, since unreacted substrate, cofactors and the products of the enzymatic reactions will not be reexposed to the column.

Closed systems may be employed when buffer recycling is possible, that is when the buffer contains high concentrations of all necessary cosubstrates, when complete consumption of injected substrate occurs within the reactor, and when products of the enzymatic reaction do not inhibit the immobilized enzyme. A closed system for immobilized oxidase enzymes is shown in Figure 4.10.

Both open and closed continuous flow systems rely on the fixed time, or endpoint method for the determination of substrate concentrations. At a fixed and constant flow rate, the injected volume of substrate will spend a fixed time on the column, and this time is related to the volume of the column (that volume not occupied by stationary phase) and the mobile-phase flow rate.

Indicator reactions that are chemical in nature may be introduced either into the mobile phase or at the end of the column by the method of postcolumn reagent addition. Postcolumn addition of reagents dilutes the column eluent, so that, when possible, the addition of indicator reagents to the mobile phase is preferable.

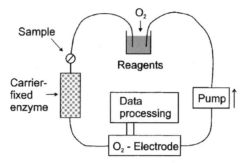

Figure 4.10. Diagram of a closed enzyme reactor system.

The conditions under which chemical indicator reactions are used often necessitates the use of postcolumn addition, however. Figure 4.11 shows an experimental setup for urea assays using an immobilized urease reactor.[30] The postcolumn addition of sodium hydroxide allows the NH_4^+ produced by the reactor to be detected as NH_3 at an ammonia gas-sensing electrode placed in a flow cell.

If the indicator reaction is also enzyme catalyzed, then two methods may be used for the conversion of primary reaction product into the detected species. The first involves the use of a second reactor column containing immobilized indicator enzyme between the primary column and the detector, in a dual-reactor system. The second, and more common approach, employs a single reactor containing coimmobilized primary and indicator enzymes. For linear conversion of primary product, it must be remembered that the quantity of indicator enzyme present must be sufficient to for the complete conversion of all primary product, so that an excess of immobilized indicator enzyme must be present.

These systems are designed for the endpoint determination of substrate concentrations, and do not provide a straightforward means for making kinetic measurements. Stopped-flow enzyme reactor systems have been designed for automated kinetic assays. A diagram of a stopped-flow reactor that uses a postcolumn chemical indicator reaction is shown in Figure 4.12.[31] In this system, the flow rate of the

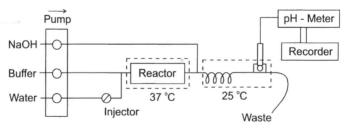

Figure 4.11. Enzyme reactor system for urea based on immobilized urease and potentiometric detection.[30]

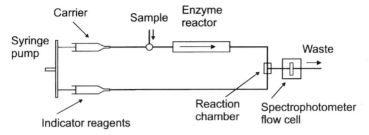

Figure 4.12. Stopped-flow enzyme reactor with absorbance detection.

mobile phase through the reactor dictates the residence time of the analyte on the column.

4.5. THEORETICAL TREATMENT OF PACKED-BED ENZYME REACTORS[32,33]

Packed-bed enzyme reactors, those employing enzymes immobilized onto a particulate phase that is subsequently packed into a column, may be characterized by their column capacity, C, and the degree of reaction P. The parameter C is defined in Eq. 4.23,

$$C = k E_t \beta \tag{4.23}$$

where k is the decomposition rate constant for the enzyme–substrate complex (either k_2 or k_{cat}), E_t is the total number of moles of enzyme immobilized, and the value of β is a constant for a given reactor, and is equal to the ratio of reactor void volume to total reactor volume (i.e., β is always less then unity). The degree of reaction, P, varies between zero (no product formed) and unity (complete conversion of substrate).

An equation equivalent to the Michaelis–Menten equation has been derived for immobilized enzymes in packed-bed reactor systems, and is given in Eq. 4.24:

$$P[S]_0 = K'_m \ln\{1 - P\} + C/Q \tag{4.24}$$

where Q is the volume flow rate of the mobile phase. In general, this equation predicts that for a given column capacity, the degree of reaction, P, is inversely related to the mobile-phase flow rate, Q. That is, the faster the analyte plug flows through the reactor, the less likely will be its complete conversion into product.

Equation 4.24 has been experimentally tested using a carboxymethylcellulose stationary phase onto which the enzyme ficin (E.C.3.4.22.3) has been immobilized. Ficin catalyzes the hydrolysis of a variety of ester substrates, but the substrate N-benzoyl-L-arginine ethyl ester (BAEE) has been chosen for this series of experiments (Eq. 4.25):

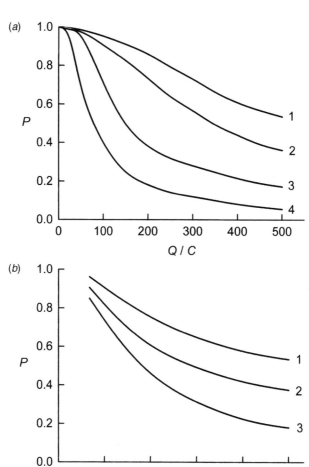

$$(4.25)$$

Figure 4.13. Theoretical (a) and experimental (b) dependence of degree of reaction on flow rate. (a) $K'_m = 2.5$ mM: curve 1, $[S]_0 = 0.1\,K'_m$; curve 2, $[S]_0 = K'_m$; curve 3, $[S]_0 = 4\,K'_m$; curve 4, $[S]_0 = 10\,K'_m$. (b) $[S]_0 = 0.5$, 5.0, and 15 mM for curves 1, 2, and 3, respectively.[32] [Reprinted, with permission, from M. D. Lilly, W. E. Hornby, and E. M. Crook, *Biochem. J.* **100**, 1966, 718–723. "The Kinetics of Carboxymethylcellulose-Ficin in Packed Beds". © 1966 by The Biochemical Society, London.]

The effect of flow rate on degree of conversion has been determined by using a constant initial substrate concentration $[S]_0$. Figure 4.13 shows the dependence of P on Q/C as predicted by Eq. 4.24, and the experimental dependence of P on Q found for the immobilized ficin reactor. Good agreement is observed between the theoretical and the experimental results, and that complete conversion of substrate into product requires low flow rates and a high column capacity. The theoretical curves use constant $[S]_0$ values that are a multiple of K_m', in order to take changes in K_m' with flow rate into account.

As noted earlier in this chapter, the apparent K_m values of immobilized enzymes vary with the thickness of the diffusion layer surrounding the particles. In packed-bed enzyme reactors, the thickness of this layer varies with the mobile phase flow rate. Faster flow rates produce smaller diffusion layers and therefore K_m' values that more closely approximate the true K_m of the enzyme. This effect has also been observed with the ficin–CM–cellulose reactor, and plots of K_m' against flow rate Q obtained at different mobile phase flow rates are shown in Figure 4.14.

For packed-bed enzyme reactors, these results show that low flow rates ensure quantitative conversion of substrate into detectable products. The variation of K_m' with flow rate indicates that lower flow rates produce higher K_m' values, so that the linear region of the saturation kinetic curve extends to higher substrate concentrations at lower flow rates. This effect becomes significant when complete conversion does not occur during the residence time of the analyte.

A more recent study involving glucose oxidase immobilized by physical entrapment in hydroxyethyl methacrylate-based hydrogels has shown that Eq. 4.24 can

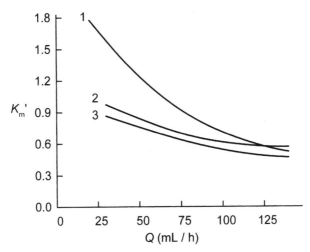

Figure 4.14. Effect of flow rate on K_m' of immobilized ficin for BAEE at ionic strength 0.04 M (curve 1), 0.1 M (curve 2), and 0.4 M (curve 3).[32] [Reprinted, with permission, from M. D. Lilly, W. E. Hornby, and E. M. Crook, *Biochem. J.* **100**, 1966, 718–723. "The Kinetics of Carboxymethylcellulose-Ficin in Packed Beds". © 1966 by The Biochemical Society, London.]

also be applied to this two-substrate immobilized enzyme.[34] Linear plots of $P[S]_0$ against $\ln(1 - p)$ were obtained, and K'_m values were calculated from the slopes of these plots. However, the cosubstrate (dissolved oxygen) concentration was not measured or controlled, and because of this, the measured glucose K'_m values *increased* from 2.95 mM at $Q = 0.33$ mL/min to 11.3 mM at $Q = 1.40$ mL/min. This illustrates the importance of using an appropriate kinetic model to understand the behavior of packed-bed enzyme reactors.

SUGGESTED REFERENCES

O. Zaborsky, *Immobilized Enzymes*, CRC Press, Cleveland, OH, 1973.

L. B. Wingard, Jr., E. Katchalski-Katzir, and L. Goldstein, Eds., *Applied Biochemistry and Bioengineering, Volume 1: Immobilized Enzyme Principles*, Academic Press, New York, 1976.

L. B. Wingard, Jr., E. Katchalski-Katzir, and L. Goldstein, Eds., *Applied Biochemistry and Bioengineering, Volume 3: Analytical Applications of Immobilized Enzymes and Cells*, Academic Press, New York, 1981.

P. W. Carr and L. D. Bowers, *Chemical Analysis, Volume 56: Immobilized Enzymes in Analytical and Clinical Chemistry*, John Wiley and Sons, Inc., New York, 1980.

G. G. Guilbault, *Analytical Uses of Immobilized Enzymes*, Pergamon Press, New York, 1984.

R. L. Lundblad, *Chemical Reagents for Protein Modification, 2nd ed.*, CRC Press, Boca Raton, FL, 1991.

Y. Lvov and H. Moehwald, Eds., *Protein Architecture: Interfacing Molecular Assemblies and Immobilization Biotechnology*, Marcel Dekker, New York, 2000.

REFERENCES

1. S. G. Caravajal, D. E. Leyden, G. R. Quinting, and G. E. Marciel, *Anal. Chem.* **60**, 1988, 1766–1786.

2. W. F. Line, A. Kwong, and H. H. Weetall, *Biochim. Biophys. Acta* **242**, 1971, 194–202.

3. H. H. Weetall, *Nature (London)* **223**, 1969, 959–960.

4. E. C. Hatchikian and P. Monsan, *Biochem. Biophys. Res. Commun.* **92**, 1980, 1091–1096.

5. S. Y. Shimizu and H. M. Lenhoff, *J. Solid Phase Biochem.* **4**, 1979, 75–94.

6. J. Kohn and M. Wilchek, *Appl. Biochem. Biotechnol.* **9**, 1984, 285–305.

7. J. K. Inman and H. M. Dintzis, *Biochemistry* **8**, 1969, 4074–4082.

8. P. R. Coulet, R. Sternberg, and D. R. Thevenot, *Biochim. Biophys. Acta* **612**, 1980, 317–327.

9. R. P. Patel, D. V. Lopiedes, S. P. Brown, and S. Price, *Biopolymers* **5**, 1967, 577–582.

10. K. L. Carraway and D. E. Koshland, Jr., *Methods Enzymol.* **25**, 1972, 616–623.

11. R. A. Zingaro and M. Uziel, *Biochim. Biophys. Acta* **213**, 1970, 371–379.

12. G. P. Royer, S. Ikeda, and K. Aso, *FEBS Lett.* **80**, 1977, 89–94.

13. A. F. S. A. Habeeb and R. Hiramoto, *Arch. Biochem. Biophys.* **126**, 1968, 16–26.

14. I. H. Silman, M. Albu-Weissenberg, and E. Katchalski, *Biopolymers* **4**, 1966, 441–448.

15. H. Ozawa, *J. Biochem.* **62**, 1967, 531–536.

16. A. J. Lomant and G. Fairbanks, *J. Mol. Biol.* **104**, 1976, 243–261.

17. Y. K. Cho and J. E. Bailey, *Biotechnol. Bioeng.* **21**, 1979, 461–476.

18. E. Ishikawa, S. Hashida, T. Kohno, and K. Tanaka, in *Nonisotopic Immunoassay*, T. T. Ngo, Ed., Plenum Press, New York, 1988. pp. 27–55.

19. A. Rosevear, J. F. Kennedy, and J. M. S. Cabral, *Immobilized Enzymes and Cells*, IOP Publishing, Philadelphia, 1987, pp. 83–97.

20. M. D. Trevan and S. Grover, *Trans. Biochem. Soc.* **7**, 1979, 28–30.

21. Y. Degani and T. Miron, *Biochim. Biophys. Acta* **212**, 1970, 362–364.

22. H. E. Klei, D. W. Sundstrom, and D. Shim, in *Immobilized Cells and Enzymes: A Practical Approach*, J. Woodward, Ed., Oxford University Press, New York, 1985, pp. 49–54, and references cited therein.

23. O. R. Zaborski, *Immobilized Enzymes*, CRC Press, Cleveland, OH, 1973, pp. 49–60.

24. L. Goldstein and E. Katchalski, *Fresenius' Z. Anal. Chem.* **243**, 1968, 375–396.

25. R. Axen, J. Carlsson, J.-C. Janson, and J. Porath, *Enzymologia* **41**, 1971, 359–364.

26. P. V. Sundaram and W. E. Hornby, *FEBS Lett.* **10**, 1970, 325–327.

27. R. Fernandez-Lafuente, J. M. Guisan, S. Ali, and D. Cowan, *Enz. Microb. Technol.* **26**, 2000, 568–573.

28. G. G. Guilbault, *Handbook of Enzymatic Methods of Analysis*, Marcel Dekker, New York, 1976, pp. 510–517.

29. T. Ngo, *Int. J. Biochem.* **11**, 1980, 459–465.

30. G. G. Guilbault and J. Das, *Anal. Biochem.* **33**, 1970, 341–355.

31. M. D. Joseph, D. Kasprzak, and S. R. Crouch, *Clin. Chem.* **23**, 1977, 1033–1036.

32. M. D. Lilly, W. E. Hornby, and E. M. Crook, *Biochem. J.* **100**, 1966, 718–723.

33. G. Johansson, *Appl. Biochem. Biotechnol.* **7**, 1982, 99–106.

34. A. Giuseppi-Elie, N. F. Sheppard, S. Brahim, and D. Narinesingh, *Biotechnol. Bioeng.* **75**, 2001, 475–484.

PROBLEMS

1. Outline the steps necessary for the preparation of a packed-bed enzyme reactor for cholesterol, using the enzyme cholesterol oxidase and polystyrene stationary-phase particles. In the absence of any indicator reaction, what kind of detector could be used to directly measure the hydrogen peroxide produced in the reactor? Suggest how the system might be modified so that a flow-through absorbance detector could be used.

2. Suggest how the covalent immobilization of urease onto particles of an ethylene–maleic anhydride copolymer might be expected to change (a) its apparent pH optimum, and (b) its apparent K_m for urea.

3. Figure 4.15 shows a new type of enzyme reactor that has been devised for assays of choline and acetylcholine, using the enzymes acetylcholinesterase and choline

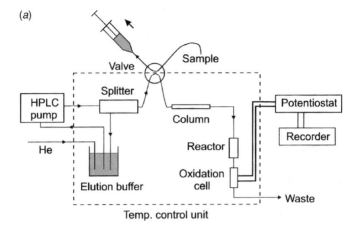

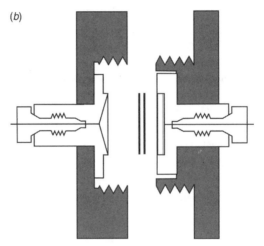

Figure 4.15

oxidase. The reactions catalyzed by these enzymes are shown below:

$$\text{Acetylcholine} + H_2O \xrightarrow{\text{acetylcholinesterase}} \text{Choline} + \text{Acetic Acid}$$

$$\text{Choline} + O_2 \xrightarrow{\text{choline oxidase}} \text{Betaine} + H_2O_2$$

In this reactor, the enzyme solution is retained between the two 0.04-μm pore membranes, and the membranes are fitted into the flow cell. A pump and injector are placed at the upstream end of the reactor. Injected acetylcholine and choline react with the enzyme(s), producing hydrogen peroxide. The H_2O_2 is detected downstream of the reactor, at a platinum working electrode poised at $+700$ mV versus SCE, so that H_2O_2 is converted to O_2.

(a) Is this a chemical or physical immobilization method? Would this type of immobilization be expected to affect the apparent K_m or pH optima values of the enzymes? Why or why not?

(b) Sketch the shape of the response current versus [acetylcholine] calibration curve, for very low concentrations up to values well above the K'_m of acetylcholinesterase.

(c) This reactor may be used to examine the properties of reversible acetylcholinesterase inhibitors, by injecting samples containing a fixed concentration of the suspected inhibitor along with increasing concentrations of acetylcholine. Sketch the Lineweaver-Burk plots of 1/current versus 1/[acetylcholine] that would be obtained in the absence and in the presence of a reversible, competitive acetylcholinesterase inhibitor.

4. At pH 6, the enzyme fumarase (50 kDa/subunit) converts fumarate to malate, and has a soluble K_m for fumarate of 5.7 μM. The dissociation rate constant for the enzyme–substrate complex into enzyme plus malate has been measured as $1.45 \times 10^3 \text{ s}^{-1}$.

(a) Give two methods by which fumarase can be covalently immobilized onto controlled-pore glass beads.

(b) The glass beads from (a) were found to possess 0.10-g enzyme/g total mass. A column 25 cm long × 5 mm i.d. was packed with 3.0 g of this product. The total volume of the column, 5.0 mL, was found to have a void volume of 2.0 mL. Calculate the capacity of this packed-bed enzyme reactor, in mol/min.

(c) Calculate the maximum flow rates that may be employed with this reactor, to achieve at least 95% conversion of fumarate into malate, when $[S]_0 = 0.01 K'_m$, $[S]_0 = 0.10 K'_m$, and $[S]_0 = K'_m$.

CHAPTER 5

Antibodies

5.1. INTRODUCTION

The enzymatic methods described in Chapters 2–4 are important bioanalytical tools for the quantitation of those biochemicals participating in enzyme-catalyzed reactions. A vast number of biochemicals that are of analytical interest, however, are not amenable to enzymatic assay techniques. These species are not involved in metabolic processes for which a sufficiently selective enzyme exists, and may not even be species that are commonly found in living systems. For many of these analytes, whether macromolecular or of relatively low molecular weight, *immunoassays* are the methods of choice.

Immunoassays rely on the selective binding properties of *antibodies*, large glycoproteins of molecular weight 150 kDa, that possess two identical binding sites per molecule and are thus called bivalent. They are produced in living organisms (excluding plants) via the immune response, in response to an *immunogen*, which may be defined as any agent capable of eliciting an immune response. Immunogenic compounds have high molecular weights (> 1 kDa), have chemical complexity (e.g., polyethylene glycol is not immunogenic) and are foreign to the individual organisms.[1] Contact with an immunogen triggers a chain of events that leads to the activation of lymphocytes (white blood cells) and the synthesis of antibodies that selectively bind the immunogen. A schematic representation of the immune response is shown in Figure 5.1.

Once antibodies have been generated, their selective binding properties toward the high molecular weight immunogens allow their interaction with other species that possess certain structural similarities. *Antigens* are species that are capable of binding selectively to components of the immune response, such as antibodies (humoral immunity) and lymphocytes (cellular immunity). Antigens are not necessarily capable of generating the immune response themselves; thus, all antigens are not immunogens, while all immunogens are also antigens. *Haptens* are low molecular weight compounds that can only elicit an immune response when they are chemically bound to a high molecular weight compound, such as a carrier protein.

Bianalytical Chemistry, by Susan R. Mikkelsen and Eduardo Cortón
ISBN 0-471-54447-7   Copyright © 2004 John Wiley & Sons, Inc.

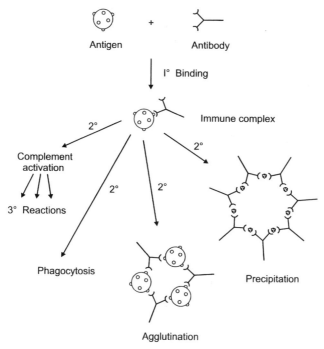

Figure 5.1. Simplified representation of the immune response.

Immunoassay methods exploit the selective recognition and binding properties of antibodies for the recognition and quantitation of antigens or haptens.

5.2. STRUCTURAL AND FUNCTIONAL PROPERTIES OF ANTIBODIES

When the red blood cells are removed from whole blood, and the resulting plasma is allowed to clot, the fibrinogen may be removed to yield *serum*. Serum contains a variety of proteins, some of which are called *globulins* because their solubility properties are different from the other serum proteins. Antibodies are a subclass of serum globulins that possess selective binding properties. Antibodies are also called *immunoglobulins* (Ig).

All immunoglobulins have a number of structural features in common.[2] They possess two "light" polypeptide chains, each with an approximate molecular weight of 25 kDa, and two "heavy" polypeptide chains of ~ 50 kDa each. These four chains are bound together in a single antibody molecule by disulfide bonds, and form a Y-shape with a central axis of symmetry (Fig. 5.2). The two halves of a natural immunoglobulin are identical.

The N-terminal ends of the light polypeptide chains (L) occur near the top of the "Y" structure, in the so-called F_{ab} fragments. These are the antigen-binding

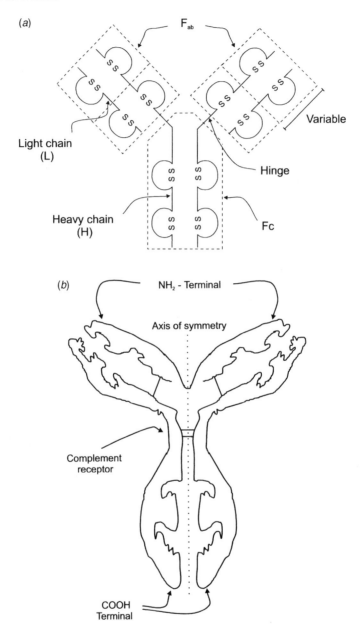

Figure 5.2. Structural diagrams representing antibody molecules.

fragments of the antibody, and have been cleaved and used in immunoassays based on primary antigen–antibody interactions in the same manner as a whole antibody molecule is used. It is the amino acid sequence of these N-terminal ends that determine the specific antigen-binding properties of the molecule. The sequence of the

TABLE 5.1. Properties of Common Immunoglobulins[a]

Immunoglobulin Class	Adult Serum Concentration	Functional Properties
IgG	12 mg/mL	Only Ig that crosses the placenta
IgE	0.3 μg/mL	Major class involved in the allergic response
IgM	1.0 mg/mL	Can cause lysis of bacteria by activating other serum components
IgA	1.8 mg/mL	Found in secretions (tears, saliva)
IgD	30 μg/mL	Found on lymphocytes

[a] See Ref. 3.

heavy chains (H) in the so-called F_c fragments (crystallizable fragments) determines the antibody class: the most abundant classes are called IgG, IgM, IgA, IgE, and IgD. Antibodies in different classes may have exactly the same antigen-binding properties, but exhibit different functional properties. The abundances and functional properties of the five major classes of antibodies[3] are listed in Table 5.1.

Antiserum is that part of the serum containing antibodies. It may be prepared from whole serum by a series of ammonium sulfate precipitation steps, where successively lower concentrations of ammonium sulfate generate a precipitate that is centrifuged and resuspended in buffer.[4] The antiserum is then dialyzed or affinity-purified, and assayed for total protein.

The selectivity of antigen–antibody (and hapten–antibody) interactions is analogous to the selectivity of substrate–enzyme interactions. The antigen-binding site of an antibody has a structure that allows a complementary fit with structural elements and functional groups on the antigen. The portion of the antigen that interacts specifically with the antigen-binding site on the antibody is called the antigenic determinant, or *epitope*. The epitope has a size of $\sim 0.7 \times 1.2 \times 3.5$ nm, which is equivalent to ~ 5–7 amino acid residues. The complementary site on the antibody is called the *paratope*, and this has about the same size.

Antigens may be classified according to their binding characteristics of valency (meaning the total number of sites) and their determinacy (meaning the number of different types of sites).[5] There are four classes of antigens:

1. Only a single epitope on the surface that is capable of binding to an antibody: *unideterminate* and *univalent*. Haptens are unideterminate and univalent.

2. Two or more epitopes of the same kind on one antigen molecule: *unideterminate* and *multivalent*.

3. Many epitopes of different kinds, but only one of each kind on one antigen molecule: *multideterminate* and *univalent*. Most protein antigens fall into this category.

4. Many epitopes of different kinds, and more than one of each kind per antigen molecule: *multideterminate* and *multivalent*. Proteins containing multiple identical subunits, as well as polymerized proteins and whole cells, fall into this category.

Binding interactions between antigen and antibody involve electrostatic, hydrophobic, and van der Waals interactions, as well as hydrogen bonding. These are all relatively weak, noncovalent forces, and for this reason, complementarity must extend over a relatively large area to allow the summation of all available interactions. Antigen–antibody association constants are often as high as $10^{10}\,M^{-1}$.

5.3. POLYCLONAL AND MONOCLONAL ANTIBODIES

Polyclonal antibodies are isolated directly from serum, by a similar procedure to the one outlined earlier in this chapter. The immunogen used to generate polyclonal antibodies may consist of a small molecule (a hapten) covalently bound to a carrier protein. The resulting antiserum will contain a mixture of antibodies that bind to different epitopes of the hapten–carrier conjugate, as well as antibodies generated in response to other immunogens present in the organism. This results in an antiserum possessing a wide range of selectivities and affinities, and may result in significant *cross-reactivities*, or interferences, when employed in immunoassays.

Monoclonal antibodies[6] are a homogeneous population of identical antibody molecules, having identical paratopes and affinity for a single antigenic epitope. They are biochemically synthesized in a procedure first described by Milstein and Kohler in the 1970s, work for which these researchers won the Nobel Prize.

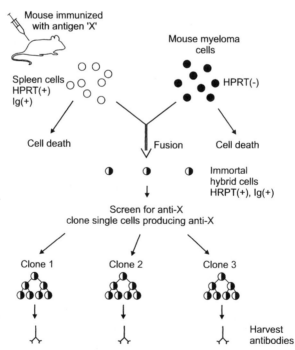

Figure 5.3. The generation of hybridoma cell lines for monoclonal antibody production.

Figure 5.3 shows an outline of Milstein and Kohler's procedure for monoclonal antibody production.

An animal immunized with an antigen (Ag) increases its population of antibody (Ab)- producing lymphocytes. These cells are short lived and can exist in cell cultures only for a few days. Malignant plasma cells are separately cultured. These cells are essentially immortal, and can be cultured for years, but they are poor producers of Abs. Some malignant plasma cells are deficient in the enzyme hypoxanthine phosphoribosyltransferase (HPRT), and will die unless HPRT is supplied to the culture medium. These are the malignant cells that are used for the production of hybridomas.

Hybridomas (or fused cell culture lines) are produced by fusing lymphocytes isolated from the spleen of an immunized animal with the HPRT deficient malignant plasma cells. The lymphocytes. The unfused cells die, since spleen cells survive only a few days, while the malignant plasma cells lack HPRT. Fused cells, however, are HPRT positive as well as immortal; these hybridomas, or hybrid cells, retain the Ab-producing characteristics of the spleen cells. Hybrid cell lines that produce a specific antibody are then cloned from the single cells and cultured. Each clone created in this manner produces antibodies of a single epitope specificity!

While monoclonal antibodies are initially almost prohibitively expensive to produce, they offer very distinct advantages over polyclonal antisera for analytical applications. Of primary importance is the immortality of the hybridoma lines: monoclonal Abs produced by this cell line will show relatively little batch-to-batch variation, and the repetitive use of animals for antibody production becomes unnecessary. In addition, the single-epitope selectivity of the monoclonal Ab tends to minimize cross-reactions, except with epitopes possessing very similar structural characteristics to the original antigen's epitope.

5.4. ANTIBODY–ANTIGEN INTERACTIONS

The interactions of antibodies with their selective binding partners, antigens or haptens, are classified as *primary* or *secondary* binding interactions. Primary interactions involve the specific recognition and combination of an antigenic determinant (epitope) with the binding site (paratope) of its corresponding antibody. All Ab–Ag interactions begin with a primary interaction, which occurs very rapidly (on a millisecond time scale), and is macroscopically invisible.

The equilibrium relationships involved in the primary binding interaction can be studied using haptens, which, due to their univalent, unideterminate nature, prevent secondary binding interactions from occurring. Free hapten exists in equilibrium with antibody-bound hapten as shown below (Eq. 5.1):

$$\text{Hapten} + \text{Antibody} \rightleftharpoons \tag{5.1}$$

Hapten
(univalent, unideterminate
antigen)

Antibody

Secondary antibody–antigen reactions occur as a result of antigen multivalency, and result in agglutination or precipitation of a polymeric antigen–antibody network. A visible product is formed over a time scale of minutes to hours (Eq. 5.2):

$$\text{Antigen A (multivalent)} + \text{Anti-A} \rightleftharpoons \tag{5.2}$$

Both primary and secondary Ag–Ab interactions may be exploited for analytical purposes, but quantitative bioanalytical methods make extensive use of primary binding interactions. The intrinsic association constant that characterizes antibody–epitope(hapten) binding is called the affinity:

$$Ab + H \rightleftharpoons Ab : H \tag{5.3}$$

$$Affinity = K = [Ab : H]/[Ab][H] \tag{5.4}$$

Typical affinity values[7] for antibody–hapten interactions range from 10^5 to $10^{12}\,M^{-1}$. Because of the very high affinities involved, the reversal of Ab—H binding may be accomplished only under extreme conditions: at high or low pH values, at high ionic strength, or in the presence of chaotropic ions (e.g., perchlorate) or denaturants (urea, guanidinium) that interfere with hydrogen bonding.

Affinity values are experimentally determined by maintaining a constant total Ab concentration, and varying the H concentration. Quantitation of $[H]_{free}$ and $[H]_{bound}$ then allows the construction of a Scatchard plot, where $[H]_{bound}/\{[H]_{free}[Ab]_{total}\}$ is plotted against $[H]_{bound}/[Ab]_{total}$. The slope of this plot is the negative of the affinity, $-K$. Scatchard plots for monoclonal and polyclonal antibodies are shown in Figure 5.4. Note that polyclonal antibodies yield a curved Scatchard plot due to their range of epitope selectivities and affinities.[8]

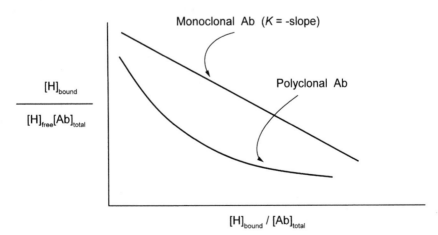

Figure 5.4. Scatchard plots for monoclonal and polyclonal antibodies.

For polyclonal antibodies, experimental K values represent an average affinity, which is usually represented as K_0.

5.5. ANALYTICAL APPLICATIONS OF SECONDARY ANTIBODY–ANTIGEN INTERACTIONS

5.5.1. Agglutination Reactions

Agglutination occurs when an antibody interacts with a multivalent, particulate antigen (i.e., an insoluble particle), resulting in cross-linking of the antigen particles by the antibody. This eventually leads to clumping, or agglutination, and will not occur with haptens. Agglutination may occur when a unideterminate, multivalent antigen (i.e., possessing many copies of single epitope "A") interacts with a single antibody (Anti-A); it may also occur if a multideterminate, univalent antigen (with epitopes A, B, C, etc.) interacts with at least two distinct antibodies (Anti-A plus Anti-B, e.g.). Whether or not agglutination occurs depends on the relative concentrations of antigens and antibodies.

The *agglutination test* is used to qualitatively test for the presence of antibodies in serum, and is also used in blood grouping. As an example, we will consider a test for the presence of antibodies to the bacterium *Brucella abortus*.[9] In this test, a series of test tubes are prepared, each containing the same quantity of the bacterial suspension, which is the antigen. To each tube is added increasingly dilute solutions of serum. If the antibody is present, agglutination will be observed over a range of serum dilutions, as shown in Figure 5.5. Two dilution zones are apparent, one called the "prozone" and the other, the "agglutination zone". In the prozone, no difference is apparent after serum is added and allowed to incubate for several minutes. In the agglutination zone, a visible precipitate forms, and settles to the bottom of the tube. The high-dilution end of the agglutination zone is called the *titer* of the serum, and is used as a semiquantitative expression of the Ab concentration for the comparison of sera. Note that tests made on sera at only one dilution may give misleading results, because of the lack of agglutination in the prozone and at high serum dilutions.

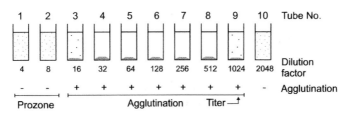

Figure 5.5. Agglutination test for *Brucella abortus* antibodies.[9] [Reprinted, with permission, from E. Benjamini, G. Sunshine, and S. Leskowitz, *"Immunology: A Short Course,"* 3rd ed., Wiley-Liss, New York, 1996, pp. 115–118. ISBN 0-471-59791-0. Copyright © 1996 by Willey-Liss, Inc.]

TABLE 5.2. The Major Human Blood Groupings

Blood Group	Erythrocyte Epitopes Present	Antibodies Present in Serum
AB	Both A and B	Neither Anti–A nor Anti–B
A	Epitope A	Anti-B only
B	Epitope B	Anti-A only
O	Neither A nor B	Both Anti-A and Anti-B

Agglutination occurs only over a certain range of serum dilutions because of either antigen or antibody excess. In the prozone, a large excess of antibody is present, so that each antibody behaves univalently and cross-linking does not occur (Eq. 5.5):

$$\tag{5.5}$$

At very high serum dilutions, a large excess of antigen exists, so that not enough antibody is present to cross-link the antigen and cause agglutination. In this case, the antigen behaves in a univalent manner (Eq. 5.6):

$$\tag{5.6}$$

Only at the intermediate dilutions do the proportions of antibody and antigen allow significant, visible agglutination to occur.

Blood typing is one common application of the agglutination test.[10] Human erythrocytes may possess either or both of epitopes A and B on their surfaces. Individuals possessing only epitope A have Anti-B in their serum, while individuals possessing only epitope B on their erythrocytes have circulating Anti-A. Some individuals have neither epitope, and both antibodies present, while others have both epitopes, and neither antibody. Table 5.2 lists the four major blood groupings, and the epitopes and antibodies present in their serum.

5.5.2. Precipitation Reactions

Precipitation reactions are similar in principle to agglutination reactions, the difference being that the antigen is a soluble, molecular species rather than a suspended particle such as a bacterium or erythrocyte. At a certain Ab/Ag ratio, the cross-linked polymeric network of antibody and antigen loses its solubility and precipitates. This is called the *precipitin* reaction.[11] Figure 5.6 shows how the quantity of precipitin produced varies with the quantity of antigen added, for a fixed total antibody concentration.

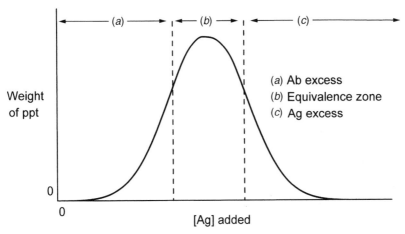

Figure 5.6. Precipitin formation as a function of added antigen.

Three zones are apparent in the curve shown in Figure 5.6. At low antigen concentrations, there is a zone of antibody excess, where antibody behaves univalently. In the equivalence zone, antigen and antibody are present in roughly equal proportions, and the largest quantity of precipitin forms. At higher antigen concentrations, the antigen behaves univalently, and its excess does not allow the formation of a highly cross-linked, insoluble network of precipitin.

The precipitin reaction can be induced in a gel, allowing qualitative tests for the presence of certain antibodies in serum. The gels used are often agar, and are cast on a flat glass plate, such as a microscope slide. The *double-diffusion method*[12] uses such a cast gel, with two wells cut into the gel, one at each end of the slide. The unknown serum is placed in one well, and the test antigen solution is placed in the other. Both solutions diffuse into the gel, and set up two opposing concentration gradients between the wells. If the suspected antibody is present in the serum, then at some distance between the wells, the Ab/Ag concentration ratio will be optimal for precipitin formation. As shown in Figure 5.7, a precipitin line forms at Ab/Ag equivalence. Provided that the diffusion rates of different antigens in the gel are known, a multiple test may be conducted on a single slide, with the positions of the precipitin lines indicating which antibodies are present in the serum sample.

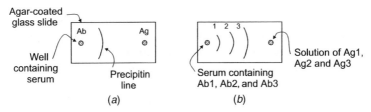

Figure 5.7. The double-diffusion method for (*a*) a single Ab and (*b*) multiple antibodies. In (*b*) the antigens diffuse at different rates with $1 > 2 > 3$.

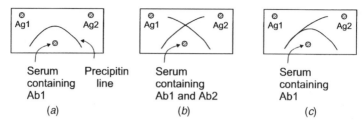

Figure 5.8. Double-diffusion test of antigen identity: (*a*) identity, (*b*) nonidentity, and (*c*) partial identity.

The double-diffusion method is an easier qualitative test than the precipitin reaction in solution, because it requires only one gel, rather than a series of test tubes with varying serum dilutions.

A double-diffusion method also exists to test for antigen identity.[13] In this test, three wells are cut into the agar gel on the microscope slide, as shown in Figure 5.8. The central well contains a known antiserum, while Ag1 is the unknown. Ag2 is a standard antigen that is suspected of being identical with Ag1. A distinctive semicircle of precipitin forms around the antibody well if Ag1 and Ag2 are identical. If the two antigens are completely different, then two independent precipitin arcs are formed. Partial identity indicates that one antigen possesses some, but not all, of the epitopes present on the other; in Figure 5.8, Ag2 possesses some of Ag1's epitopes, but is not identical to Ag1.

The *single radial immunodiffusion test* (SRID)[14] is a quantitative assay for antigen, and is also based on the precipitin reaction. This test is performed in a gel that is cast with a constant, uniform concentration of antibody distributed throughout. Wells cut into the gel contain standard antigen solutions at different concentrations. The precipitin produced after Ag diffuses into the gel forms a ring around the wells, as shown in Figure 5.9, and the ring d^2 (diameter^2) is directly proportional to [Ag] in the well. Values of d are corrected by subtraction of the well diameter. A calibration curve of d^2 versus [Ag] is used for antigen quantitation.

Precise quantitative work with the SRID test requires that the wells be cut reproducibly, and have vertical walls to produce distinct rings. To facilitate measurement, once the precipitin forms, the slide can be rinsed with PBS overnight to remove free Ab from the gel, then stained with a general protein stain such as Coomassie Blue. With this staining technique, detection limits of $\sim 10^{-7}$ M antigen have been achieved.

Secondary reactions such as agglutination and precipitation are generally used to identify or screen for the presence of certain suspected antibodies or antigens. They require antigens to be either multivalent, or multideterminate, or both, to allow crosslinking to occur. Secondary reactions cannot be used to quantitate or identify haptens, since haptens are always unideterminate and univalent.

Primary antigen–antibody interactions can be used to quantitate both antigens and haptens. Methods exploiting primary interactions have been developed for both antigen–hapten and antibody quantitation. These methods involve combining

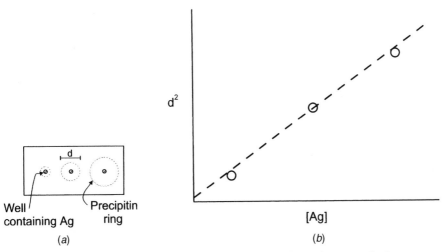

Figure 5.9. The single radial immunodiffusion test for antigen quantitation.

antibody with antigen, and then determining the concentrations of free and bound reactants. Competition between a labeled analyte (a reagent) and an unlabeled (unknown) analyte allows quantitation. These types of assays are the subject of Chapter 6.

SUGGESTED REFERENCES

E. Harlow and D. Lane, *Antibodies: A Laboratory Manual*, Cold Spring Harbor Laboratory, Cold Spring Harbor, New York, 1988.

I. R. Tizard, *"Immunology: An Introduction,"* CBS (Saunders) College Publishing, New York, 1984.

J. T. Barrett, *"Textbook of Immunology,"* 4th ed., C. V. Mosby Co., St. Louis, MO, 1983.

REFERENCES

1. C. R. Young: "Structural Requirements for Immunogenicity and Antigenicityy," in *"Molecular Immunology,"* M. Z. Atassi, C. J. Van Oss, and D. R. Absolom, Eds., Marcel Dekker, New York, 1984.

2. E. Benjamini, G. Sunshine, and S. Leskowitz, *Immunology: A Short Course*, Wiley-Liss, New York, 1996, pp. 57–75.

3. D. R. Davies and H. Metzger, *Ann. Rev. Immunol.* **1**, 1983, 87–117.

4. L. Hudson and F. C. Hay, *Practical Immunology*, 3rd ed., Blackwell Scientific Publications, London, 1989, p. 2.

5. E. Benjamini, G. Sunshine, and S. Leskowitz, *"Immunology: A Short Course,"* Wiley-Liss, New York, 1996, pp. 48–50.

6. E. Benjamini, G. Sunshine, and S. Leskowitz, *"Immunology: A Short Course,"* Wiley-Liss, New York, 1996, pp. 133–135.

7. D. Hawcroft, T. Hector, and F. Rowell, *"Quantitative Bioassay,"* John Wiley & Sons, Inc., New York, 1987, p. 91.

8. E. T. Maggio, *"Enzyme-Immunoassay,"* CRC Press, Boca Raton, FL, 1980, pp. 12–14.

9. E. Benjamini, G. Sunshine, and S. Leskowitz, *"Immunology: A Short Course,"* Wiley-Liss, New York, 1996, pp. 115–118.

10. A. Nowotny, *"Basic Exercises in Immunochemistry,"* Springer-Verlag, New York, 1979, p. 219.

11. E. Benjamini, G. Sunshine, and S. Leskowitz, *"Immunology: A Short Course,"* Wiley-Liss, New York, 1996, pp. 121–122.

12. G. Mancini, D. R. Nash, and J. F. Heremans, *Immunochemistry* **7**, 1970, 261–264.

13. A. J. Crowle, *"Immunodiffusion,"* 2nd ed., Academic Press, New York, 1973, pp. 247–303.

14. A. J. Crowle, *"Immunodiffusion,"* 2nd ed., Academic Press, New York, 1973, pp. 226–232.

PROBLEMS

1. Two polyclonal antibody preparations have been prepared according to the method outlined in this chapter. The preparations were tested for antibody concentration by measuring their titer. Preparation A possesses a titer of 1:1024, while preparation B has a titer of 1:128. Which is the more concentrated preparation of antibodies?

2. Explain why Scatchard plots for antigen binding are linear for monoclonal antibodies, and curved for polyclonal antibodies.

3. Which of the following combinations may be expected to yield precipitin near 1:1 antibody:antigen equivalence:

    (a) A univalent, multideterminate antigen (epitopes A, B, C, and D) and Anti-A,

    (b) A multivalent, unideterminate antigen (epitope A) and Anti-A,

    (c) A univalent, unideterminate antigen (epitope A) and Anti-A, and

    (d) A multivalent, multideterminate antigen (epitopes A, B, C, and D) and Anti-A.

4. An agglutination test for a bacterium was performed on serial dilutions of a freshwater sample. Using dilution factors of 1:1 to 1:512, the results showed the appearance of an agglutination zone and a zone of antibody excess, but no prozone was observed. Was the bacterium present? Why was no prozone observed?

5. In Figure 5.7, the double-diffusion method shows a precipitin line present as an arc nearer to the Ab well than to the Ag well, and curving toward the Ab well. These results are expected if the diffusion coefficient of Ag through the gel is much faster than that of Ab. What would be the appearance of the precipitin line if the Ab diffused much faster than the Ag (e.g., if Ag was a very high molecular weight species)?

CHAPTER 6

Quantitative Immunoassays with Labels

6.1. INTRODUCTION

Quantitative immunoassays constitute an enormous group of assay techniques designed for the selective quantitation of trace levels of low and high molecular weight species in complex biochemical media. In these immunoassay methods, *the ligand (antigen or hapten) is almost always the analyte*; while the qualitative identification of unknown antibodies is often of clinical importance, precise quantitative results are not normally required. In this chapter, the term ligand is used synonymously with hapten and antigen, because secondary binding reactions are not involved. Quantitative immunoassay methods exploit the primary binding of antibody and ligand, that is, the recognition and combination of antibody paratope with antigen epitope, with intrinsic affinity K. Secondary binding interactions do not generally occur, even with macromolecular antigens, because a large excess of either antibody or ligand is usually present. In practice, however, the possibility of secondary interactions should be considered during the development of new immunoassays.

The classification of immunoassay methods is based on (a) whether they are homogeneous, with no separation step needed prior to measurement, or heterogeneous, where a separation step is required; (b) which species, antibody or antigen, is labeled; and (c) the type of label employed.

The ideal label for immunoassay methods has the following properties. It is inexpensive, safe, and simple labeling procedures exist. It is covalently linked to the assay reagent at multiple sites, for high sensitivity. The labeled species is stable. Labeling has a minimal effect on the binding behavior, that is, the labeled and unlabeled reagents behave identically with respect to antibody–antigen binding. The label is easily detected using inexpensive instrumentation that is readily automated. Finally, the label should have properties that enable the differentiation of the free and bound forms without requiring a separation step.

Bianalytical Chemistry, by Susan R. Mikkelsen and Eduardo Cortón
ISBN 0-471-54447-7   Copyright © 2004 John Wiley & Sons, Inc.

While no real labels meet all of these needs, the properties of some of the more recently introduced labelling systems are approaching the ideal. Radioisotopes, once the only type of label used for immunoassays, have clearly been overwhelmed by current applications of fluorescent labeling methods, enzyme labels, and even coenzyme and prosthetic group labels. A variety of alternative labels has also been investigated, including red blood cells, latex particles, viruses, metals, and free radicals. Table 6.1 shows a representative listing of labels used in modern immunoassays.[1]

TABLE 6.1. Immunoassay Labels[a]

Immunoassay Type	Label Type	Label	Property Measured
Heterogeneous	Fluorophore	Fluorescein	Fluorescence intensity
		Europium chelate	Time-resolved fluorescence
		Phycobiliproteins	Fluorescence intensity
	Chemiluminescent	Acridinium ester	Ester hydrolysis
		Phenanthridinium ester	Ester hydrolysis
	Enzyme	Alkaline phosphatase	Enzyme activity
		β-Galactosidase	Enzyme activity
		Peroxidase	Enzyme activity
		Urease	Enzyme activity
	Cofactor	ATP	Kinase activity
		NAD	Dehydrogenase activity
	Lysing agent	Mellitin	Release of liposome-trapped enzyme (activity)
	Secondary label	Biotin	Binds avidin-enzyme and streptavidin-enzyme conjugates (activity)
Homogeneous	Fluorophore	Fluorescein	Fluorescence polarization Fluorescence quenching Fluorescent energy transfer
	Chemiluminescent	Acridinium ester	Ester hydrolysis
		Isoluminol	Light emission
	Prosthetic group	FAD	Glucose oxidase activity
	Enzyme	Glucose-6-phosphate dehydrogenase	
	Enzyme activity	Malate dehydrogenase	Enzyme activity
		Peroxidase	Aggregation and enzyme activity
		Hexokinase	Substrate channeling
	Electroactive	Ferrocenes	Oxidation current
	Spin label	Nitroxides	Broadening of ESR[b] signal
	Substrate	Galactosyl-umbelliferone	Enzymatic hydrolysis

[a] See Ref. 1.
[b] Electron spin resonance = ESR.

6.2. LABELING REACTIONS

Reagents used for the labeling of protein antigens and antibodies are similar to those used for the immobilization of enzymes by covalent and cross-linking methods (cf. Chapter 4). Homobifunctional and heterobifunctional cross-linking reagents are common, using NHS ester, maleimido, aldehyde, thiocyanate, and iso-thiocyanate groups to provide selectivity. Glycoproteins such as antibodies may also be bound to labels containing primary amine groups through the initial oxidation of sugar residues by periodate; the resulting aldehyde groups form Schiff bases with primary amines, which may be subsequently reduced to secondary amines under mild conditions with sodium borohydride.[2] Following the coupling reactions, labeled macromolecular reagents are usually purified by gel filtration chromatography: Porous stationary phases retain small solutes, but allow high molecular weight species to elute unretarded in aqueous buffer mobile phases. Dialysis or affinity chromatography may also be used for conjugate purification.

The preparation of labeled haptens may occur under more extreme conditions, since protein denaturation is only a factor if the label is an enzyme. The first critical step in hapten labeling is the introduction of a reactive group onto the hapten, which may be done by the alkylation of O or N substituents with haloesters,[3] followed by hydrolysis (Eq. 6.1):

$$RNHR' + I(CH_2)_3-\overset{\overset{\displaystyle O}{\|}}{C}-OCH_3 \longrightarrow RR'N(CH_2)_3-\overset{\overset{\displaystyle O}{\|}}{C}-OCH_3 \longrightarrow RR'N(CH_2)_3-\overset{\overset{\displaystyle O}{\|}}{C}-OH \quad (6.1)$$

A second reaction converts hydroxyl groups into carboxylates using succinic anhydride (Eq. 6.2):[4]

$$R-OH \; + \; \text{[succinic anhydride]} \longrightarrow R-O-\overset{\overset{\displaystyle O}{\|}}{C}-(CH_2)_2-\overset{\overset{\displaystyle O}{\|}}{C}-OH \quad (6.2)$$

Alternatively, ketone groups may be used to generate reactive carboxylates (Eq. 6.3):[5]

$$\underset{R_1 \;\; R_2}{\overset{\overset{\displaystyle O}{\|}}{C}} \; + \; H_2N-O-CH_2-\overset{\overset{\displaystyle O}{\|}}{C}-OH \longrightarrow \underset{R_1 \;\; R_2}{\overset{\displaystyle N-O-CH_2-\overset{\overset{\displaystyle O}{\|}}{C}-OH}{\overset{\|}{C}}} \quad (6.3)$$

Following the introduction of a carboxylic acid group onto the hapten, a variety of reagents may be used to activate these groups toward nucleophiles such as primary amine groups. The mixed-anhydride method uses isobutylchloroformate[6] to generate a mixed anhydride in the presence of a base such as triethylamine (Eq. 6.4):

$$R-\overset{\overset{\displaystyle O}{\|}}{C}-OH \; + \; R'-O-\overset{\overset{\displaystyle O}{\|}}{C}-Cl \; \overset{Et_3N}{\longrightarrow} \; R-\overset{\overset{\displaystyle O}{\|}}{C}-O-\overset{\overset{\displaystyle O}{\|}}{C}-OR' \quad (6.4)$$

The product of the mixed-anhydride reaction is reactive toward primary amine groups as well as hydroxyl groups, but is susceptible to hydrolysis. Yields of 20–30% conjugate have been obtained using a 10- to 20-fold excess of isobutylchloroformate.

Carboxylic acids may also be activated using carbodiimide and carbodiimide–hydroxysuccinimide reagents, as described in Chapter 4 for enzyme immobilization. For the carbodiimide reaction alone, average yields of 20% conjugate have been reported, due to a competing side reaction that converts the intermediate O-acylisourea to a stable N-acylurea. The NHS esters may be generated *in situ* and reacted directly with the label; for the progesterone + β-galactosidase reaction, a 10% loss in activity was reported, and 26% of the active enzyme was found to be immunoreactive. However, the stability of the NHS ester allows its isolation and purification; the purified NHS ester of progesterone, following reaction with enzyme, yielded a 100% immunoreactive enzyme–hapten conjugate, and the steroid/enzyme molar ratio was found to be 2:1.[7] Excess carbodiimide coupling reagent has been found to deactivate many enzymes when the in situ method is used, and the isolation of the NHS ester intermediate allows direct control over stoichiometry in the coupling step.

Conjugates are generally characterized by determining the hapten/label or antibody/label molar ratio, and by examining the characteristics of the label to determine whether conjugation has resulted in property changes. For example, the specific activity of enzyme labels is determined, and the molar absorptivity and quantum yields of fluorescent labels are compared before and after coupling. Enzymatic labels may be protected from deactivation by including a competitive inhibitor in the reaction mixture during coupling; the presence of this species can protect reactive groups at the active site from modification.[8] Immunoreactivity, the fraction of the hapten–label or antibody–label conjugate that can be bound by excess antibody or antigen, respectively, is often used to characterize a labeled reagent, and binding affinities may also be determined.

The above reactions are given as examples of the methods used to link haptens and antibodies to detectable labels; a comprehensive listing of conjugation procedures is beyond the scope of this book.

6.3. HETEROGENEOUS IMMUNOASSAYS

Many immunoassays require a separation step prior to quantitation, in order to separate the bound and free fractions of the labeled species. Consider an immunoassay in which a labeled ligand, Ag^*, competes with unlabeled analyte, Ag, for a limited quantity of antibody binding sites. This competitive equilibrium is represented by Eqs. 6.5 and 6.6,

$$Ab + Ag \rightleftharpoons Ab{:}Ag \tag{6.5}$$

$$Ab + Ag^* \rightleftharpoons Ab{:}Ag^* \tag{6.6}$$

We will assume that the presence of the label has no effect on the affinity of the antibody for the ligand. After equilibration, the reaction mixture contains Ag, Ag^*, Ab:Ag, and $Ab:Ag^*$. Because the quantity of Ab present is low relative to the quantities of Ag and Ag^*, essentially no free Ab remains. If the label itself is not affected by antibody binding, then separation of the bound and free fractions is necessary to determine the extent of binding of the labeled species. Separation may be accomplished by adding a precipitation step, or by conducting the assay with antibodies immobilized onto solid supports or particles.

Precipitation methods employ salts or solvents to precipitate the bound fraction, that is, Ab:Ag and $Ab:Ag^*$, from the reaction mixture. If this bound fraction is large or unstable enough to precipitate spontaneously, a simple centrifugation or filtration step is added. More often, ammonium sulfate or ethanol is used to promote precipitation. Precipitation may be enhanced by using a second antibody, for example, anti-IgG prepared by innoculating an animal with the IgG from a different species. The high molecular weight Ab2:Ab1:Ag and $Ab2:Ab1:Ag^*$ that result are easier to precipitate quantitatively. Polyethylene glycol (PEG) may also be added to yield lower particle weight precipitates, increasing the speed and specificity of the precipitation process.

Modern immunoassay methods employ antibodies immobilized onto solid supports. Antibodies bind strongly and spontaneously to glass and some plastics. They can thus be attached to beads, tubes, columns, or the wells of microtiter plates. These plates, shown in Figure 6.1, consist of 8 rows and 12 columns of wells, and are usually injection molded from polystyrene, to which antibodies and other proteins adsorb irreversibly. Standard dimensions are 86-mm width, 128-mm long, and 7-mm diameter wells, with a volume between 0.3 and 0.4 mL.

Immobilization of the antibody onto a solid support results in all bound species (Ab:Ag and $Ab:Ag^*$) also being immobilized. Unbound material is easily removed

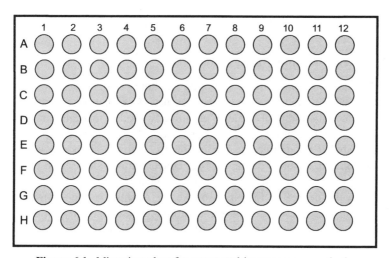

Figure 6.1. Microtiter plate for automated immunoassay methods.

(a) as a centrifugation supernatant, if beads or particles are the solid phase; (b) by elution or rinsing, if the solid phase is a tube or column of particles; or (c) by decantation and rinsing, if the solid phase is a microtiter plate. These methods require that the covalent binding or adsorption of the antibody to the solid phase is *reproducible* and *irreversible*.

A novel solid phase used by three commercial immunoassay kit manufacturers employs antibody covalently bound to cellulose particles that contain an iron(III) oxide core.[9] These small, paramagnetic particles do not spontaneously sediment, and have a large surface area. The particles are retained magnetically in a tube during decantation and rinsing steps, allowing easy separation of the bound fraction after equilibration with Ag and Ag*.

6.3.1. Labeled-Antibody Methods

Labeled-antibody methods rely on the presence of a large excess of the labeled antibody. This antibody is incubated with standard concentrations of added antigen (the analyte), and the concentration of bound antibody is determined. The resulting calibration curve is linear, as shown in Figure 6.2(a), and linearity extends upward to concentrations where Ab* is no longer in excess. Even for very low antigen concentrations, if an excess of Ab* is present, some complex will form; in principle, even a single Ag molecule can be detected.[10]

Sandwich assays employ two antibodies, and are useful for ligands possessing two or more distinct epitopes. The primary, or unlabeled antibody, is immobilized onto a solid support, and must be present in excess quantities relative to the total [Ag] present in standards and unknowns. Following incubation with antigen, to allow quantitative capture Ag, and a rinse step, the labeled antibody is added in excess, and a second incubation proceeds. The labeled antibody reacts only with the antigen that has been retained on the solid phase by the primary antibody. After a second rinse, the quantity of immobilized Ab:Ag:Ab* "sandwich" is then determined, and is equal to the quantity of antigen in the sample.

6.3.2. Labeled-Ligand Assays

In labeled-ligand immunoassays, a limiting amount of antibody is allowed to react with excess labelled antigen (Ag*) in the presence of varying concentrations of analyte Ag. Generally, the total quantity of antibody used is that amount required to produce 50% binding of the labeled ligand in the absence of analyte, or unlabeled ligand. Unlabeled Ag inhibits the binding of Ag* with the antibody, through competitive equilibria. A series of standard Ag concentrations are used to construct an inhibition curve, as shown in Figure 6.2(b), and an unknown [Ag] can then be determined from the degree of inhibition of Ab:Ag* formation. The detection limits obtained with this method depend on the affinity of the antibody for the antigen, and the experimental uncertainty in quantitating the bound or free fraction of labeled antigen.[10] For a K_0 value of 10^{12} M^{-1}, antigen detection limits of 10^{-14} M have been achieved.

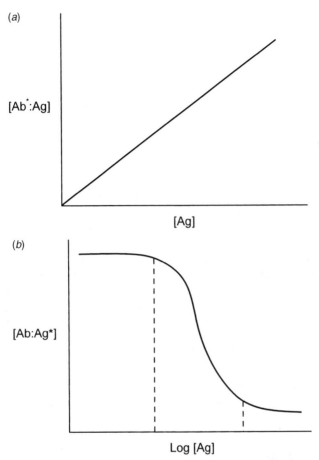

Figure 6.2. Antigen calibration curves for (a) labeled antibody methods, and (b) labeled ligand methods, where the dashed lines show the analytically useful region of the curve.

For example, a separation-based radioimmunoassay (RIA) has been developed for human placental lactogen (HPL),[11] which is a protein hormone secreted by the placenta. The HPL levels have been used as indicators of normal placental function during pregnancy. It is antigenic, and antibodies may be produced by immunizing an animal with the human protein. The HPL is labeled by iodination with ^{125}I at its tyrosine residues, to give a labeled antigen, HPL*. This labeled species is a low-energy γ-ray emitter. Figure 6.3 shows an outline of the assay procedure. A fixed amount of HPL* and the unknown HPL are incubated with a fixed, insufficient quantity of anti-HPL antibodies. Competition for the Ab-binding sites ensues, and after a fixed incubation period, the bound fraction is precipitated by the addition of ethanol. After centrifugation, the radioactivity in either the precipitate or the supernatant is measured.

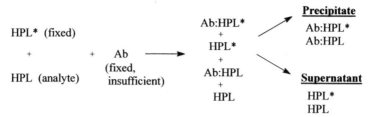

Figure 6.3. Outline of RIA for HPL.

Standard curves are constructed by plotting radioactivity against log[HPL], as shown in Figure 6.4. If precipitate radioactivity is measured, then the measured signal decreases as HPL concentration increases, since fewer HPL* molecules will have access to the binding sites on the antibody. If supernatant radioactivity is quantitated, then the inverse will be true. Both curves exhibit a sigmoid shape, and the position of the midpoint of the curve on the log[HPL] axis depends on antibody affinity K_0.

For quantitative purposes, these sigmoid curves can be linearized using the logit transformation. First, the measured signals are normalized to values that lie between zero and one. The logit transformation is then performed, using $\text{logit}(y) = \ln\{y/(1-y)\}$. If y is the fraction of the label in the supernatant, then a plot of $\text{logit}(y)$ against log[HPL] will be linear with a negative slope, as detailed in Chapter 16. If the radioactivity of the precipitate is measured, then the logit transformation will yield a straight line with a positive slope.

6.3.3. Radioisotopes

Because radioisotopes were the first labels used in immunoassay methods, there exists a vast array of radioimmunoassays. Radioisotopes are easily detected at low levels, and labeling procedures are simple. The label has virtually no effect on antibody–antigen binding. However, radioisotopes are costly, hazardous, and require inconvenient monitoring and disposal procedures. In addition, isotopic decay necessitates the regular replacement of the labeled component.

In general, ^{125}I is used as a label for large-protein antigen labeling.[12] It has a half-life of 60 days, is a low-energy γ-emitter, and thus requires only inexpensive instrumentation for detection. The isotope ^{3}H is commonly used for hapten labeling. It has a 12-year half-life and is a β-emitter, thus requiring a scintillation counter for detection. A further disadvantage of ^{3}H is that it has a low specific activity, and yields poor detection limits relative to ^{125}I. Some specialized assays employ isotopes of Co, Fe, and Se.

The RIA detection limits for antigens are $\sim 10^{-12}\,M$, and are equalled only by the enzymatic labels, since they are capable of catalytic signal amplification.

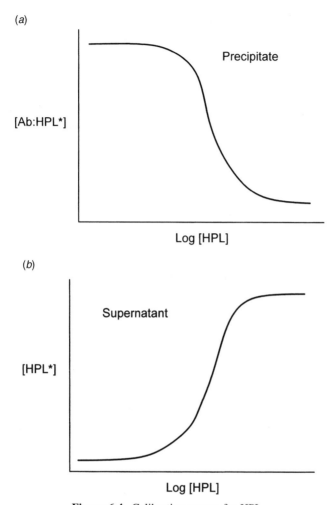

Figure 6.4. Calibration curves for HPL.

6.3.4. Fluorophores

Commonly used fluorescent labels[13] include fluorescein, rhodamine, and umbelliferone derivatives (Fig. 6.5). Recently, a group of proteins isolated from algae have been used as immunoassay labels due to their high molar absorptivity (at least 10 times that of fluorescein) and their quantum yields > 0.8; these are the phycobiliproteins, called phycoerythrin, phycocyanin, and allophycocyanin. Fluorescent labels are safe, and require no licensing for their use. Antigen detection limits of $\sim 10^{-10} M$ are normal with fluorescent labels, two orders of magnitude higher than those of radioactive labels, as a result of scattering, quenching, and background fluorescence from biological samples, especially those containing significant quantities of protein. Fluorescent labels offer many options for signal

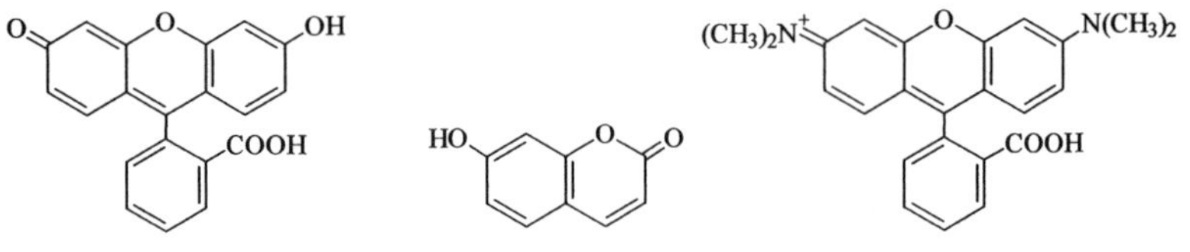

Figure 6.5. Common fluorescent labels used in immunoassays.

generation in heterogeneous immunoassays,[14] and may be classified as indirect, competitive, or sandwich methods.

6.3.4.1. Indirect Fluorescence. This method is normally used for antibody quantitation or screening. Antigen is immobilized onto a solid support, and antiserum is added. If specific antibodies are present, these antibodies will be immobilized on the support. Following a rinse step, fluorophore-labeled antibodies to the particular antibody class are added (e.g., fluorescein–anti-IgG). These labeled antibodies are immobilized only if primary antibodies to the immobilized antigen are present on the solid support. Fluorescence intensity thus increases with antibody concentration.

6.3.4.2. Competitive Fluorescence. Antibodies to the antigen or hapten (analyte) are immobilized onto a solid support. Fluorophore-labeled and unlabeled antigen compete for a limited number of antibody binding sites. The support is rinsed and fluorescence is measured. Higher analyte concentrations result in fewer immobilized labeled species, yielding reduced fluorescence intensity.

6.3.4.3. Sandwich Fluorescence. This method, also called a two-site fluorometric immunoassay, is not readily adapted to hapten assays, because the antigen must be large enough to possess at least two distinct epitopes, to allow the independent binding of two different antibodies. Immobilized primary antibody is allowed to quantitatively bind antigen (analyte) in the sample. Fluorophore-labeled second antibody is then added, to form a support–antibody–antigen–antibody–fluorophore "sandwich". The fluorophore is immobilized only if antigen is bound to the primary antibody. Fluorescence intensity increases linearly with the amount of antigen present in the sample.

6.3.4.4. Fluorescence Excitation Transfer. This modified-sandwich method (shown in Eq. 6.7) has both antibodies labeled, but with different fluorophores (F1 and F2) that are chosen so that the emission spectrum of one overlaps with the excitation spectrum of the other. Internal quenching thus occurs when F1 and F2 are held in close proximity. If F1 and F2 are close, the emission of F1 is quenched by F2 absorption. Following a rinse step, the emission of F1 is measured, so that, for increasing antigen concentrations, emission intensity decreases.

$$\text{—Ab—F1} \xrightarrow{\text{Ag}} \begin{matrix} \text{—Ab—Ag} \\ \text{F1} \end{matrix} \xrightarrow{\text{Ab-F2}} \begin{matrix} \text{—Ab—Ag—Ab} \\ \text{F1} \quad \text{F2} \end{matrix} \qquad (6.7)$$

Support

This method has been developed into a "two-site sandwich assay" for digoxin, using phycoerythrin as F1 and fluorescein as F2.[15] Phycoerythrin is isolated from red algae, has a quantum yield of 0.80, and absorbs 30 times more light than does fluorescein. Phycoerythrin-labeled primary antibody is immobilized onto microtiter plate wells. Following incubation with the digoxin sample to allow quantitative binding, fluorescein-labeled second antibody is added. Fluorescein quenches the emission from phycoerythrin, as shown in Figure 6.6, so that greater quantities of immobilized digoxin yield reduced emission intensities. This assay has a detection limit of 0.5-mg/L digoxin, with a 6% coefficient of variation:

6.3.4.5. Time-Resolved Fluorescence. This instrumental technique may be applied to most immunoassays that require fluorescence intensity measurements,

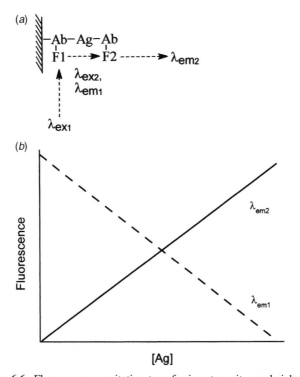

Figure 6.6. Fluorescence excitation transfer in a two-site sandwich assay.

but do not require antibodies to the fluorophore itself. Its popularity stems from the rejection of interfering fluorescent signals and scattered excitation radiation. If a mixture of short- and long-lived fluorophores is excited with a short (<1 μs) pulse of light, then the excited molecules will emit fluorescence that is either short- or long lived. In both cases, fluorescence will decay exponentially with time, but short-lived fluorescence will decay to nearly zero very quickly. If emission measurements are taken after 100–200 μs, only long-lived fluorescence will be measured. Short-lived fluorescence and scattered light may thus be eliminated from measured signals. The labeling fluorophore must possess a long-lived excited state, and only a few species, notably europium chelates, have suitable fluorescence properties.[16] Proteins labeled with p-isothiocyanatophenyl–EDTA at lysine residues will chelate Eu^{3+} with formation constants in excess of $10^{12}\,M^{-1}$. Typically, a 1 μs excitation pulse at 337 nm is followed by a 200-μs delay, following which emission at 613 nm is integrated between 200 and 600 μs. Background long-lived fluorescence from polystyrene (used for microtiter plates) is extremely low, but other solid-phase supports, such as nitrocellulose and 3M paper, exhibit significant long-lived background signals.

6.3.5. Chemiluminescent Labels

Chemiluminescent labels may be employed in sandwich or competitive antigen assays. In sandwich assays, a solid support holds a primary antibody, and incubation with ligand yields a species that is detectable following a second incubation step with a labeled second antibody. Luminol has been tested as an immunoassay label; it may be coupled to proteins through its primary amino group. Luminol reacts with hydrogen peroxide and hydroxide in a microperoxidase-catalyzed reaction, which yields light at 430 nm (Eq. 6.8):

$$+ \ 2\,H_2O_2 \ + \ OH^- \longrightarrow \qquad + \ N_2 \ + \ 3\,H_2O \ + \ h\nu \qquad (6.8)$$

Luminol   3-Aminophthalate

Following the second incubation and rinse steps in the sandwich immunoassay, immobilized Ab:Ag:Ab-Luminol is detected by the addition of OH^- and microperoxidase to the support, then integrating the light produced upon H_2O_2 addition for a defined period (5–10 s), as shown in Figure 6.7. Since this is a noncompetitive, labeled-antibody method, the integrated light intensity increases linearly with antigen concentration.

The luminol reaction is not particularly effective as a labeling system, since the fluorescence quantum yield of luminol is much lower if luminol is chemically bound to another species.

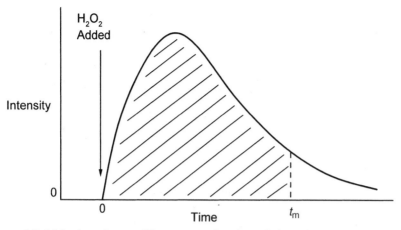

Figure 6.7. Light intensity at 430 nm as a function of time, for Ab:Ag:Ab-Luminol sandwich assay. Integration occurs between the time of H_2O_2 addition and t_m.

A recently introduced chemiluminescent labeling system employs arylacridinium esters[17] as labels that are readily coupled to protein tyrosine residues. In the presence of hydroxide and H_2O_2, the ester is oxidatively cleaved, yielding 10-methylacridone plus light (Eq. 6.9). This reaction does not require an enzymatic catalyst, and quantum yields are high even when the label is attached to proteins, since quenching by hapten or protein is minimized by the release of the excited acridone. Nearly all light is given off within 10 s after oxidation, and so high S/N values are achievable. Furthermore, antibodies can be labeled to such an extent that specific activities of 10^{15} photons are obtained per gram of antibody. However, the phenyl ester bond is inherently unstable to hydrolysis, so that labeled reagents have a limited shelf-life. Research toward more stable acridinium esters is underway.

$$\begin{array}{c} \underset{\substack{\\ C=O \\ | \\ OR}}{\overset{\substack{CH_3 \\ | \\ N^+}}{\text{acridinium ester}}} \quad \xrightarrow[H_2O_2]{OH^-} \quad \underset{\substack{\\ O}}{\overset{\substack{CH_3 \\ | \\ N}}{\text{10-methylacridone}}} \quad + \quad h\nu \end{array} \qquad (6.9)$$

Chemiluminescent labels may also be used in labeled-antigen (competitive) assays. The antigen (analyte) competes with the labeled analyte for immobilized antibody, and, following a rinse step, reagents are added to generate chemiluminescence from the labels.

Immunoassay kits using luminol and arylacridinium ester labels have been developed by nine companies for a variety of thyroid, steroid and pituitary hormones, viruses, digoxin, and creatine kinase. One assay for total and free thyroxin has detection limits of ~ 20 pM in serum, while another for total and free triiodothyronine has a 3-pM detection limit in serum. These analytes are used as clinical indicators of thyroid gland malfunction. While research has shown that detection limits using chemiluminescent labels should equal those obtained using enzymatic or radioactive labels, practical applications show limited precision as a result of emission kinetics. The luminescence reactions proceed so rapidly (500,000 photons/s) that the reproducibility of capturing a consistent fraction of photons requires a photomultiplier that possesses excellent precision over a dynamic range of many orders of magnitude. Tests using a thyroxine–isoluminol conjugate yielded coefficients of variation in integrated response of 13–18% over 15 replicate measurements.

6.3.6. Enzyme Labels

Enzymes are currently the most widely used and investigated labels for immuno-assays, because a single enzyme label can provide multiple copies of detectable species. This *catalytic amplification* results in immunoassay detection limits that rival those of radioimmunoassay without the storage and disposal problems associated with radioisotopes. Enzyme immunoassays label either ligands or antibodies with enzyme, and enzyme *activity* in bound or free fractions is measured. Heterogeneous immunoassays employing enzymatic labels have been named enzyme-linked immunosorbent assays (ELISAs). ELISA methods usually employ antibody immobilized onto the wells of polystyrene microtiter plates, and may be

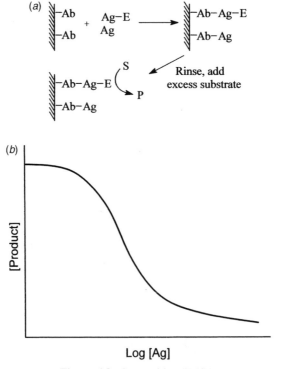

Figure 6.8. Competitive ELISA.

competitive or noncompetitive. In a competitive ELISA, enzyme-labeled antigen competes with free antigen (the analyte) for a fixed, insufficient quantity of immobilized antibody binding sites, as illustrated in Figure 6.8. After incubation, the support is rinsed to remove unbound species, and the enzyme substrate is added in saturating concentration. The conversion of substrate to product may be measured continuously, in a kinetic assay, in which the rate of conversion decreases with increasing free antigen concentration. More often, a fixed-time approach is used; after a given incubation time, the reaction is stopped by the addition of strong acid or base that denatures the enzyme. Product quantitation then yields a calibration curve in which product concentration decreases with increasing free antigen concentration.

Noncompetitive ELISA methods are based on sandwich assays, as shown in Figure 6.9. An immobilized primary antibody is present in excess, and quantitatively binds antigen. An enzyme-labeled second antibody is then allowed to react with the bound antigen, forming a sandwich that is detected by measuring enzyme activity bound to the surface of the support. Noncompetitive ELISAs yield calibration curves in which enzyme activity increases with increasing free antigen concentration.

Both competitive and noncompetitive ELISA methods employ microtiter technology for automation. Robotics are used for reagent addition and timing, while

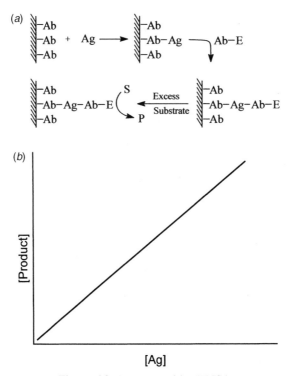

Figure 6.9. Noncompetitive ELISA.

automated microtiter plate readers are available for the measurement of absorbance values in each well. A typical ELISA uses some rows for serial dilutions and replicates of a standard antigen solution, while others are used for serial dilutions and replicates of the unknowns. Several unknown dilutions are employed to ensure that at least one dilution value will yield results that allow interpolation on the calibration curve. Furthermore, the greatest sensitivity in a competitive ELISA occurs when equal concentrations of free and enzyme-labeled antigen compete for immobilized antibody. With noncompetitive ELISAs, the antigen concentration must be low enough for complete (100%) antigen binding to occur.

An ideal labeling enzyme should have a high turnover number (k_{cat}), a product that is easily measured, and a substrate that does not interfere with measurement. In addition, it should be inexpensive, stable, and resistant to interferences that may be present in biological samples. The enzyme, and its substrate and product should not normally be present in samples for assay.

Five enzymes are commonly employed as labeling enzymes in immunoassays.[18] The most widely used is horseradish peroxidase (E.C.1.11.1.7), because of its high specific activity (4500 U/mg at 37 °C) and because this 40-kDa enzyme is relatively nonspecific for its secondary substrate. This means that a variety of reduced dyes may be used as substrates that are converted to their highly absorbing oxidized forms. Pyrogallol is often used as a substrate for this reaction (Eq. 6.10):

$$2 \quad \text{[structure]} \quad + \text{ H}_2\text{O}_2 \longrightarrow \text{[structure]} \quad + \quad 3 \text{ H}_2\text{O} \tag{6.10}$$

Alkaline phosphatase (E.C.3.1.3.1) is also used in many enzyme-amplified immunoassays. It catalyzes the hydrolysis of an orthophosphoric monoester, to produce an alcohol and orthophosphate. This 100 kDa enzyme has a pH optimum near 9, and exhibits activities of 1000 U/mg (37 °C). With p-nitrophenylphosphate as substrate, the p-nitrophenol produced may be monitored at 450 nm (Eq. 6.11):

$$\text{[structure]} \quad + \text{ H}_2\text{O} \longrightarrow \text{[structure]} \quad + \quad \text{HPO}_4^{2-} \tag{6.11}$$

β-Galactosidase (E.C.3.2.1.23) catalyzes the hydrolysis of terminal nonreducing β-D-galactose residues in β-D-galactosides. This enzyme has a molecular weight of 540 kDa, a pH optimum of 7.0 and a specific activity (37 °C) of 600 U/mg. A synthetic substrate, o-nitrophenyl-β-D-galactoside, yields readily detectable o-nitrophenol upon hydrolysis (Eq. 6.12):

$$\begin{array}{ll} o\text{-Nitrophenyl-} + \text{ H}_2\text{O} \longrightarrow o\text{-Nitrophenol} + \text{ D-Galactose} \\ \beta\text{-D-galactose} \end{array} \tag{6.12}$$

Glucose-6-phosphate dehydrogenase (E.C.1.1.1.49) is also commonly used as a label. This enzyme has a molecular weight of 104 kDa, and pure preparations show optimum activity at pH 7.8, with 400-U/mg specific activity at 37 °C. The dehydrogenase catalyzes the oxidation of D-glucose-6-phosphate with concomitant reduction of NADP$^+$ (Eq. 6.13). Absorbance or fluorescence of the resulting NADPH may be used for quantitation.

$$\text{D-Glucose-6-phosphate} \quad + \quad \text{NADP}^+ \longrightarrow \text{D-Glucono-}\delta\text{-lactone-6-phosphate} \quad + \quad \text{NADPH} \quad (6.13)$$

Urease (E.C.3.5.1.5) catalyzes the hydrolytic decomposition of urea into ammonium, bicarbonate, and hydroxide ions. It has a molecular weight of 483 kDa, and preparations with specific activities of 10,000 U/mg at 37 °C and pH 7.0 are commercially available.

Heterogeneous immunoassays may be used with conventional microtiter systems in clinical laboratories, but have also been adapted for doctor's office and home diagnostic test devices. For example, a device for detecting antibodies to the human immunodeficiency virus (HIV), operates as shown in Figure 6.10.[19]

Preparations of glucose oxidase (GO), recombinant HIV antigen, and human IgG, all immobilized on latex beads (0.9 mm diameter), are embedded in a glass fiber paper wick at successive positions along the wick (A, B, and C, respectively). A whole blood sample is added at position A, and it spreads and interacts with the HIV-latex beads. A glucose oxidase–peroxidase substrate solution is then allowed to flow along the wick from the sampling end, and a protein A–horseradish peroxidase conjugate is added at position B. Protein A binds immunoglobulins in a region far removed from the antigen-binding site. The flow of substrate solution results in hydrogen peroxide release at position A, where GO is immobilized. This is carried in the substrate solution to position B, where it reacts with the protein A–peroxidase conjugate that is captured by anti-HIV antibodies bound to the HIV-latex beads. A reduced dye present in the substrate solution is converted to an insoluble colored product, which forms a colored deposit at site B. Unreacted protein A–peroxidase conjugate flows downstream to site C, where it is captured by the anti-IgG–latex beads. The same insoluble colored deposit is therefore also formed at site C. The flow of substrate solution along the wick is maintained by positioning a sponge, or absorbent wick, at the far end. Results are obtained in < 15 min, with color at sites B and C representing a positive test for anti-HIV, while color at only site C represents a negative test. The detection limit of this device is similar to a conventional

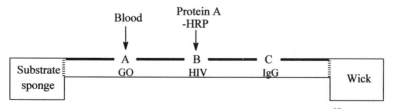

Figure 6.10. Immunoassay test device for HIV antibodies.[19]

ELISA method for anti-HIV, but it is designed to be used as a qualitative, disposable system.

6.4. HOMOGENEOUS IMMUNOASSAYS

Homogeneous immunoassay methods have been slower to appear than heterogeneous methods. They rely on labeled antigen species that show large signal changes upon antibody binding, so that separation of the bound and free fractions of the label is unnecessary. Fluorophores and enzymes represent the majority of labels used in homogeneous immunoassays,[20] and representative examples of these are given below.

6.4.1. Fluorescent Labels

6.4.1.1. Enhancement Fluorescence. In this method, the antigen or ligand is labeled with a fluorophore. The fluorescence intensity increases on antibody binding, due to either the change in the environment of the fluorophore (hydrophilic to hydrophobic, e.g.) or due to energy transfer from fluorophores such as tryptophan and tyrosine in the antibody. An increase in unlabeled antigen concentration results in a decrease in binding of labeled antigen, and therefore a decrease in fluorescence intensity. One reported enhancement fluorescence immunoassay for thyroxine (T_4) employs fluorescein-labeled T_4 as the hapten-label conjugate. The mechanism for fluorescence enhancement upon antibody binding is thought to involve the intramolecular quenching of fluorescein fluorescence by the iodine atoms on T_4; the quenching is relieved upon antibody binding, when the iodothyronine epitope interacts with the antibody paratope, so that, in the absence of free T_4 (the analyte), a maximum in fluorescence intensity is observed. The addition of T_4 results in reduced fluorescence intensity. This homogeneous assay shows a fourfold increase in fluorescence intensity on complete binding of labeled antigen. Applications of this method are rarely reported, because enhancement effects are often too small to be analytically useful.

6.4.1.2. Direct Quenching Fluorescence. This homogeneous method employs the decrease in fluorescence intensity that occurs upon binding of the fluorophore-labeled ligand by antibody as the analytical signal. The decreased fluorescence from the label occurs as a result of quenching of the label's fluorescence by the antibody. At high antigen (analyte) concentrations, less of the labeled antigen is bound to the antibody, so that the fluorescence intensity increases. The change in fluorescence signal is generally small with this method, especially when the antigen is large and partially shelters the fluorescent label. However, a useful assay for gentamicin has been developed using this principle. The fluorescence of fluorescein-labeled gentamicin decreases by $\sim 40\%$ when it is bound to anti-gentamicin; this quenching is inhibited in the presence of free gentamicin, so that fluorescence intensity increases with analyte concentration.

6.4.1.3. Indirect Quenching Fluorescence.

6.4.1.3. Indirect Quenching Fluorescence. With this method, a second antibody is generated to the fluorescent label itself. Following incubation of fluorophore-labeled and unlabeled antigen with the primary antibody, antibody to the fluorescent label is added. Because of steric hindrance, this second antibody will only bind the fluorescent label if the labeled antigen is free, that is, not bound to the primary antibody. Upon binding of the second antibody, the fluorescence of the label is quenched. Increased free antigen concentrations yield increased free labeled antigen concentrations, and therefore decreased fluorescence intensity (Eq. 6.14):

$$
\begin{array}{ccccc}
\text{Ab} + \text{Ag--F}^* & \longrightarrow & \text{Ab:Ag--F}^* & + & \text{Ag--F}^* \\
& & \downarrow \text{Ab}_F & & \downarrow \text{Ab}_F \\
& & \text{Ab:Ag--F}^* & & \text{Ag--F:Ab}_F
\end{array}
\tag{6.14}
$$

This principle has been applied to assays for T_4, human placental lactogen, and human serum albumin. The T_4 assay involves fluorescein-labeled T_4 complexed with a low-affinity antifluorescein antibody. When anti-T_4 is added, it reacts with the fluorescein-labeled T_4 and pulls it away from the antifluorescein antibody. This results in an increase in fluorescence intensity. In the presence of free T_4, less of the labeled T_4 is bound to the anti-T_4 antibody, so that more antifluorescein–fluorescein-T_4 complex is present, and fluorescence intensity is reduced. One important advantage of this method over other homogeneous immunoassay methods is that relatively impure antigen may be used for the labeling reaction. In principle, fluorophore-labeled impurities in the labeled-antigen reagent will be quenched by the antifluorophore antibodies, so that they will not contribute to background fluorescence.

6.4.1.4. Fluorescence Polarization Immunoassay.

6.4.1.4. Fluorescence Polarization Immunoassay. Polarized excitation radiation yields polarized fluorescence emission due to the fixed relationship between molecular orientation and the absorption and emission of radiation. First reported in 1973,[21] fluorescence polarization immunoassay (FPIA) methods rely on differences in the speed of rotation, or tumbling, with molecular weight. Ideally, a small fluorophore-labeled antigen or hapten will rotate very quickly in solution, and will exhibit a maximum in emission intensity measured perpendicular to the incident polarized excitation beam. Upon binding to the large antibody molecule, slow rotation yields a significantly diminished perpendicular emission (Eq. 6.15). Competitive homogeneous assays therefore yield increased perpendicular emission intensities with increased free antigen concentration, since more fluorophore–antigen conjugate exists free in solution. This principle has been successfully applied to a variety of commercial immunoassays marketed by Abbott Laboratories.

$$
\begin{array}{ccccc}
\text{Ab} + \text{Ag--F}^* & \longrightarrow & \text{Ab:Ag--F}^* & + & \text{Ag--F}^* \\
& & \text{High polarization} & & \text{Low polarization}
\end{array}
\tag{6.15}
$$

A phenobarbitone assay, for example, using a fluorescein label, has a free labeled antigen molecular weight of ~ 500, which increases to > 150 kDa upon antibody binding. A similar competitive assay for digoxin yields a detection limit of 0.2 µg/mL.

6.4.1.5. Fluorescence Excitation Transfer. Similar to the heterogeneous assay described earlier, the excitation transfer immunoassay employs two fluorophores, F1 and F2. When held in close proximity, F2 quenches F1 emission, but when F2 and F2 are far apart, essentially no quenching occurs. In the examples that have been published, F1 is fluorescein isothiocyanate, which, when bound through a thiourea linkage to an antibody or antigen, absorbs at 490 nm and emits at 525 nm. F2 is tetraethylrhodamine, which exhibits an absorbance maximum at 525 nm. A direct assay for morphine uses rhodamine-labeled antimorphine and fluorescein-labeled morphine. A maximum quenching of 72% (or a minimum fluorescence intensity of 18%) was observed for the rhodamine–antimorphine–morphine–fluorescein conjugate, compared to the free fluorescein–morphine conjugate in solution. In the bound complex, the fluorophore and quencher are within 5–10 nm of each other, facilitating dipole–dipole coupled excitation energy transfer. Unlabeled morphine competes with the fluorescein-labeled morphine for antibody binding sites, and results in increased fluorescence intensity (or decreased quenching).

6.4.2. Enzyme Labels

Both competitive and noncompetitive methods have been incorporated into homogeneous enzyme-labeled immunoassay kits that ultimately relate enzyme activity to analyte concentration.[22] The competitive-binding assays are called enzyme-multiplied immunoassay technique (EMIT), substrate-labeled fluorescein immunoassay (SLFIA), apoenzyme reactivation immunoassay (ARIS), and cloned enzyme donor immunoassay (CEDIA), while a noncompetitive method is called enzyme inhibitory homogeneous immunoassay (EIHIA).

6.4.2.1. Enzyme-Multiplied Immunoassay Technique. EMIT is a homogeneous method for the quantitation of haptens, especially hormones, therapeutic drugs, and drugs of abuse. This method is a competitive assay, in which hapten and enzyme-labeled hapten compete for a fixed, insufficient quantity of antibody (Eq. 6.16).

$$\text{Enzyme-H} + \text{H} + \text{Ab} \longrightarrow \quad\quad\quad \text{(6.16)}$$

Free enzyme-labeled hapten exhibits high enzyme activity, but if antibody is bound to the hapten moiety, enzyme activity is drastically reduced due to steric hindrance or structural changes at the active site that result from antibody binding. Enzyme activity therefore increases with increasing free antigen concentration. With these assays, glucose-6-phosphate dehydrogenase or lysozyme are used as the labeling enzyme. Detection limits of ~ 1 ng/mL are typical.

A second type of EMIT has been developed using the enzyme malate dehydrogenase as the enzymatic label. Research has shown that thyroxine competitively inhibits malate dehydrogenase. A conjugate prepared with thyroxine covalently bound close to the enzyme's active site shows very low specific activity that can be restored by binding of the thyroxine to *anti*-thyroxine antibody. In this very specific assay for thyroxine, enzyme activity increases upon antibody binding, so that in a competitive assay for free thyroxine, activity decreases with increasing free thyroxine concentration.

6.4.2.2. Substrate-Labeled Fluorescein Immunoassay. This assay, also called enzyme-release fluorescence immunoassay, employs a fluorophore-labeled antigen in which fluorescence is either quenched by the antigen or is nonexistent as a result of the conjugation. Following incubation with antibody, the resulting mixture is incubated with a hydrolytic enzyme that cleaves the fluorescent label from free labeled antigen only. Cleavage does not occur with antibody-bound labeled antigen, because the presence of the antibody prevents access to the enzyme's active site. Increased free antigen concentrations yield increased free labeled antigen, and therefore increased fluorescence intensity (Eq. 6.17). This method has been adapted into several commercial assay kits, including >10 therapeutic drugs such as the antiasthmatic drug theophylline as well as the antibiotics gentamicin and tobramycin. In the gentamicin assay, gentamycin coupled to β-galactosylumbelliferone is used as the labelled antigen, which is not fluorescent. Free β-galactosylumbelliferone reacts with β-galactosidase to form a product that emits at 453 nm following excitation at 400 nm. When the labeled antigen is bound to *anti*-gentamicin antibody, steric hindrance prevents the enzymatic cleavage of the conjugate into a fluorescent product. This assay can easily detect gentamicin levels of 1-μg/mL serum, using serum volumes of 1 μL.

$$Ab + Ag{-}F \longrightarrow Ab{:}Ag{-}F \ + \ Ag{-}F$$
$$\downarrow E \qquad\qquad \downarrow E \qquad\qquad (6.17)$$
$$Ab{:}Ag{-}F \qquad Ag + F^{*}$$

6.4.2.3. Apoenzyme Reactivation Immunoassay (ARIS). This assays employ a prosthetic group, FAD, as an antigen label for competitive immunoassay in solution and in the new film and paper strip formats for dry immunochemistry. In solution, competition between free and FAD-labeled antigen for antibody binding sites yields a mixture of free and bound FAD species. The free species are able to combine with the glucose oxidase apoenzyme to form an active holoenzyme, while the bound antigen–FAD conjugates are sterically hindered from combining

with the apoenzyme. Enzyme activity thus increases as free antigen concentration increases. A commercial ARIS assay kit for theophylline has a detection limit of 5 ng/mL.

6.4.2.4. Cloned Enzyme Donor Immunoassay.

This technique, abbreviated CEDIA, has arisen as a result of recombinant DNA technology, and a fundamental understanding of the enzyme β-galactosidase. This enzyme is active only in its native tetrameric structure; the four identical subunits comprising the tetramer are inactive if the quaternary structure of the holoenzyme is absent. It is now known that if any of the amino acids between positions 10 and 60 from the N-terminal end of the β-galactosidase subunit are absent, an active tetramer cannot be formed.

Given this information, researchers have devised "donor" and "acceptor" components of the β-galactosidase subunit. The enzyme donor is conjugated to antigen, and becomes the indirect label in a competitive immunoassay. If this labeled antigen is not bound to antibody, then the combination of donor and acceptor result in a complete subunit that can combine with three other complete subunits to form an active tetramer. If the labeled antigen is antibody-bound, steric effects prevent donor–acceptor combination, so that the generation of active enzyme is not possible from this potential subunit. The concepts in this assay are shown in Eq. 6.18:

$$(6.18)$$

The CEDIA assays show increased β-galactosidase activity with increased free antigen concentration. Commercial CEDIA assays for digoxin and other antigens show detection limits of the order of nanograms per milliliter of serum, and so are competitive with other enzyme-labeling methods.

6.4.2.5. Enzyme Inhibitory Homogeneous Immunoassay.

This non-competitive, homogeneous immunoassay method uses an enzyme-labeled antibody, where the enzyme substrate is an insoluble species. The assay has been found most suited to macromolecular antigens, such as α-fetoprotein (where high levels indicate abnormal pregnancies) and ferritin, an iron transport protein. The enzymes employed in the commercial assays are α-amylase and dextranase, both of which hydrolyse high molecular weight substrates into low molecular weight constituents.

The EIHIA method is based on the inability of the enzyme to access substrate in the presence of excess antigen, as shown in Eq. 6.19:

$$(6.19)$$

In effect, antibody–antigen binding blocks the active site of the enzyme, and does not allow its approach to the insoluble substrate. The higher the molecular weight of the antigen, the greater is the inhibitory effect on enzyme activity in the enzyme–antibody–antigen complex. The assay kits for α-fetoprotein and ferritin have detection limits of 1 and 5 ng/mL, respectively, and show very good correlations with conventional RIA and ELISA methods.

6.5. EVALUATION OF NEW IMMUNOASSAY METHODS[23,24]

This section highlights some aspects of validation especially important for immunoassay methods. Some background in data analysis is assumed; methods generally used for evaluation and validation of new assays are described later in this book (Chapter 16) and can be consulted to clarify some of the concepts used in this section.

The sample matrix plays a more important role in immunoassays than in many other types of analysis. By matrix, we mean everything in the sample that is not the species of interest. In immunoassays, the matrix has many more components than a typical reagent blank used for colorimetry. The assay reactions are often very sensitive to the presence of proteins, the ionic strength, and the presence of lipids and enzymes. Matrix components may interact by protecting the analyte from adsorbing to glass or plastic, and so may even increase recoveries. Protein concentration is a major contributor to the matrix effect. For this reason, most immunoassay kits provide standards prepared in human or animal serum. Animal sera are often used for zero standards, and may be used throughout the curve. Therapeutic drug levels are often measured in nonserum samples such as cerebro spinal fluid, urine or ultrafiltrates; separate standardizations in these matrices are required, since errors arise when serum standard curves are used. Control runs must be performed to assess nonspecific binding of the assay reagents; these controls include all assay components except antibody, where an inert protein is substituted to maintain exactly the same matrix.

During methods evaluation, standard curves should be plotted in several ways, since different plots reveal different reaction characteristics. For example, the bound/free versus concentration plot is very steep at the low concentration end, and may be used to calculate the detection limit of the assay. The sigmoidal plots of bound/total against log concentration clearly show regions of insensitivity (low and high concentration ends) that might not be clearly apparent in the logit–log curve.

The low detection limits of immunoassays depend on the typically high affinities of antibodies for haptens and antigens, as well as the detection limits of the labels used. The detection limit is usually defined as the concentration that yields a signal that is equal to the mean of the blank signal plus two or three standard deviations. This establishes the confidence range for the zero response. For this calculation, the bound/free versus concentration plot is used. While detection limits allow comparisons of different immunoassay methods at the lower concentration end, they say nothing about assay reliability; for this reason, both detection limits and precision profiles should be compared.

Accuracy can be defined as the fundamental ability any assay to measure the true concentration of an analyte. Pure standards in a valid matrix are often difficult to obtain, and the biological reactivity of less pure standards may not parallel their immunoreactivity in the assay. On the other hand, standards may be pure, but the assay antibody may react with similar molecules or fragments; this cross-reactivity is discussed in detail below. Accuracy must be confirmed by comparison with an established method(s) using real samples.

Cross-reactivity (an indicator of assay specificity) has critical importance for immunoassay methods in which a particular analyte is assayed in the presence of very similar species; for example, in the monitoring of a therapeutic drug in serum where various metabolites of the drug are also present. For the sake of uniformity, cross-reactivity is usually reported as the mass or concentration of interferent

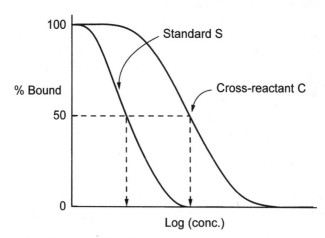

Figure 6.11. Cross-reactivity of an interferent in a competitive immunoassay.

required to displace 50% of the label, as shown in Figure 6.11.

$$\%C = 100 \times \frac{[S]_{50}}{[C]_{50}}$$

Cross-reactivity is evaluated by comparing the ability of each potential cross-reacting material to displace label. The percent cross-reactivity is equal to 100 times the concentration of analyte at 50% response divided by the concentration of interferent at 50% response.

Cross-reactivity depends on the selectivity of an antibody for a particular epitope, and can be controlled to some extent by the careful design of the immunogen used to produce antibodies. Both the site of attachment and the nature of the hapten linkage to the carrier protein are critical for the selectivity of the assay. Additionally, the hapten–carrier conjugate should be prepared in the same way as the hapten–label conjugate used in the assay, so that similar epitopes are available for antibody binding. Antibodies raised against a hapten conjugated to a carrier protein through a linking group sill often recognize the linking group; for example, the hapten 2,4-dinitrophenol can be linked to a carrier protein by the reaction of 2,4-dinitrofluorobenzene (Sanger's reagent) with the primary amine groups of lysine residues, yielding an antibody that preferentially binds 2,4-dinitrophenyllysine over 2,4-dinitrophenol.[25]

Heterology is the term used to describe the use of different hapten derivatives for the preparation of the immunogen and the preparation of the labeled hapten. There are three types of heterology that are recognized as being important to assay selectivity.[26] Hapten heterology occurs when different haptens are used to produce the immunogen and the labeled species; for example, an estradiol assay that produces the immunogen by linking estradiol to the carrier protein through the 11-hydroxy group, while the enzyme-labeled species was produced by linking esterone to peroxidase through the 11-hydroxy group, using the same linking reagent. Bridge heterology occurs when, for example, a succinyl linking group is used in the immunogen, and a glutaryl group is used for the hapten-label conjugate. Finally, site heterology occurs when the same linking group is connected to different sites on the hapten, such as the 11-hydroxy and 17-hydroxy groups on estrogen.

In the absence of nonspecific interferences, any heterologous combination of immunogen and labeled hapten result in better detection limits, often as much as 100-fold lower than those achievable with no heterology. The reason is that heterology reduces the association constant for the labeled hapten with the antibody, and free hapten will displace the labeled hapten more readily. Lower concentrations of free hapten are thus more readily detected.

While heterology improves the detection limits of an assay, it significantly increases the cross-reactions that occur with similar analytes. It has been found that antibodies with the best selectivity are produced when immunogens are prepared by coupling haptens at a site far from the regions where interfering species have slight structural differences. In an assay for morphine (Fig. 6.12), for example, the immunogen and the enzyme–hapten conjugate were prepared in exactly the

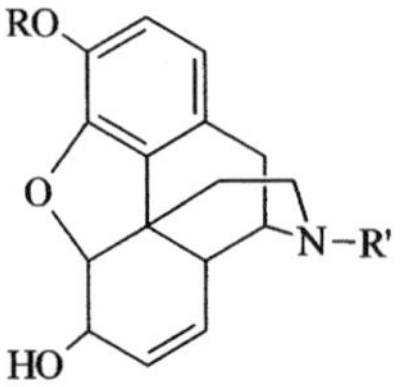

Name	R	R'
Morphine	H	CH$_3$
Codeine	CH$_3$	CH$_3$
Normorphine	H	H
Norcodeine	CH$_3$	H

Figure 6.12. Structures of morphine and interfering species.

same way for each assay tested, while the linking sites were varied from one assay to the next.[27]

The selectivity of each assay was determined by measuring the responses to the interfering species and to morphine at a fixed, equal concentrations. The structures of the conjugates, and the selectivities of the assays are shown in Figure 6.13.

These site heterology experiments with the morphine assay show that substitution at the R' position with linking groups yielded the best selectivity for morphine over codeine. The reason is that the R' group is the same for morphine and codeine, and the R' group is far from the R group, which is where morphine and codeine are structurally different. The antibodies raised to this conjugate thus have a paratope

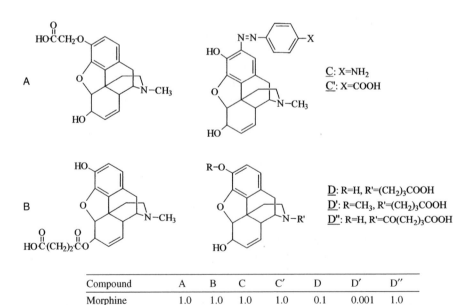

Compound	A	B	C	C'	D	D'	D''
Morphine	1.0	1.0	1.0	1.0	0.1	0.001	1.0
Codeine	1.0	1.0	0.83	0.1	0.0003	1.0	0.048

Figure 6.13. Selectivity of morphine assay in the presence of codeine.

complementary to this "protruding" region of the immunogen, and show high selectivity for morphine.

The cross-reactivity of polyclonal antibodies raised to an atrazine–protein conjugate (linked at the chlorine group) was evaluated, using a peroxidase-atrazine labeled species linked in the same manner, in order to determine the reliability of a newly available competitive immunoassay kit for environmental atrazine monitoring.[28] A wide variety of similar compounds (Fig. 6.14) were screened, and the concentrations given beside the compounds indicate the apparent concentration of atrazine that would be erroneously recorded from the responses obtained to 1-μg/mL concentrations of each of these interferents.

Cross-reactivity was highest with those compounds most closely resembling atrazine, that is, those possessing the 4-ethylamino and 6-isopropylamino groups. The substituent at the 2-position had little effect on response, and this is to be expected since this is the site used to link atrazine to both the carrier protein for the immunogen and the peroxidase label. The responses indicate that the relative response of the immunoassay is related to antibody recognition and binding to the N-alkyl side chains on the triazine ring.

Nonspecific interferences increase or decrease the apparent result of an assay, without reacting specifically with the antibody or antigen. Components such as

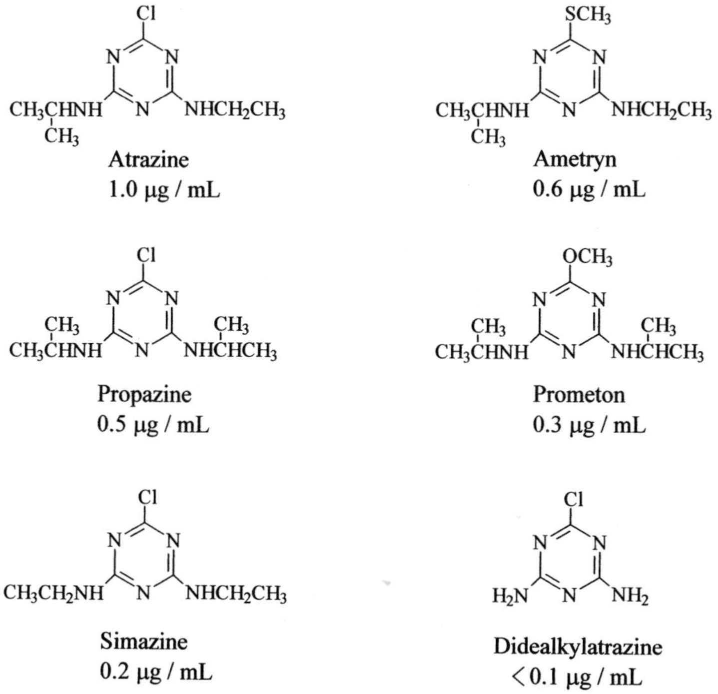

Figure 6.14. Chemical structures and immunoassay responses of 1-μg/mL concentrations of triazine interferents in an atrazine immunoassay kit.

lipids, pigments, and endogenous enzymes may adsorb or bind nonspecifically to the solid phase or to reagents, or may affect the detection step. Blocking reagents[29] are commonly used to prevent the nonspecific adsorption of proteins onto microtiter plates following antibody immobilization; albumin or other proteins are used to saturate areas of the support not containing antibody, so that further binding is prevented. A unique interference is present in some samples that contain antibodies to one of the reactants in the immunoassay; for example, endogenous *anti*-insulin is present in the serum of patients treated with beef or pork insulin, and invalidates the measurement of insulin levels by some immunoassays.[30] Interferences in thyroxine assays include human serum albumin and some genetic variants, which have high affinity for thyroxine.[31,32]

SUGGESTED REFERENCES

T. T. Ngo, Ed., *Nonisotopic Immunoassay*, Plenum Press, New York, 1988.

T. T. Ngo and H. M. Lenhoff, Eds., *Enzyme-Mediated Immunoassay*, Plenum Press, New York, 1985.

E. T. Maggio, *Enzyme-Immunoassay*, CRC Press, Boca Raton, Florida, 1980.

R. M. Nakamura, Y. Kasahara, and G. A. Rechnitz, Eds., *Immunochemical Assays and Biosensor Technology for the 1990s*, American Society for Microbiology, Washington, DC, 1992.

C. P. Price and D. J. Newman, Eds., *Principles and Practice of Immunoassay*, Macmillan (Stockton Press), New York, 1991.

D. Wild, Ed., *The Immunoassay Handbook*, 2nd ed., Nature Publishing Group, New York, 2001.

REFERENCES

1. R. M. Nakamura, Y. Kasahara, and G. A. Rechnitz, Eds., *Immunochemical Assays and Biosensor Technology for the 1990s*, American Society for Microbiology, Washington DC, 1992.

2. P. K. Nakane, *Ann. N. Y. Acad. Sci.* **254**, 1975, 203–211.

3. J. B. Gushaw, M. W. Hu, P. Singh, J. G. Miller, and R. S. Schneider, *Clin. Chem.* **23**, 1977, 1144.

4. B. F. Erlanger, *Pharmacol. Rev.* **25**, 1973, 271–280.

5. B. F. Erlanger, F. Borek, S. M. Beiser, and S. Lieberman, *J. Biol. Chem.* **228**, 1957, 713–727.

6. B. F. Erlanger, F. Borek, S. M. Beiser, and S. Lieberman, *J. Biol. Chem.* **234**, 1959, 1090–1094.

7. D. Exley and R. Abuknesha, *FEBS Lett.* **79**, 1977, 301–304.

8. G. L. Nicolson and S. J. Singer, *J. Cell Biol.* **60**, 1974, 236–248.

9. J. L. Guesdon and S. Avrameas, *Immunochemistry* **14**, 1977, 443–447.

10. R. M. Nakamura, A. Voller, and D. E. Bidwell, in *Handbook of Experimental Immunology, Volume 1: Immunochemistry*, 4th ed., D. M. Wier, Ed., Oxford (Blackwell Scientific), New York, 1986. pp. 27.1–27.20.

11. D. Hawcroft, T. Hector, and F. Rowell, *Quantitative Bioassay*, John Wiley & Sons, Inc., New York, 1984, pp. 158–159.

12. T. M. Jackson and R. P. Ekins, *J. Immunol. Methods* **87**, 1986, 13–20.

13. I. Hemmila, *Clin. Chem.* **31**, 1985, 359–370.

14. D. S. Smith, M. H. H. Al-Hakiem, and J. Landon, *Ann. Clin. Biochem.* **18**, 1981, 253–274.

15. D. Hawcroft, T. Hector, and F. Rowell, *Quantitative Bioassay*, Wiley, New York, 1984, p. 172.

16. E. P. Diamandis and T. K. Christopoulos, *Anal. Chem.* **62**, 1990, 1149A–1157A.

17. D. Hawcroft, T. Hector, and F. Rowell, *Quantitative Bioassay*, Wiley, New York, 1984, pp. 183–185.

18. A. Johannsson, in *Principles and Practice of Immunoassay*, C. Price and D. Newman, Eds., Macmillan, New York, 1991, pp. 300–303.

19. R. M. Nakamura, Y. Kasahara, and G. A. Rechnitz, Eds., *Immunochemical Assays and Biosensor Technology for the 1990s*, American Society for Microbiology, Washington, DC, 1992.

20. C. Price and D. Newman, in *Principles and Practice of Immunoassay*, C. Price and D. Newman, Eds., Macmillan, New York, 1991, pp. 393–416.

21. W. B. Dandliker, R. J. Kelly, J. Dandliker, J. Farquhar, and J. Levin, *Immunochemistry* **10**, 1973, 219–227.

22. P. Khanna, in *Principles and Practice of Immunoassay*, C. Price and D. Newman, Eds., Macmillan, New York, 1991, p.327.

23. R. M. Nakamura, Y. Kasahara, and G. A. Rechnitz, Eds., *Immunochemical Assays and Biosensor Technology for the 1990s*, American Society for Microbiology, Washington, DC, 1992.

24. R. Ekins, in *Principles and Practice of Immunoassay*, C. Price and D. Newman, Eds., Macmillan, New York, 1991, pp. 96–153.

25. H. N. Eisen and G. W. Siskind, *Biochemistry* **3**, 1964, 996–1008.

26. D. S. Kabakoff, in *Enzyme-Immunoassay*, E. T. Maggio, Ed., CRC Press, Boca Raton, FL, 1980, p. 79.

27. B. K. Van Weemen and A. H. W. M. Schuurs, *Immunochemistry* **12**, 1975, 667–670.

28. D. A. Goolsby, E. M. Thurman, M. L. Clark, and M. L. Pomes, in *Immunoassays for Trace Chemical Analysis* (ACS Symp. Ser. 451), M. Vanderloon, L. H. Stanker, B. E. Watkins, and D. W. Roberts, Eds., American Chemical Society, Washington, DC, 1990, pp. 86–99.

29. A. M. Campbell, *Monoclonal Antibody and Immunosensor Technology*, Elsevier, New York, 1991, pp. 68–69.

30. M. Rendell, R. G. Hamilton, H. M. Drew, N. F. Adkinson, Jr., *Am. J. Med. Sci.* **282**, 1981, 18–26.

31. C. E. Petersen, C. E. Ha, M. Mardel, and N. V. Bhagavan, *Biochem. Biophys. Res. Commun.* **214**, 1985, 1121–1129.

32. J. R. Stockigt, V. Stevens, E. L. White, and J. W. Barlow, *Clin. Chem.* **29**, 1983, 1408–1410.

PROBLEMS

1. In the binding reaction between each of the following pairs of reactants, indicate which component would normally be regarded as the ligand: (a) estrogen and anti-estrogen antibody, (b) goat anti-human α-fetoprotein and human α-fetoprotein, (c) mouse antibody and rabbit anti-mouse Ig, and (d) nicotinic acetylcholine receptor and acetylcholine?

2. In a heterogeneous, competitive, labeled-ligand immunoassay: (a) which component is bound to the solid phase, (b) which component(s) is/are present at limited concentration, and (c) which component(s) is/are present in excess?

3. From the list of immunoassay labels given in Table 6.1, classify each label type as giving either (a) a single detectable event, (b) a continuous signal, or (c) catalytic signal amplification.

4. A new strategy has recently been developed for the competitive sandwich immunoassay of small molecules, using automated ELISA technology and polystyrene microtiter plates. The method is based on the formation and quantitation of the complex Ab:H—H:Ab*, where Ab and Ab* are the immobilized and labeled antibodies, respectively, and H—H is a synthetic bis-analogue of the hapten H. The bis-analogue competes with monomeric hapten (the analyte) for immobilized antibody binding sites. This assay has been tested for the model analyte testosterone (A), using a competing bis(testosterone) species (B) for the competitive sandwich immunoassay, and antibodies (both Ab and Ab*) that were raised to the immunogen shown in (C), shown in Eq. 6.20.

(6.20)

(a) Which of the four species, Ab, Ab*, H—H, and H, must be present at limited concentration, and which must be present in excess?

(b) In addition to the species Ab:H—H:Ab*, which other species involving Ab and any of Ab*, H—H, and H will exist on the polystyrene microtiter plates following the final incubation with Ab* and rinsing away of excess reagents?

(c) Three potential interferents are shown below in structures D, E, and F (Eq. 6.21). Which of these is likely to present the greatest cross-reactivity in the competitive sandwich immunoassay, and which is expected to present the least cross-reactivity? Why?

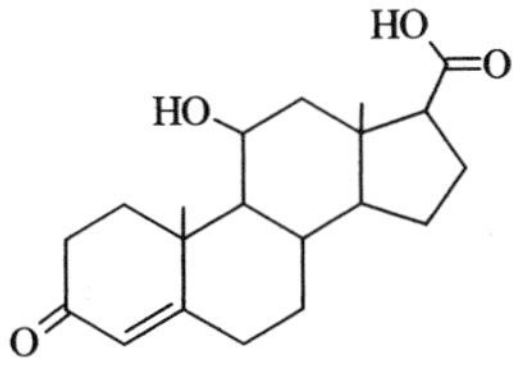

D 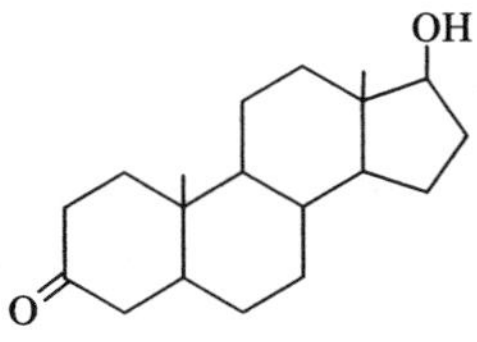   E   (6.21)

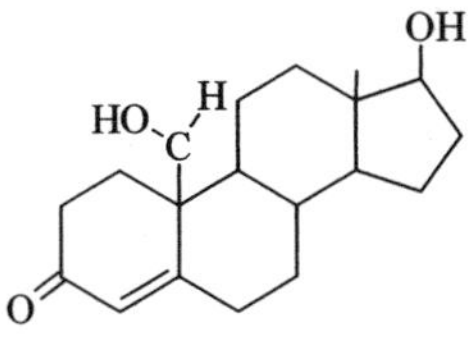

F

(d) Draw a plot of label signal versus log[testosterone] that would be expected in this type of assay.

5. Biotin deficiency results from a number of inborn errors of metabolism, and biotin quantitation therefore represents a general screening method for these diseases. Recently, a new method was proposed for biotin quantitation in serum. The method has been called an ELLSA, and relies on the following competitive reaction:

$$\text{Biotin}_s + \text{Biotin}_f + \text{Avidin-E} \longrightarrow \text{Biotin}_s : \text{Avidin-E} + \text{Biotin}_f : \text{Avidin-E}$$

where Biotin_s and Biotin_f are the adsorbed and free (analyte) forms of biotin, respectively, Avidin-E is an Avidin–alkaline phosphatase conjugate that is added in limited quantity. Following an incubation period, reagents not bound to the microtiter plates are rinsed away, and the remaining enzyme activity is quantitated using a fixed time method after the addition of excess substrate.

(a) Sketch the plot of enzyme activity versus log[Biotin_f] that would be expected for this type of assay.

(b) A synthetic substrate (**I**, below) was used for enzyme activity measurements. This substrate undergoes dephosphorylation in the presence of alkaline phosphatase to yield an unstable intermediate (**II**) that decomposes in a chemiluminescent reaction (Eq. 6.22):

$$\text{(6.22)}$$

If the K_m of alkaline phosphatase for compound **I** is 0.10 mM, what concentration of this substrate should be used for the enzyme activity determination?

(c) Show the plot of light intensity versus time that would be obtained by the continuous measurement of light emitted in a control well, where no free biotin had been added to the reaction mixture. How does this differ from a typical absorbance versus time plot that would be obtained if the substrate yielded a colored (absorbing) product?

6. In a heterogeneous sandwich immunoassay employing fluorescence excitation transfer (Eq. 6.7 and Fig. 6.6), draw the plots of emission intensity versus log[Ag] that would be expected if emission from (a) F1 and (b) F2 were measured.

CHAPTER 7

Biosensors

7.1. INTRODUCTION

The real-time and *in situ* quantitation of biologically and environmentally important analytes has long been a goal of analytical research. The need for continuous monitoring of substrates in fermentation broths, pesticides, and environmental contaminants in natural waters, and biochemicals or pharmaceutical metabolites in living organisms has led to extensive activity in the field of biosensor research. Biosensors are devices, ideally small and portable, that allow the selective quantitation of chemical and biochemical analytes. They consist of two components: the transducer and the chemical recognition element. Chemical recognition is accomplished by exploiting the natural selectivities of biochemical species such as enzymes, antibodies, chemoreceptors, and nucleic acids. In the presence of the analyte, these agents, immobilized at the surface of the transducer, cause a change in a measurable property in the local environment near the transducer surface. The transducer monitors this property, and converts the chemical recognition event into a measurable electronic signal. Transducers may measure electrochemical, optical, thermal, or adsorption processes that change in the presence of the analyte.

The first biosensor was introduced by Clarke in 1962,[1] and consisted of an enzyme, glucose oxidase (GO) trapped at the surface of a platinum electrode by a semipermeable dialysis membrane that allowed substrates and products to freely diffuse to and from the enzyme layer, as shown in Figure 7.1. The rate of the enzymatic reaction is proportional to the substrate concentration in the external solution. The conversion of glucose and molecular oxygen to gluconolactone and hydrogen peroxide may be monitored electrochemically, by reoxidizing hydrogen peroxide to molecular oxygen at the surface of the platinum electrode. The generated current depends on the local H_2O_2 concentration, and therefore on the bulk glucose concentration.

The magnitude of the response current of the amperometric glucose biosensor shown in Figure 7.1 depends on four main factors. First, mass-transport kinetics determine the rates at which substrates can be supplied to, and products removed from, the reaction layer in which enzyme is trapped. Second, enzyme kinetics

Bianalytical Chemistry, by Susan R. Mikkelsen and Eduardo Cortón
ISBN 0-471-54447-7   Copyright © 2004 John Wiley & Sons, Inc.

Working electrode (Pt)

Auxilliary electrode Reference electrode

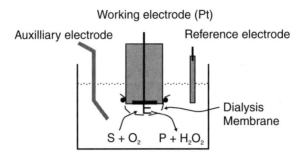

Dialysis
Membrane

S + O$_2$ P + H$_2$O$_2$

Figure 7.1. Amperometric glucose biosensor using GO for chemical recognition.

determine the rate at which H$_2$O$_2$, the detected species, is generated within this reaction layer. Third, electrochemical or heterogeneous reaction kinetics determine the rate at which H$_2$O$_2$ produced enzymatically can be converted to O$_2$ to generate the measured signal. Final, some H$_2$O$_2$ will escape across the semipermeable membrane to the bulk of the solution without undergoing oxidation; the fraction oxidized is called the conversion efficiency, and measured signals are directly related to this factor. It is important to remember that all of these processes occur only within the thin reaction layer at the surface of the transducer; steady-state concentrations of reactants and products exist within this layer when a constant transducer signal is achieved. Enzyme-based biosensors do not generally alter concentrations in the bulk of the analyte solution.

The glucose biosensor shown in Figure 7.1 represents the first successful combination of a chemical recognition agent with a transducer, but it possesses a number of deficiencies that preclude its use for *in vivo* or *in situ* glucose quantitation. An obvious difficulty for *in vivo* use is its size, but even with miniaturized versions of this sensor, difficulties associated with immunological responses to the membrane material exist. In addition, the response currents achieved in both *in vivo* and *in situ* environments will depend not only on analyte (glucose) concentration, but also on the concentration of dissolved oxygen, which may be very low, or variable, leading to unreliable results. Much modern biosensor research is directed toward "reagentless" biosensors, where the concentration of a single analyte species determines the response.

Since enzymes are the chemical recognition agents most often involved in the sensing process, this chapter begins with a theoretical description of the responses of enzyme-based biosensors, and identifies some parameters of importance for optimizing response. A representative survey of the types of transducers and chemical recognition agents currently being investigated is then followed by a discussion of biosensor evaluation methods.

7.2. RESPONSE OF ENZYME-BASED BIOSENSORS[2]

The theoretical description of the response of enzyme-based biosensors assumes that within the enzyme layer, steady-state concentrations of substrates and products exist. Because the reaction is effectively occurring at a two-dimensional (2D) surface, rather than in a bulk, or three-dimensional (3D) medium, the rates of the various processes involved in signal generation are represented as fluxes, and the magnitudes of these fluxes depend on distance from the transducer surface (i.e., concentration gradients exist). The flux, j, is equal to the homogeneous reaction rate multiplied by the thickness of the enzyme reaction layer, l, as given in Eq. 7.1.

$$j = (\text{rate}) \times l \tag{7.1}$$

For a homogeneous reaction rate given in moles per cubic centimeter second [mol/ $(cm^3 s)$] and an enzyme layer thickness given in centimeters (cm), flux units of mol/ $(cm^2 s)$ are obtained. The importance of the transducer surface area in determining the magnitude of the response is thus immediately apparent. The 2D nature of the reaction results in concentration gradients for both substrate and product: At the surface of the transducer, substrate concentrations are lower, and product concentrations are higher than those in the bulk solution. These concentration gradients exist over a distance called the diffusion layer thickness, δ. The diffusional transport of substrate across this layer is represented by the reaction given in Eq. 7.2, and occurs at a rate proportional to the bulk-to-surface concentration difference, Eq. 7.3:

$$S^* \xrightarrow{k_S} S_0 \tag{7.2}$$
$$j_S = k_S([S]^* - [S]_0) \tag{7.3}$$

where j_S is the substrate flux, $[S]^*$ is the bulk and $[S]_0$ is the surface substrate concentration, and k_S is the mass transport rate constant, which is equal to the substrate diffusion coefficient, D_S, divided by the diffusion layer thickness, Eq. 7.4:

$$k_S = D_S/\delta \tag{7.4}$$

Once substrate has arrived in the reaction layer, simple one-substrate Michaelis–Menten kinetics (Eq. 7.5 for oxidoreductase enzymes, where E_O and E_R represent oxidized and reduced enzyme, respectively) yield fluxes for the formation of the enzyme–substrate complex (j_C) and the decomposition of this complex to form product (j_P), Eqs. 7.6 and 7.7:

$$E_O + S \underset{k_{-1}}{\overset{k_1}{\rightleftharpoons}} ES \xrightarrow{k_2} E_R + P \tag{7.5}$$

$$j_C = l(k_1[E_O][S] - k_{-1}[ES]) \tag{7.6}$$

$$j_P = lk_2[ES] \tag{7.7}$$

These fluxes are simply the homogeneous reaction rates multiplied by the thickness of the enzymatic reaction layer, as given in Eq. 7.1.

For amperometric transducers, it is the regeneration of E_O by reaction with an oxidant (O_2 or another mediating species M^{2+}) that results in the species actually measured by the transducer. This reaction, Eq. 7.8, results in a flux for the regeneration of E_R given by Eq. 7.9:

$$E_R + M^{2+} \xrightarrow{k_E} E_O + M \tag{7.8}$$

$$j_E = lk_E[E_R][M^{2+}] \tag{7.9}$$

Recalling that at all distances from the transducer surface, the total enzyme concentration is equal to the sum of the concentrations of the three enzyme species ($[E]_T = [E_O] + [ES] + [E_R]$), and setting $j = j_S = j_C = j_P = j_E$ (as a direct result of the steady-state assumption), the flux j is described by Eq. 7.10:

$$1/j = (K'_m/\{lk_2[E]_T([S]^* - j/k_S)\}) + 1/\{lk_2[E]_T\} + 1/\{lk_E[E]_T[M^{2+}]\} \tag{7.10}$$

As written in the inverse form, the flux is separated into three component terms. The first describes substrate transport and ES complex formation, the second describes the decomposition of the ES complex, and the third term describes the regeneration of E_O by oxidant. Ideally, the decomposition of ES and the regeneration of E_O are rapid with respect to complex formation, so that $[S]_0 = 0$. In this case, only the first term of Eq. 7.10 is significant, and it can be simplified to Eq. 7.11:

$$1/j = K'_m/\{lk_2[E]_T[S]^*\}) + 1/\{k_S[S]^*\} \tag{7.11}$$

The first term of Eq. 7.11 is related to enzyme kinetics, while the second term gives the flux dependence on substrate transport to the reaction layer. It can be seen that the flux increases with increasing substrate (analyte) concentration, with increasing enzyme concentration in the reaction layer, and with increasing reaction layer thickness.

The current measured by an amperometric electrode is directly proportional to the flux described in Eq. 7.11, with proportionality constants n (electrons in the stoichiometric electrochemical reaction), F (Faraday's constant, 96,487 C/mol), A (electrode area) and B (fractional collection efficiency):

$$i = nFABj \tag{7.12}$$

By substituting for the flux, we obtain

$$i = nFAB\{lk_2[E]_T k_S/(K'_m k_S + lk_2[E]_T)\}[S]^* \tag{7.13}$$

It can be seen from Eq. 7.13 that the measured signal is directly proportional to analyte concentration. This relationship will hold as long as $[S]_0 = 0$, that is, at analyte concentrations low enough for complete and rapid conversion to product to occur within the reaction layer. At higher analyte concentrations, $[S]_0 > 0$, and the measured current plateaus at a limiting value. This expression was derived using a number of assumptions that allowed simplification; for example, uniform concentrations of substrate and product within the reaction layer (no gradients) and the rapid decomposition of the ES complex and rapid regeneration of E_O were assumed. More detailed derivations have been performed that do not rely on these simplifications.[3,4]

Many enzyme-based biosensors do not rely on the consumption of one of the products for the generation of a signal. For example, optical fiber transducers rely on local absorbance or fluorescence measurements, where signal magnitudes are related to steady-state product concentrations. Similarly, potentiometric transducers measure, but do not consume, product. These biosensors are described by solving the steady-state flux equations for the local product concentration [P] within the reaction layer (the solution of these equations is greatly simplified by assuming uniform concentrations within the reaction layer). The analytically useful regions of these curves again occur for $[S]^* < K'_m$, and [P] is then given by Eq. 7.14:

$$[P] = (1 + lk_2[E]_T/k_S K'_m)[S]^* \tag{7.14}$$

For optical transducers, the measured signals are directly proportional to [P], so that, once again, reaction layer thickness and mass-transport kinetics determine the sensitivity of the biosensor, and signals are directly proportional to analyte concentration. For potentiometric transducers, signals are proportional to log[P], and therefore to $\log[S]^*$.

While enzyme-based biosensors employ only one of many possible chemical recognition agents, they are the devices that are currently the most thoroughly understood and described. Other recognition agents, such as antibodies and chemoreceptors, rely on primary binding interactions for the preconcentration of detectable species at the surface of a transducer. The following section shows examples of the diversity of possible sensing configurations.

7.3. EXAMPLES OF BIOSENSOR CONFIGURATIONS

7.3.1. Ferrocene-Mediated Amperometric Glucose Sensor[5]

This enzyme-based biosensor uses glucose oxidase (GO) as a chemical recognition element, and an amperometric graphite foil electrode as the transducer. It differs from the first reported glucose biosensor discussed in the introduction to this chapter in that a mediator, 1,1'-dimethylferricinium, replaces molecular oxygen as the oxidant that regenerates active enzyme. The enzymatic reaction is given in Eq. 7.15, and the electrochemical reaction that provides the measured current is shown in Eq. 7.16.

$$\beta\text{-D-Glucose} + 2(C_5H_4CH_3)_2Fe^+ \xrightarrow{\text{GO}} \text{Gluconolactone} + 2(C_5H_4CH_3)_2Fe + 2H^+ \tag{7.15}$$

$$(C_5H_4CH_3)_2Fe \longrightarrow (C_5H_4CH_3)_2Fe^+ + e^- \tag{7.16}$$

While GO is very selective for its primary substrate, β-D-glucose, a number of low molecular weight oxidants may be used in place of molecular oxygen. The ferrocene species used in this sensor is coadsorbed to the surface of the graphite foil in its reduced form with GO, and the sensor surface is then covered with a dialysis membrane. The resulting sensor is used with reference and counterelectrodes, and the potential is maintained at +0.16 V versus SCE to generate the ferricinium form of the mediator. At this potential, any H_2O_2 generated by the competing reaction of reduced enzyme with molecular oxygen remains undetected, and the low applied potential ensures that electroactive interferences present in biological fluids, such as ascorbic acid, will not be oxidized. Figure 7.2 shows a diagram of the sensor, and the glucose calibration curves that are recorded in nitrogen-saturated, air-saturated, and oxygen-saturated solutions.

The calibration curves in Figure 7.2(b) show that very little difference in electrochemical signal is observed for solutions in which oxygen is absent and for those in which oxygen is present at ambient levels (0.2 mM). Significant interference

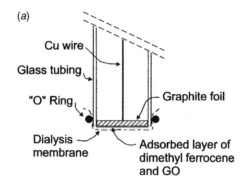

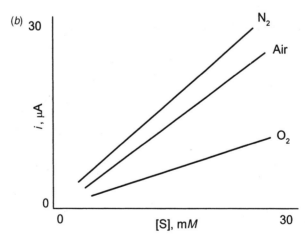

Figure 7.2. (a) Ferrocene-mediated glucose sensor; (b) calibration curves recorded in N₂, air, and O₂-saturated solutions.

from the competing reaction with oxygen occurs only at high (saturating) oxygen concentrations of ~ 1.2 mM.

These principles have been incorporated into home testing devices for diabetics, using a different ferrocene derivative. A drop of patient blood is placed onto a test strip containing (a) ferrocene and GO adsorbed onto a conductor, and (b) a counter-electrode conductor. The test strip is inserted into a portable, handheld potentiostat, and the current measured by the device is directly related to the blood glucose concentration. A similar device for the quantitation lactate is available, and another, for cholesterol quantitation, is to be introduced in the near future.

7.3.2. Potentiometric Biosensor for Phenyl Acetate[6]

Phenyl acetate has been chosen as a model analyte for the evaluation of *catalytic antibodies* as chemical recognition agents. A catalytic antibody is an antibody raised to an immunogen, where the immunogen contains a transition state analogue for a chemical reaction. In this example, the reaction is the hydrolytic cleavage of the ester linkage of phenyl acetate:

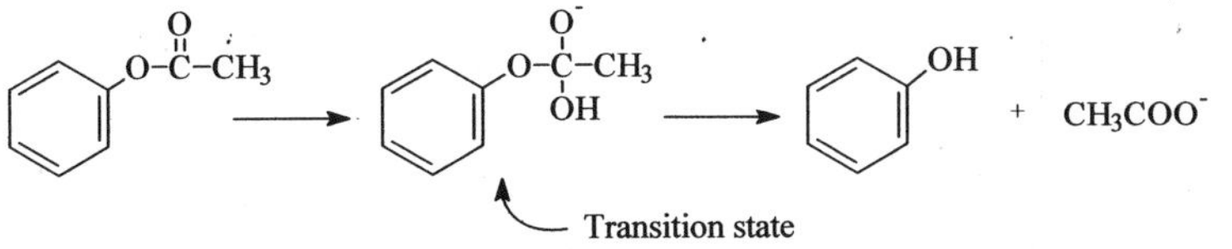

$$\text{(7.17)}$$

Antibodies raised to the transition state analogue (Fig. 7.3) will bind to the transition state of the hydrolysis reaction, lowering the activation energy and therefore catalyzing the reaction. These antibodies were trapped at the surface of a pH electrode using a dialysis membrane [Fig. 7.4(a)]. The reaction (Eq. 7.17) produces a change in local pH at the surface of the electrode, since acetic acid is one of the products. The measured pH therefore decreases as the phenyl acetate concentration increases in the external solution, since the steady-state concentration of acetic acid in the reaction layer increases [Fig. 7.4(b)].

The linear dynamic range of the sensor's response to phenyl acetate is 0.02–0.50 mM, and the detection limit has been reported as 5 μM. Studies of the sensor's selectivity have shown that a number of structurally similar compounds yield signals, particularly those containing the $RCOOC_6H_5$ group. However, the selectivity is similar to that of hydrolytic enzymes. Research has also shown that ambient

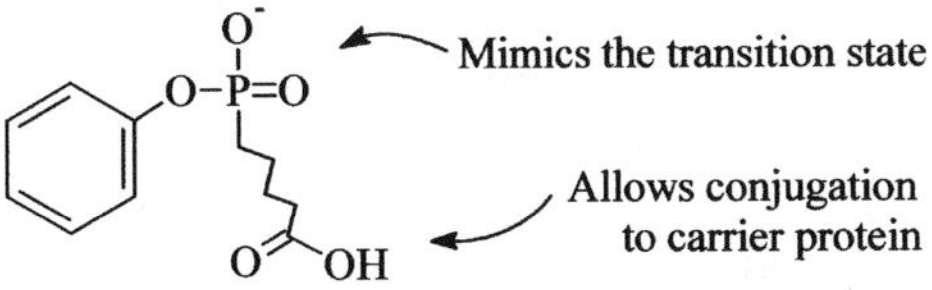

Figure 7.3. Transition state analogue hapten used for the generation of catalytic antibodies for phenyl acetate hydrolysis.

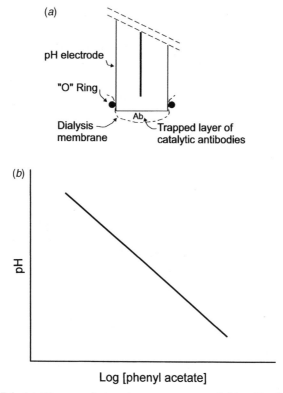

Figure 7.4. (*a*) Diagram of phenyl acetate sensor and (*b*) calibration curve.

buffer concentration is critical to the magnitude of the signals obtained, and an inverse linear dependence of log(signal) against log[buffer] was obtained over the 10^{-4}–10^{-2} M range for a Tris buffer (0.25 mM).

The phenyl acetate biosensor is not of great commercial importance, but it illustrates a novel use of antibodies as chemical recognition agents. In principle, catalytic antibodies may be raised for any reactions for which stable transition state analogues may be prepared.

7.3.3. Potentiometric Immunosensor for Digoxin[7]

This biosensor employs a potassium (K^+) ion-selective electrode as the transducer, and chemical recognition involves the *anti*-digoxin antibody and a digoxin-crown ether conjugate. A diagram of the sensor is shown in Figure 7.5.

In this device, a poly(vinyl chloride) (PVC) membrane separates an internal K^+ solution from an external layer of digoxin antibody that is trapped at the sensor surface with a dialysis membrane. The digoxin-crown ether conjugate, which selectively binds K^+, is present only in the PVC layer due to its hydrophobic properties. Competition between free digoxin (analyte) that freely crosses the

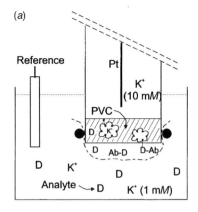

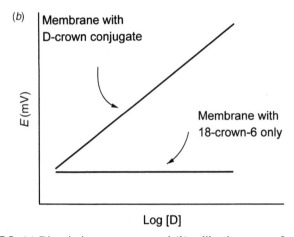

Figure 7.5. (*a*) Digoxin immunosensor and (*b*) calibration curves for digoxin.

dialysis membrane and the digoxin–crown conjugate at the PVC–aqueous interface for antibody in the reaction layer causes a change in the transport of K^+ across the PVC membrane, because the antibody–digoxin–crown conjugate is held at the interface. This biosensor incorporates the principles of competitive immunoassay, and allows the binding interaction to be monitored as it proceeds, by measuring potential as a function of time when digoxin is added to the analyte solution. The response time of this sensor is slow ($\sim$ 10 min) and this has precluded its use in practical situations.

7.3.4. Evanescent-Wave Fluorescence Biosensor for Bungarotoxin[8]

This biosensor employs a quartz optical fiber as a transducer, and the chemical recognition element is the nicotinic acetylcholine receptor (AcChR). The receptor is a membrane protein that spans a lipid bilayer; it binds acetylcholine rapidly and reversibly, and changes shape upon binding, to allow the transport of ions through a

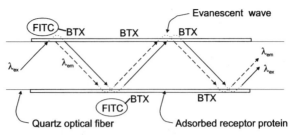

Figure 7.6. Principles of evanescent wave fluorescent biosensor for BTX.

channel created through the membrane. Concentration gradients of sodium and potassium ions exist across membranes *in vivo*, and an open ion channel allows the flow of ions, and therefore an ion current exists across the membrane. The AcChR is isolated from the electroplax tissue of electric eels and rays, where it is found at high levels. The isolated receptors bind AcCh and its analogues, as well as α-bungarotoxin (BTX, a protein isolated from snake venom) and neurotoxins such as tubocurarine. The receptor behaves like an antibody in that it has selective-binding sites, but it possesses no catalytic properties. It may be immobilized by standard protein immobilization methods.

The AcCh biosensor uses competitive binding between FITC labeled or unlabeled bungarotoxin and receptor immobilized by adsorption onto the surface of a quartz optical fiber. Once bound to the protein BTX, FITC exhibits absorption and emission maxima of 495 nm and 520 nm, respectively. In the absence of analyte (BTX), a maximal quantity of FITC–BTX binds to the immobilized receptor protein.

The fiber optic transducer exploits total internal reflection in acting as a waveguide with minimal loss of light intensity between the fiber ends. Such total internal reflection systems exhibit an *evanescent wave* at the quartz-solution boundary (i.e., along the length of the fiber) that penetrates the surrounding medium to a distance of about one-half the excitation wavelength. The evanescent wave allows excitation of FITC bound near the quartz surface, and allows capture of the light emitted by FITC at 520 nm, as shown in Figure 7.6.

As in competitive immunoassays, competition between FITC-labeled and unlabeled BTX results in signal changes—in this case, the intensity of light emitted at 520 nm decreases with increasing BTX concentration. Any analyte species that binds to the BTX site on the AcCh receptor will displace labeled BTX, and cause a signal decrease. Figure 7.7 shows curves obtained for three analytes: BTX, tubocurarine, and carbamylcholine. The curve obtained for AcCh falls between those obtained for carbamylcholine and tubocurarine. Because the binding reaction for FITC–BTX with receptor is essentially irreversible, a kinetic method is used to determine signal magnitudes. The change in fluorescence intensity with time is monitored for several minutes, for FITC–BTX solutions alone, and for solutions containing FITC–BTX and an analyte. In the presence of analyte, the fluorescence increase with time is less than in the absence of analyte. The irreversible

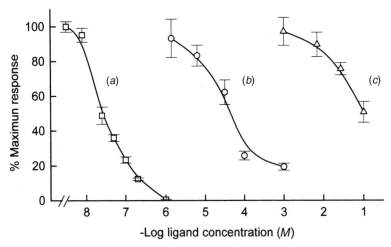

Figure 7.7. Calibration curves of relative fluorescence response against log concentration for three analytes at the fiber/AcChR/FITC-BTX sensor: (*a*) BTX; (*b*) tubocurarine; and (*c*) carbamylcholine. [Reprinted, with permission, from K. R. Rogers, J. J. Valdes, and M. E. Eldefrawi, *Analytical Biochemistry* **182**, 1989, 353–359. "Acetylcholine Receptor Fiber-Optic Evanescent Fluorosensor". Copyright © 1989 by Academic Press, Inc.]

nature of the reaction necessitates cleaning of the fiber surface and readsorption of fresh receptor between measurements, so that the sensor is not considered a reusable or continuous-use device.

7.3.5. Optical Biosensor for Glucose Based on Fluorescence Energy Transfer[9]

This glucose sensor relies on macromolecules trapped at one terminus (the distal end) of an optical fiber for signal generation. The macromolecules, FITC-labeled dextran and rhodamine-labeled concanavalin A (Rh–ConA), are trapped by a hollow dialysis fiber that fits snugly over the end of the optical fiber, as shown in Figure 7.8(*a*). A competitive equilibrium is set up between glucose and FITC–dextran, both of which bind to Rh–ConA:

$$\text{Glucose} + \text{Rh–ConA} \rightleftharpoons \text{Glucose : ConA–Rh} \qquad (7.18)$$
$$\text{(analyte)} \quad \text{(receptor)} \qquad \text{(analyte : receptor)}$$

$$\text{FITC–Dextran} + \text{Rh–ConA} \rightleftharpoons \text{FITC–Dextran : ConA–Rh} \qquad (7.19)$$
$$\text{(competitor)} \quad \text{(receptor)} \qquad \text{(competitor : receptor)}$$

FITC is excited at 490 nm and emits light at 520 nm. Rhodamine has an absorption maximum at 550 nm, and emits at 580 nm. Figure 7.8(*b*) shows the overlap between the emission spectrum of FITC and the absorption spectrum of rhodamine.

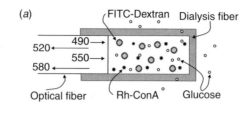

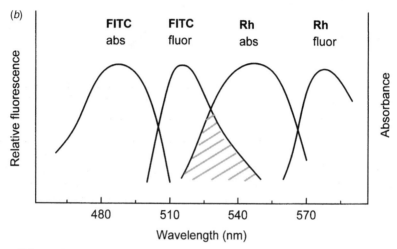

Figure 7.8. (*a*) Glucose biosensor; (*b*) spectral properties of FITC and Rh. [Reprinted, with permission, from D. Meadows and J. S. Schultz, *Talanta* **35** (No. 2), 1988, 145–150. "Fiber-Optic Biosensors Based on Fluorescence Energy Transfer". Copyright © 1988 Pergamon Journals Ltd.]

The emission intensity of FITC is monitored at 520 nm as the analytical signal. When FITC-labeled dextran is bound to Rh–ConA, the rhodamine label quenches the 520-nm emission, so that the intensity measured is at a minimum. In the presence of the analyte (glucose), which diffuses across the dialysis fiber at the tip of the optical fiber, FITC–dextran is displaced from Rh–ConA by glucose, and emission intensity at 520-nm increases. This sensor allows glucose quantitation at concentrations up to ~ 5 mM, and has a relatively slow response due to equilibration of the macromolecular reactions.

7.3.6. Piezoelectric Sensor for Nucleic Acid Detection[10]

This device employs single-stranded poly(adenylic acid) [poly(A)] as the chemical recognition agent. This species selectively recognizes its complementary polymer, poly(U), through hybridization to form a double-stranded nucleic acid. The poly(A) is immobilized onto the activated surface of a quartz piezoelectric crystal, which is a mass-sensitive transducer. Electric dipoles are generated in anisotropic materials (such as quartz crystals) subjected to mechanical stress, and these materials will then undergo dimensional changes, or oscillations, in the presence of an electric field. The frequency of oscillation is of the order of megahertz (MHz, 10^6 Hz). This resonant frequency decreases if mass is added to the surface of the crystal; in fact, it has been shown that the resonance frequency is inversely proportional to the surface mass. These transducers are sensitive to humidity, so that measurements must be made under dry or reproducibly humid conditions; *in situ*

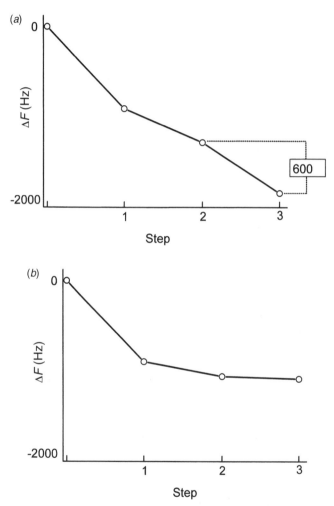

Figure 7.9. Change in resonant oscillation frequency of a piezoelectric transducer (*a*) with, and (*b*) without poly(A) immobilized to the surface. Step 1 is surface modification with copolymer, Step 2 is poly(A) immobilization, and Step 3 is hybridization to target poly(U).[10] [Reprinted, with permission, from N. C. Fawcett, J. A. Evans, L.-C. Chien, and N. Flowers, *Anal. Lett.* **21**, 1988, 1099–1114. "Nucleic Acid Hybridization Detected by Piezoelectric Resonance". Copyright © 1988 by Marcel Dekker, Inc.]

measurements in solution are not possible because of viscous damping of the oscillations.

The nucleic acid sensor was prepared by first derivatizing the quartz surface with a 3:1 styrene–acrylic acid copolymer. Poly(A) was then covalently immobilized onto pendant carboxylic acid groups by amide bond formation with the amino groups on the adenine base. Hybridization occurred during incubation with poly(U). Following each of the three steps, the sensor was rinsed and dried and the resonant frequency of oscillation was measured. Prior to any treatment, the quartz crystals exhibited a resonant frequency of 9 MHz. Because each step in the surface treatment involved the addition of mass to the crystal surface, a frequency decrease was expected after each step. Figure 7.9 shows the actual frequency changes measured for a sensor prepared as described above, as well as for a sensor prepared with no poly(A) included in the second step (i.e., a control sensor).

While this sensor employed only a model nucleic acid probe (poly(A)), it demonstrates the principle of detection of the complementary sequence, the analyte poly(U). Figure 7.9(a) shows that a frequency decrease of 600 Hz was observed following the hybridization step, Step 3, and that this decrease was not observed with a control sensor that did not possess surface-bound poly(A). The sensor exhibits a relatively small frequency change that is superimposed on a large initial resonance frequency (9 MHz), and so S/N is low; in addition, difficulties associated with *ex situ* measurements at constant humidity preclude the use of this device for practical DNA or RNA detection.

7.3.7. Enzyme Thermistors[11]

A variety of enzyme-based biosensors have been tested using thermistors as the means of signal transduction. Thermistors are similar to electrical resistors, but possess highly temperature dependent resistance values. Since many enzymatic

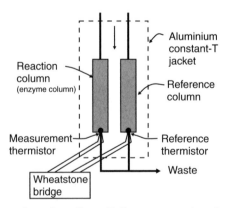

Figure 7.10. Typical configuration of a split-flow enzyme thermistor with an aluminum constant-temperature jacket.

TABLE 7.1. Thermal Biosensors[a]

Analyte	Immobilized Enzyme(s)	Conc. Range (mM)
Ascorbic acid	Ascorbate oxidase	0.05–0.6
ATP	Apyrase	1–8
Cellobiose	β-Glucosidase + glucose oxidase/catalase	0.05–5
Cephalosporin	Cephalosporinase	0.005–10
Cholesterol	Cholesterol oxidase + catalase	0.03–0.15
Creatinine	Creatinine iminohydrolase	0.01–10
Ethanol	Alcohol oxidase/catalase	0.01–2
Galactose	Galactose oxidase	0.01–1
Glucose	GO/catalase	0.001–0.8
Glucose	Hexokinase	0.5–25
Lactate	Lactate-2-monooxygenase	0.005–2
Lactate	Lactate oxidase/catalase	0.002–1
Lactose	β-Galactosidase + GO/catalase	0.05–10
Oxalic acid	Oxalate oxidase	0.005–0.5
Oxalic acid	Oxalate decarboxylase	0.1–3
Penicillin	β-Lactamase	0.01–500

[a] See Ref. 11. [Reprinted, by permission, from B. Danielsson and F. Winquist in *"Biosensors: A Practical Approach"*, A. E. G. Cass, Ed., Oxford University Press, New York, 1990. "The Practical Approach Series, Edited by D. Rickwood and B. D. James. © Oxford University Press 1990.]

reactions are accompanied by significant enthalpy changes, thermistors with immobilized enzymes can detect local temperature changes in the environment near the thermistor surface. For example, the reactions catalyzed by catalase, GO and NADH dehydrogenase are accompanied by enthalpy changes of -100, -80, and -225 kJ/mol, respectively.

Figure 7.10 shows a schematic diagram of an enzyme thermistor used to test a variety of immobilized enzymes in a flow-through biosensing device. The temperature probe is sealed inside a flow-through reaction chamber, which is encased in an aluminum constant-temperature jacket (to avoid responses due to ambient temperature fluctuations). The Wheatstone bridge allow differential measurement of both termistors, thus permitting measurements of small temperature changes (down to 10^{-4} °C).

By using devices similar to that shown in Figure 7.10, many immobilized enzymes have been tested. Table 7.1 summarizes the analytes, the enzymes, and the dynamic ranges over which the enzyme thermistors provide substrate quantitation.

7.4. EVALUATION OF BIOSENSOR PERFORMANCE

In addition to the usual evaluation parameters for analytical methods (Chapter 16), the sensitivity, detection limit, dynamic range, and precision profile, biosensors are also characterized with respect to the rapidity of their response and recovery. This additional evaluation is a result of their intended use, for the rapid and continuous monitoring of analytes in biological or environmental samples.

The response time of a biosensor has been defined as the time after analyte addition for the sensor's response to reach 95% of its final value. Similarly, the recovery time is defined as the time after analyte removal when 95% of the baseline signal has been achieved. Biosensors with response and recovery times of the order of tens of seconds or less provide useful *in vivo* or *in situ* devices. Many enzyme-based biosensors exhibit very similar response and recovery times, with parameters such as the reaction layer thickness and analyte concentration affecting the time required for 95% change to occur. Sensors that employ antibodies or chemoreceptors for chemical recognition often show fast responses, but take inordinate amounts of time to regenerate baseline readings in recovery time studies. This finding is primarily a result of the large association constants of these species for their specific binding partners, with association kinetics being rapid, and dissociation kinetics being very slow. Chaotropic reagents and denaturants are often used *ex situ*, to decrease the affinity of the recognition agent for its ligand by interfering with hydrogen bonding or by reversibly altering the tertiary structure of the protein.

Biosensors are also evaluated with respect to their stability. The reproducible generation of signal for a given analyte concentration is evaluated with respect to storage time, under various conditions of temperature and pH, and the results indicate the *storage stability* of the sensor. *Operational stability* is a similar type of evaluation, performed under conditions of continuous use. Operational stability is usually much poorer than storage stability. Both parameters are reported as the time required for the loss of a given percentage of the initial signal; for example, the time required for the signal generated by a particular analyte concentration to decrease to 50% of its initial value.

Interferences are of particular importance for devices destined for continuous use in very complex matrices. Biosensors are tested for interferences not just from species that are expected to bind to or react with the particular chemical recognition agent employed; the end use of the biosensor is considered, and components of that sample matrix are examined for potential interference. Test assays are conducted in the sample matrix, and compared with results obtained in simple buffers in order to determine analyte recovery.

In vivo examinations of glucose biosensors have been conducted using rats, dogs, and chimpanzees as test subjects.[12,13] The objective of these continuing studies is an artificial pancreas, where a glucose sensor is combined with a switching mechanism to allow automatic insulin delivery in diabetic patients. These and other studies have shown that immune responses are often generated towards membrane materials and immobilized species that are directly exposed to the living system. For this reason, a considerable worldwide research effort has been directed toward the development of biocompatible materials for use in *in vivo* sensors. In the references cited above, polyurethane–polysiloxane[12] and polyethylene oxide[13] were used as the outer, biocompatible layers for the short-term *in vivo* studies.

To date, a significant variety of enzyme-based biosensors have been commercialized for clinical testing laboratories and even for use by lay consumers. The most widely available devices measure blood glucose (diabetics) or lactate concentrations (high level athletes). Individual sensors are now being incorporated onto multisensor arrays in microfluidic devices capable of simultaneous measurement of a number of analytes; examples include arrays that have used enzymes,[14] antibodies,[15] and nucleic acids[16] as recognition agents with electrochemical[14] or optical transduction.[15,16]

SUGGESTED REFERENCES

A. P. F. Turner, I. Karube, and G. S. Wilson, Eds., *Biosensors: Fundamentals and Applications*, Oxford University Press, New York, 1987.

A. E. G. Cass, Ed., *Biosensors: A Practical Approach*, Oxford University Press, New York, 1989.

A. J. Cunningham, *Introduction to Bioanalytical Sensors*, John Wiley & Sons, Inc., New York, 1998.

V. C Yang and T. T. Ngo, Eds., *Biosensors and their Applications*, Kluwer Academic/Plenum Publishers, New York, 2000.

R. M. Nakamura, Y. Kasahara, and G. A. Rechnitz, Eds., *Immunochemical Assays and Biosensor Technology for the 1990s*, American Society for Microbiology, Washington, DC, 1992.

A. M. Campbell, *Monoclonal Antibodies and Immunosensor Technology*, Elsevier, New York, 1991.

REFERENCES

1. L. C. Clark, Jr., and C. Lyons, *Ann. N. Y. Acad. Sci.* **102**, 1962, 29–45.

2. M. J. Eddowes, in *Biosensors: A Practical Approach*, A. E. G. Cass, Ed., Oxford University Press, New York, 1990, pp. 211–237.

3. C. Bourdillon, C. Demaille, J. Moiroux, and J. M. Saveant, *Acc. Chem. Res.* **29**, 1996, 529–535.

4. P. N. Bartlett and K. F. E. Pratt, *J. Electroanal. Chem.* **397**, 1995, 61–78.

5. A. E. G. Cass, G. Davis, G. D. Francis, H. A. O. Hill, W. J. Aston, I. J. Higgins, E. V. Plotkin, D. L. Scott, and A. P. F. Turner, *Anal. Chem.* **56**, 1984, 667–671.

6. G. F. Blackburn, D. B. Talley, P. M. Booth, C. N. Durfor, M. T. Martin, A. D. Napper, and A. R. Rees, *Anal. Chem.* **62**, 1990, 2211–2216.

7. D. L. Bush and G. A. Rechnitz, *J. Membrane Sci.* **30**, 1987, 313–322.

8. K. R. Rogers, J. J. Valdes, and M. E. Eldefrawi, *Anal. Biochem.* **182**, 1989, 353–359.

9. D. Meadows and J. S. Schultz, *Talanta* **35**, 1988, 145–150.

10. N. C. Fawcett, J. A. Evans, L.-C. Chien, and N. Flowers, *Anal. Lett.* **21**, 1988, 1099–1114.

11. B. Danielsson and F. Winquist, in *Biosensors: A Practical Approach*, A. E. G. Cass, Ed., Oxford University Press, New York, 1990, pp. 191–209.

12. B. Aussedat, V. Thome-Duret, G. Reach, F. Lemmonier, J. C. Klein, Y. Hu, and G. S. Wilson, *Biosens. Bioelectron.* **12**, 1997, 1061–1071.

13. J. G. Wagner, D. W. Schmidtke, C. P. Quinn, T. F. Fleming, B. Bernacky, and A. Heller, *Proc. Natl. Acad. Sci. U.S.A.* **95**, 1998, 6379–6382.

14. I. Moser, G. Jobst, and G. A. Urban, *Biosens. Bioelectron.* **17**, 2002, 297–302.

15. K. E. Sapsford, Z. Liron, Y. S. Shubin, and F. S. Ligler, *Anal. Chem.* **73**, 2001, 5518–5524.

16. A. H. Forster, M. Krihak, P. D. Swanson, T. C. Young, and D. E. Ackley, *Biosens. Bioelectron.* **16**, 2001, 187–194.

PROBLEMS

1. A biosensor for glucose has been prepared by trapping a 0.50-mM solution of GO between an amperometric carbon electrode of 0.10-cm^2 area and a dialysis membrane, to form a reaction layer 0.10 mm thick. The substrate, glucose, has a diffusion coefficient of $\sim 5 \times 10^{-6}$ cm^2/s, and it has been found that if the sensor is rotated at a constant rate of 600 rpm, the thickness of the diffusion layer may be assumed equal to that of the reaction layer. In the GO reaction, the rate of ES complex decomposition, k_2, is equal to 780 s^{-1} and K'_m is 50 mM for glucose. The collection efficiency of the electrode is 0.45. A ferrocene species, used to regenerate the oxidized form of GO and deliver electrons to the electrode surface ($n = 2$ mol electrons/mol glucose), shows very fast kinetics for both processes, so that Eq. 7.13 is valid.

   (a) Estimate the current, in microamperes, that would be obtained for a 6-mM solution of glucose (this is the normal physiological concentration in whole blood).

   (b) Estimate the current that would be obtained if the reaction layer thickness was reduced by one order of magnitude (i.e., to 0.010 mm).

2. The mathematical descriptions of enzyme-based biosensor response assume that steady-state concentrations of substrates and products exist within the thin reaction layer at the surface of the transducer, and that the bulk concentrations of these species do not change significantly with time. Suggest one experimental situation in which these assumptions are not expected to be valid.

3. Polystyrene optical fibers can be derivatized by first nitrating and then reducing the surface to produce aminophenyl groups. These can then be diazotized to allow protein immobilization through (mainly) tyrosine residues. An evanescent-wave fluorescence biosensor for glucose is to be designed, using a bienzyme system consisting of hexokinase and glucose-6-phosphate dehydrogenase (both are <10 nm in hydrodynamic diameter) according to the following reactions:

$$\text{Glucose} + \text{ATP} \longrightarrow \text{G6P} + \text{ADP}$$

$$\text{G6P} + \text{NADP}^+ \longrightarrow \text{6-Phosphoglyceric acid} + \text{NADPH}$$

where the product NADPH fluoresces at 460 nm following excitation at 340 nm. Which of the following three sensor configurations may be expected to generate the largest fluorescence intensity signals for the same bulk glucose concentration?

(a) Simultaneous coadsorption of both enzymes to underivatized polystyrene.

(b) Nonpolymerizing covalent immobilization of glucose-6-phosphate dehydrogenase, followed by cross-linked immobilization of hexokinase.

(c) Coimmobilization of the two enzymes through tyrosine residues to form a bienzyme monolayer on the polystyrene surface.

4. Why is the operational stability of an enzyme-based thermal sensor for oxalic acid expected to be poorer than an amperometric oxalate sensor prepared using the same enzyme, oxalate oxidase?

CHAPTER 8

Directed Evolution for the Design of Macromolecular Bioassay Reagents

8.1. INTRODUCTION

Macromolecular reagents, mainly enzymes and antibodies, are widely used in bio-analytical chemistry. Proteins used as reagents are obtained mainly from natural microbial, plant, or animal sources, but in some cases an "industrial strain" of microorganism, able to produce large amounts of the desired protein, is selected. Sometimes the protein of interest is produced by a bacterial strain not suitable for industrial production. To overcome this problem, the gene coding for the protein may be cloned into an appropriate bacterial strain, and the resulting recombinant bacterium is used as the protein source. Both of these methods allow the production of proteins that are identical, at least in their primary structure, to the original "natural" protein.

From the origin of life until the present time, enzymes and other biomolecules have been naturally selected for their abilities to perform relevant functions in a living cell. This slow biological evolution process is responsible for molecular design changes. Evolution and adaptation are necessary in order to cope with environmental and biological changes along the geological time span, improving the fitness of the selected phenotypes.

Natural biological evolution occurs as a result of naturally occurring mutations, fragmentation and loss of DNA, duplication, and other genetic processes. Furthermore, sexual recombination increases the possibility of successful genetic combinations. The selection of improved variants occurs as a result of environmental conditions or selection pressures.

Natural proteins are adapted or optimized for function in specific physicochemical conditions. For example, enzymes isolated from bacteria found in Antarctic seawater have high activity at low temperature, but they are denatured at relatively low temperature. Thermophilic bacteria adapted to survive on hotspring water or in volcanic lakes harbor thermostable enzymes possessing almost no activity at low

Bianalytical Chemistry, by Susan R. Mikkelsen and Eduardo Cortón
ISBN 0-471-54447-7   Copyright © 2004 John Wiley & Sons, Inc.

temperature. The combination of thermostability and low-temperature activity in the same enzyme is not found in nature, because naturally occurring biomolecules are not adapted to cope with such extreme temperature changes. However, enzymes possessing these two characteristics have been designed or evolved in the laboratory.

Biological macromolecules have evolved their structures and functions for specific intracellular conditions. Posttranslational modifications occur to regulate structure and function in this environment. For example, the enzymatic turnover rate is one of the variables that can be controlled to regulate cellular metabolic pathways that are critical to ensure life. Biomolecules adapted to the complex intracellular environment generally do not possess ideal properties for analytical or industrial applications, where control of activity (through modification) and concentration (through biosynthesis and degradation rate) is irrelevant. The differences in the requirements for intracellular and analytical or industrial environments have created new research opportunities for the design of nonnatural biomolecules by genetic modification of their naturally occurring counterparts.

Goals for enzymatic targets of directed protein evolution have been to increase stability to denaturing agents (temperature, pH, organic solvents), improve solubility in water–organic solvent mixtures, and increase activity and/or affinity for substrate. More ambitious projects have also been undertaken to design enzymes with catalytic activity for new (but usually related) substrates.

Goals for antibody targets are related to improvement of affinity, production of antibodies with new binding selectivities, and the design of catalytically active

TABLE 8.1. Examples of Biomolecules Improved by Directed Evolution[a]

Biomolecule	Altered Function	Approach
Subtilisin E (protease)	Increased activity ($\approx$170-fold) in organic solvents (60% dimethylformamide)	Error-prone PCR
β-Lactamase (degradation of antibiotics, i.e., penicillin)	Activity toward a new substrate (cefotaxime)	DNA shuffling
p-Nitrobenzyl esterase	Thermostability and activity increase (14 °C)	Error-prone PCR/DNA shuffling
Biphenyl dioxygenases (degradation of poly-chlorinated biphenyls)	Gained activity toward substrates poorly degraded, improved activity	DNA shuffling of homologous genes
Green florescent protein	40-Fold brighter bacterial colonies	DNA shuffling
Immunoglobulin variable domain	Tolerates loss of structural disulfide bridge	Error-prone PCR
Lipase	Increased enantioselectivity in hydrolysis	Error-prone PCR

[a] The methods used to generate diversity are described later in the chapter.

antibodies. Ribozymes (ribonucleotides with catalytic activity) are also targets for directed evolution; efforts are underway to obtain ribozymes with new catalytic activities. Examples of target molecules and their new/improved functions are presented in Table 8.1.

8.2. RATIONAL DESIGN AND DIRECTED EVOLUTION

Rational protein design (or protein engineering) implies the use of knowledge to design or improve the characteristics of a biomolecule. Rational protein design is accomplished by methods such as site-directed mutagenesis and the insertion or deletion of DNA sequences to yield predictable changes in protein properties. Rational methods have been used successfully to improve enzymatic stability at high temperature and toward other denaturing physical or chemical agents.

Rational methods are based on experimental evidence that the effect of single amino acid substitutions on protein stability can be well approximated as additive, distributed, and large independent interactions. Moreover, by comparison of homologous enzymes from thermophilic and nonthermophilic microorganisms, it is known that introduction of disulfide bridges as well as increased numbers of proline residues increase protein stability.

Rational design methods have been applied to a glucose dehydrogenase enzyme to improve thermal stability, EDTA tolerance, and substrate selectivity.[1] Pyrroloquinoline quinone (PQQ) enzymes are found in Gram-negative bacteria. This family of enzymes uses PQQ as the prosthetic group. PQQ–glucose dehydrogenase (PQQGDH) is a monomeric membrane-bound protein with a MW of 87 kDa. An improved version of this enzyme would have an important application in glucose sensors; the enzyme commonly used in these devices, GO, cannot be used under anaerobic conditions unless an oxidant is present to perform the function of O_2.

Yoshida et al.[1] used homologous recombination of PQQGDH genes from *Escherichia coli* and *Acinetobacter calcoaceticus,* and previous knowledge to generate several chimera proteins. Their goals were to improve the PQQ binding constant (assayed as EDTA tolerance since a divalent ions is required for holoenzyme formation), to increase temperature stability, and to improve substrate selectivity. Their procedure is shown schematically in Figure 8.1. Seven genetic variants were produced from the original *E. coli* gene.

Site-directed mutagenesis was used to substitute the His residue in position 775 to Asn (His775Asn). This mutation improves the substrate specificity but simultaneously causes a decrease in thermal stability. Some regions of *A. calcoaceticus* that are responsible for EDTA resistance and for thermal stability were inserted into the *E. coli* gene by homologous recombination.

The chimeric genes obtained were used to transform *E. coli* cells, with each single clone expressing one variant of PQQ–glucose dehydrogenase. The enzyme was isolated and assayed in a buffer containing PQQ to form the holoenzyme, and a colorimetric indicator. Enzymatic activity was assayed for different substrates, and the effect of EDTA and thermal stability were investigated. When

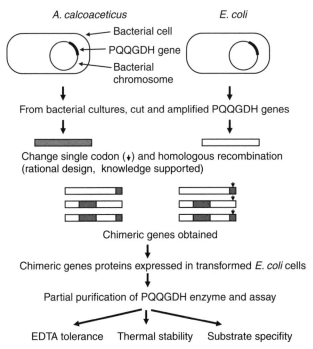

Figure 8.1. Steps followed for the construction (rational design) of an improved PQQGDH enzyme.

compared with the original *E. coli* enzyme, the thermal stability of the best chimeras was eight times greater, and the selectivity toward glucose was improved. The EDTA tolerance was as high as the original enzyme from *A. calcoaceticus*.

In contrast to rational design methods, where changes in protein properties may be predicted from sequence changes, random mutation methods have been introduced to empirically improve protein properties in the absence of a priori knowledge of structure–function relationships. Error-prone polymerase chain reaction (PCR), for example, can be used to introduce approximately random mutations in amplified DNA sequences. The protein products of these mutations may then be screened to select successful or improved variants.[2]

Directed evolution may exploit both rational design and random mutation methods. Generally, a three-step cycle is reiterated until the desired properties are observed. In the first step, genetic diversity is generated. The second step involves the physical or chemical linking of the genotype (nucleic acid) to the phenotype (the translated protein). In the final step, the successful variants are identified and selected. An example of a typical directed evolution protocol is shown in Figure 8.2.

The number of iterations of the three-step cycle required to achieve the desired changes in properties depends on two main factors. First, the so-called

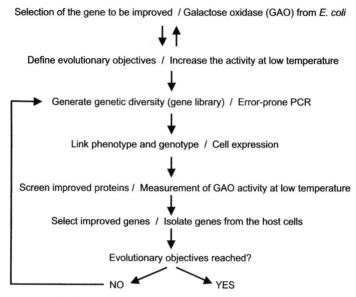

Selection of the gene to be improved / Galactose oxidase (GAO) from *E. coli*

Define evolutionary objectives / Increase the activity at low temperature

Generate genetic diversity (gene library) / Error-prone PCR

Link phenotype and genotype / Cell expression

Screen improved proteins / Measurement of GAO activity at low temperature

Select improved genes / Isolate genes from the host cells

Evolutionary objectives reached?

NO   YES

Figure 8.2. A simplified directed evolution protocol; on the right an example of each step is shown.

"evolutionary distance" between the initial protein (or gene) and the target molecule in important because it relates to the total number and identity of required mutations. This parameter is often difficult to estimate because the extent of potential improvement is generally unknown. The second factor is the extent of genetic diversity generated in the first step of the cycle; if this value is high, fewer cycles should be required to obtain the desired properties. Several authors have reported success with as few as two to three cycles, causing three to four amino acid substitutions in the evolved enzymes.[3,4]

8.3. GENERATION OF GENETIC DIVERSITY

The generation of a large genetic library (or repertoire) is fairly straightforward by methods such as error-prone PCR and DNA shuffling as described below. Theoretical studies have modeled the probability, P, that a given ligand–epitope will be recognized by at least one antibody in a repertoire, or genetically diverse population of antibodies. In this model, P depends on the size of the repertoire (N_{ab}) and the desired threshold value of the dissociation constant K_d (or more specifically pK_d). The parameter P is the probability that an antibody will recognize any given epitope with an affinity of at least K_d.

$$P = 100 \times \left(1 - e^{-N_{ab}(pK_d)}\right) \tag{8.1}$$

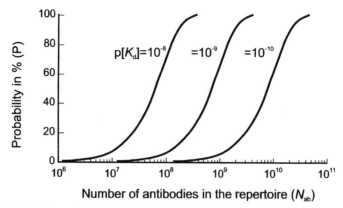

Figure 8.3. Probability of finding at least one antibody than recognizes any given epitope as a function of repertoire size and threshold K_d value.[5] [Reprinted, with permission, from A. D. Griffiths and D. S. Tawfik, *Current Opinion in Biotechnology* **2**, 2000, 338–353. "Man-Made Enzymes—From Design to In Vitro Compartmentalization". © 2000 Elsevier Science Ltd.]

This models the relationship between repertoire size and repertoire completeness, and shows that at least 10^8–10^{10} different antibodies are needed in the repertoire if a threshold K_d of 10^{-8}–10^{-10} is required, depending of the probability selected (Fig. 8.3).[3] Therefore, for one initial or starting sequence, the first step in the directed evolution cycle must be capable of generating 10^8–10^{10} nucleic acid sequences that code for different proteins. This library or repertoire size is needed for directed evolution according to this model. However, good results have been obtained experimentally (mainly with enzymes) using smaller libraries, of $\sim 10^4$ members.[6,7]

Although early work in this area used physical and chemical mutagenesis, there are two main methods that are now used to generate diversity: *error–prone PCR* and *DNA shuffling*. Both techniques require the initial selection and isolation of the gene or genes of interest. Restriction endonuclease enzymes can be used to cleave the selected gene per second from chromosomal genetic material; this is followed by purification and eventually amplification. Often the starting gene is isolated from the corresponding messenger RNA (mRNA), and the enzyme reverse transcriptase is used to obtain the corresponding complementary DNA (cDNA).[8]

8.3.1. Polymerase Chain Reaction and Error-Prone PCR

PCR is normally used to increase the concentration of a given DNA sequence for analysis. For this primary use, the reagent concentrations and other conditions are carefully controlled to avoid the introduction of mistakes (mutations) during the copying process.

In this reaction, a DNA sequence (template) up to 10,000 bases is used. DNA polymerase from a thermophilic bacterium is used. For example Taq-polymerase (from *Thermus aquaticus*) is used to catalyze the extension of two primers by the sequential incorporation of complementary nucleotides, as shown in Figure 10.9. Taq-polymerase has high thermal stability, retaining activity even after many heating cycles at 95 °C. Primers are single-stranded, short DNA sequences, frequently 15–25 bases long, complementary to the two sequences at the 5' ends of the region of interest in the double-stranded template DNA. In this region, the Taq-polymerase begins to copy the template by extending the primers in the 3' direction, until the template is totally copied or the process is stopped by a temperature increase that denatures the DNA. The number of copies obtained is equal to $2^n - 1$, where n is the number of cycles. Substrate exhaustion normally occurs after 20–30 cycles.

Error-prone PCR uses altered reaction conditions in order to increase the rate of production of copying errors. Taq-polymerase, when used *in vitro*, has an intrinsic error rate of about one noncomplementary nucleotide for every 10,000–100,000 bases (an error rate of 1 in 10^4 or 10^5). In Table 8.2, conditions are described to increase the rate of mutations, where the error rate is increased by increasing the concentration of divalent cations (Mg^{2+} or Mn^{2+}) and using increasingly unbalanced initial concentrations of nucleotides. Note that the error rate is tuned to between 2 and 15 errors for every 1000 base pairs, an increase of one to three orders of magnitude over the *in vitro* Taq-polymerase mutation rate.

Some protocols generate random DNA mutations, but random mutations do not introduce a random incorporation of new amino acids in the expressed protein, which is the target molecule to be improved by directed evolution. During the translation process, every consecutive 3-base sequence (codon) codes for an amino acid or for a stop codon, where protein translation ends. Leucine and arginine are coded by six different codons, whereas only one codon exists for tryptophan. Most other amino acids are coded by two to four codons. The implications of multiple codons

TABLE 8.2. Random Mutagenesis of Gene-Sized Fragments Using Error-Prone PCR[a]

Error (%)	[dATP]	[dCTP]	[dGTP]	[dTTP]	[MnCl$_2$]	[MgCl$_2$]
0.2	0.35	0.4	0.20	1.35	0.5	2.50
0.3	0.20	0.2	0.18	1.26	0.5	2.04
0.4	0.22	0.2	0.27	1.86	0.5	2.75
0.5	0.22	0.2	0.34	2.36	0.5	3.32
0.6	0.23	0.2	0.42	2.90	0.5	3.95
0.8	0.23	0.2	0.57	4.00	0.5	5.20
1.0	0.12	0.1	0.36	2.50	0.5	3.28
1.5	0.12	0.1	0.55	3.85	0.5	4.82

[a] All the concentrations are in millimolar (mM). A typical experiment is done with nanograms of template. The [dNTP] is used to calculate the corresponding MgCl$_2$ concentration.[9] [Reprinted, with permission, from M. Fromant, S. Blanquet, and P. Plateau, *Analytical Biochemistry* **224**, 1995, 347–353. "Direct Random Mutagenesis of Gene-Sized DNA Fragments Using Polymerase Chain Reaction". Copyright © 1995 by Academic Press, Inc.]

for error-prone PCR are not trivial: Random DNA mutations produce a bias in amino acid changes, where the probability of incorporation is higher for amino acids coded by several codons, like leucine and arginine. The dNTP concentrations in the starting PCR mixture can be adjusted to partially compensate for this, for example, by reducing the total dGTP + dCTP concentration.

8.3.2. DNA Shuffling

In this method of generating genetic diversity, the gene of interest is cleaved with DNAse I into many short double-stranded fragments (10–50 base pairs) that are then purified and recombined in a PCR-like process without primers. When the genes are partially recombined, terminal primers are added and full-length sequences are amplified. This process is usually used after an error-prone PCR step, where different mutants of the gene are generated. Alternatively, the conditions of DNA shuffling can be adjusted to introduce mutations, and to avoid regeneration of the original sequence.[10] Using a powerful derivative technique called "family shuffling", homologous genes pooled from different organisms are recombined. Figure 8.4 shows a comparison of DNA shuffling and family shuffling.

Family shuffling uses genes that have evolved in different microbial species from a common ancestral protein, therefore possessing sequences with a high degree of identity. This is a requirement for the process; > 50% sequence identity is needed. Proteolytic enzymes like subtilisin have been shuffled, using a large family of 26

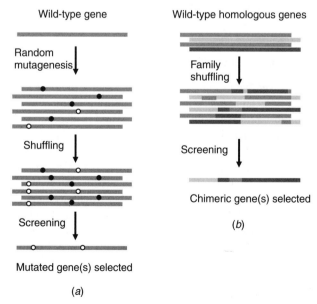

Figure 8.4. Comparison of DNA shuffling (*a*) and family shuffling (*b*). White and black dots represent advantageous and deleterious mutations, respectively.

subtilisin sequences, and improved chimeras have been obtained, with properties that significantly outperform the best parental variant.[11] New methods that mimic sexual recombination have also been proposed, and some of these have been reviewed recently.[12]

8.4. LINKING GENOTYPE AND PHENOTYPE

Directed evolution methods to design new or improved biomolecules are possible because individual nucleic acid molecules can be bound to their translated protein product. This kind of linkage, between genotype and phenotype, allows the examination of protein properties, selection of the best phenotypes, and exploitation of the attached nucleic acid through sequencing or reverse transcription–cloning for large scale genotype–phenotype production. Genotype–phenotype linkages may occur by compartimentalization or may be strong, noncovalent interactions or covalent bonds. The strategies that have been developed to generate the linkage are either cell dependent (*in vivo*) or cell free (*in vitro*) display.

In vitro methods have the advantage that the size of the library to be screened can be as large as 10^{12}–10^{13}, whereas the size achievable using expression into cells (*in vivo* methods) have between 10^8 and 10^{10} different DNA members. Proteins that interfere or are detrimental to metabolic cell processes (toxins, inhibitors of protein synthesis) cannot be selected using *in vivo* methods.[13]

In vivo methods for creating genotype–phenotype linkages are well established and exploit the cellular machinery necessary to process some proteins. Cell surface display of proteins coupled with fast cell-sorting systems can increase the size of the library that can be displayed and selected or screened.

8.4.1. Cell Expression and Cell Surface Display (*in vivo*)

In this method, evolved genes are incorporated into a bacterial plasmid using standard molecular biology techniques. Plasmids containing different evolved genes are used to transform different bacterial cells; the plasmid is incorporated into the bacterial cytoplasm and its genetic code is translated into proteins. Cells are plated on agar, and single colonies (resulting from single cells) are isolated and replated. Each colony produces only one genetic (evolved) variant. The protein is expressed without a direct linkage (bond) with the plasmid, and may be secreted or remain in the cytoplasm, as shown in Figure 8.5. The number of clones that can be screened using plates or liquid multiwell assays is usually limited to 10^5 clones.

To overcome the limited number of clones that can be screened using cell expression methods, cell surface display techniques have been developed, in which multiple copies of the evolved protein are displayed on the cell surface, allowing rapid identification and selection. Display methods generally produce a physical connection between gene and protein product, allowing the use of selection processes rather than screening.[14] With cell–surface display, the protein product is physically attached to the cell containing its genetic code.

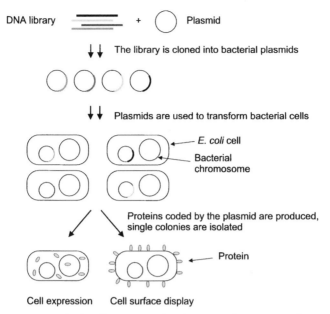

DNA library + Plasmid

The library is cloned into bacterial plasmids

Plasmids are used to transform bacterial cells

E. coli cell

Bacterial chromosome

Proteins coded by the plasmid are produced, single colonies are isolated

Protein

Cell expression   Cell surface display

Figure 8.5. *In vivo* methods allow connection between phenotype and genotype using cellular machinery. Synthesized proteins remain in the cytoplasm (or are secreted) in the nondisplay method. Cell surface display is a more recently introduced technique that uses special plasmids in order to produce proteins capable of transport into and retention by the cell membrane.

8.4.2. Phage Display (*in vivo*)

Phage display begins with the ligation of each member of the genetic library to a section of viral DNA that codes for one of the coat proteins of the virus. The most commonly used virus is the filamentous M13 phage, a common lab strain, and the coat protein is the gene III protein (gIIIp). The population of viruses is used to infect a bacterial culture; the resultant daughter phages display the evolved proteins and contain the genetic code responsible for their displayed phenotypic variants.

A variation of this method involves the construction of a plasmid containing the evolved variant DNA linked to gIIIp DNA. The plasmid is introduced to *E. coli*, which are cultured before M13 phage is added to the culture. This infection is necessary to allow the production of all the phage proteins, and therefore the production of phages in the bacterial cell occurs at the expense of the host cellular machinery. When phages are assembled in the bacterial cell, the viral capsids display the evolved proteins on their surfaces and at the same time contain the corresponding gene, as shown in Figure 8.6. Phage display is often used when molecular affinity molecules (e.g., antibodies) are being selected. An excellent description of this display system has been written by Johnsson and Ge.[15]

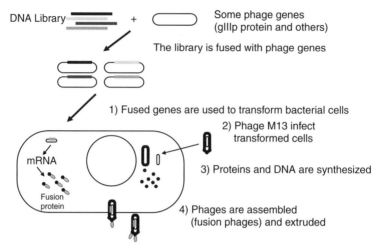

Figure 8.6. Phage display method. When the phages are extruded, binding to immobilized ligands can be used to select evolved variants.

8.4.3. Ribosome Display (*in vitro*)

Ribosome display uses the unmodified DNA library, which is transcribed and translated *in vitro*. The link occurs between the evolved protein and mRNA, and is achieved by stalling the translating ribosome at the end of the mRNA, which lacks a stop codon. Without a stop codon, the protein is not released by the ribosome, and the complex formed by the mRNA, protein (usually correctly folded) and ribosome is used directly for selection against an immobilized target. This method is shown schematically in Figure 8.7(*a*).

Later the selected mRNA complexes are separated from the immobilized target, and are dissociated with EDTA. Using reverse transcription PCR, the corresponding DNA is synthesized using the purified mRNA as the template.[16]

8.4.4. mRNA–Peptide Fusion (*in vitro*)

Also called mRNA display, this method was derived from ribosome display technology. Puromycin-tagged mRNA and several additional steps must be used to achieve covalent coupling between the protein product and its mRNA, as shown in Figure 8.7(*b*). Puromycin, which mimics aminoacyl–tRNA (tRNA = transfer RNA), can be attached covalently to mRNA by the ribosomal machinery.

8.4.5. Microcompartmentalization (*in vitro*)

In this two-phase technique, the DNA library, together with enzymes, cofactors, and monomers needed for the transcription and translation processes are encapsulated in a water–oil emulsion. Each aqueous compartment has a size at or below that of a

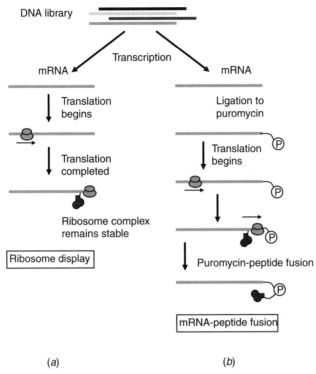

Figure 8.7. Ribosome display is used to display binding molecules; the noncovalent complex (ribosomes, protein, and mRNA) is stabilized by changing the medium composition and lowering the temperature (*a*). In mRNA–peptide fusion, the link between genotype (mRNA) and phenotype (protein) is covalent (*b*).

bacterial cell (2.6-μm diameter), and contain, on average, a single genetic variant. Advantages of this method when compared with cell expression methods include a higher number of useful compartments per milliliter of solution, and a fully controllable expression system.[17] One disadvantage is the lack of a direct genotype–phenotype linkage, which would be problematic if conditions favored coalescence of microcompartments.

8.5. IDENTIFICATION AND SELECTION OF SUCCESSFUL VARIANTS

Screening is usually defined as the identification of variants with the desired properties, whereas selection is the enrichment of desired variants in a new, smaller molecular library. Screening is limited in practice to 10^5–10^7 sequences, since it requires the assay of each variant, whereas selection (e.g., "panning" using immobilized ligands) may be used on much larger libraries for enrichment.

Screening commonly relies on the detection of an optical change, such as an absorbance or luminescence increase, via activation of a reporter gene, such as the β-galactosidase or luciferase gene. In eukaryotic cells, the so-called "two hybrid"[18] system has been described; it is used to screen and select for protein–protein interactions. In the two-hybrid system, transcription of a reporter gene is activated through the interaction of two proteins, which reconstitute a transcription factor composed of DNA-binding and DNA-activation domains; commonly used reporters genes are LEU2 or HIS3 from yeast. When protein–protein interaction occurs, transcription is activated, conferring to the cell the capacity to grow on minimal media (without leucine or histidine), allowing selection. Some high-throughput techniques are based on cell culture in low volume, high-density micro-titer plates (6144 wells with volumes of 1–2 μL or less).[19]

Selection is based upon the functional properties of the evolved protein (pheno-type). Binding to antigen immobilized in the wells of a microtiter plate or on the surfaces of magnetic beads are frequently used approaches. Flow cytometry sorting systems, some of which can sort up to 10^8 cells/h, represent a promising new class of selection technique.

8.5.1. Identification of Successful Variants Based on Binding Properties

Gene libraries can be enriched for successful variants using a technique called "panning", in which a ligand bound to the surface of a support material is exposed to the product mixture, containing linked phenotype–genotype species. The ligand may be bound to the surfaces of microtiter plate wells, magnetic beads, membranes, or chromatographic stationary phase materials. In principle, successful variants bind to the solid-supported ligand, while unsuccessful variants are readily removed during a rinsing step.

In practice, panning methods can also select for undesirable properties. One problem with the use of immobilized ligands is that the immobilization process may alter their structure and binding properties. Moreover, nonspecific binding can occur, such as adsorption to the support material. Both of these problems lead to selection of undesired properties.

One approach that can overcome immobilization-related problems involves the use of antigen (or ligand) covalently bound to biotin (Ag-B). The Ag-B is added to the solution at the desired concentration to allow solution-phase binding, so that the selection process is based mainly on the differential affinity of each antibody or other binding molecule. The mixture is then added to strepavidin-coated beads or wells, where the formation of the biotin–strepavidin noncovalent complex allows the selection of high affinity antigen-binding molecules.

Enzymes have been successfully enriched from libraries by selecting variants that bind to transition state analogs, or by covalent trapping with surface bound sui-cide inhibitors,[20] but this approach does not usually select the most active enzy-matic variants.

8.5.2. Identification of Successful Variants Based on Catalytic Activity

When improved catalytic activity is the goal of the directed evolution process every variant is often screened following spatial distribution (e.g., single bacterial colonies in microtiter wells), and this generally involves the measurement of an optical change in each well. This allows the screening of 10^2–10^6 variants, and is a reliable but labor-intensive method.

One selection strategy for enrichment of catalytic activity is based on the measurement of single-turnover activity.[21] Catalytic antibodies capable of hydrolyzing glycosidic bonds have been studied using this procedure, as shown in Figure 8.8. In this example, reaction of the glycosidic bond with the evolved catalytic antibodies, expressed by phage display, results in the production of a highly reactive quinone methide species at or near the active site. Reaction with any nearby nucleophile (e.g., an amine group on the protein) results in covalent immobilization of the phage-displayed catalytic antibody. The immobilization allows separation of successful from unsuccessful variants by washing with buffer, removing noncovalently bound phage.

A second example involves different evolved subtiligases. Subtilisins are proteolytic enzymes; sutiligases are mutants that catalyzes the ligation of peptides. Evolved variants of subtiligase that ligate a biotinylated peptide onto their own extended N-termini were selectively captured, as shown in Figure 8.9. The reaction of the biotin-labeled peptide substrate with the evolved enzyme variant results in a

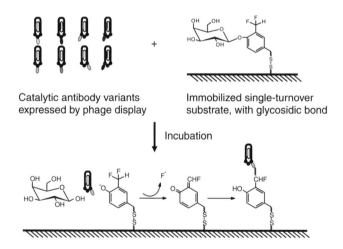

Catalytic antibody variants expressed by phage display

Immobilized single-turnover substrate, with glycosidic bond

Incubation

When the glycosidic bond is hydrolyzed, the catalytically active phage is captured

Figure 8.8. Selection for improved catalytic activity using single-turnover activity. Phage mixture contains many variants, and each fusion phage carries the DNA coding for its evolved protein that is attached to its coat protein.[21] [Reprinted, with permission, from K. D. Janda, Lee-C. Lo, Chih-H. Lo, Mui-M Sim, R. Wang, Chi-H. Wong, and R. Lerner, *Science* **275**, February 1997, 945–948. "Chemical Selection for Catalysis in Combinatorial Antibody Libraries". Copyright © 1997 by AAAS.]

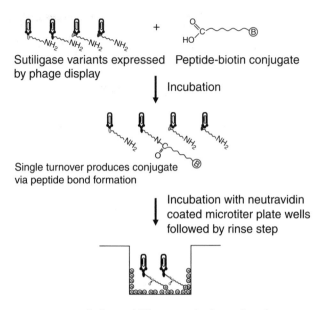

Figure 8.9. Selection for improved catalytic activity using single-turnover activity. Phage mixture contains many variants, and each fusion phage carries the DNA coding for its evolved protein that is attached to its coat protein.

covalent link between the peptide and enzyme. The biotin label allows separation of successful from unsuccessful variants via neutravidin coated microtiter plate wells.[22]

With single-turnover selection processes, only a single catalytic event is necessary for selection; therefore, the selected variants are not necessarily improved catalysts.

8.6. DIRECTED EVOLUTION OF GALACTOSE OXIDASE

Galactose oxidase (EC 1.1.3.9, GAO) from the fungus *Fusarium sp.* was used as a starting gene for directed evolution. The GAO has a molecular weight of 68 kDa, and is a monomeric glycoprotein with 1.7% carbohydrate, formed by 639 amino acid residues. It catalyzes the following reaction:

$$1° \text{ alcohol} + O_2 \longrightarrow H_2O_2 + \text{aldehyde} \qquad (8.2)$$

where the alcohol can be glycerol, allyl alcohol, D-galactose, galactopyranosides, oligosaccharides, and polysaccharides. This enzyme has been used in biosensors to measure D-galactose and lactose in blood samples, dairy samples and in fermentation process control. The goals of the directed evolution experiment were to increase stability and activity toward nonnatural substrates.

In the first step, the GAO gene was isolated from *Fusarium sp.*, and fused to a strong promoter (*lacZ*). When the promoter is induced, transcription by RNA polymerase begins, and both mRNA and GAO are produced.

Error-prone PCR was used to generate genetic diversity, at a 2–3 base substitution rate per gene. Mutated genes were cloned into a vector (plasmid pUC18), and used to transform *E. coli* cells by electroporation. Transformed cells were plated on agar, and single colonies were picked off. These were transferred to a duplicate plate, grown 12 h in the presence of IPTG to induce the *lacZ* promoter, and GAO was then produced. The cells were then centrifuged, washed, and lysed. Galactose oxidase activity was measured for each variant culture product before and after 10 min at different temperatures (55–70°C) to determine stability.

Selected improved variants were later recombined using a method similar to DNA shuffling. Three or four rounds of mutation allowed the selection of a mutant enzyme with 15-fold higher activity, and $\sim 5\,°C$ increase in stability. Substrate selectivity was unchanged with respect to the wild-type enzyme from *Fusarium*.[23] This example is one of the many possible applications of directed evolution for the generation of new bioassay reagents with new or improved characteristics.

SUGGESTED REFERENCES

M. Famulok, E.-L. Winnacker, and C.-H. Wong, Eds., "Combinatorial Chemistry in Biology", in *Current Topics in Microbiology and Immunology*, Vol. 243, Spriger-Verlag, Berlin, 1999.

F. H. Arnold, Ed., "Evolutionary Protein Design", *Advances in Protein Chemistry*, Vol. 55, Academic Press, San Diego, CA, 2001.

REFERENCES

1. H. Yoshida, K. Kojima, A. B. Witarto, and K. Sode, *Protein Eng.* **12**, 1999, 63–70.

2. M. B. Tobin, C. Gustafsson, and G. W. Huisman, *Curr. Opin. Structural Biology* **10**, 2000, 421–427.

3. W. P. Stemmer, *Nature (London)* **370**, 1994, 389–391.

4. L. You and F. H. Arnold, *Protein Eng.* **9**, 1996, 77–83.

5. A. D. Griffiths and D. S. Tawfik, *Curr. Opin. Biotechnol.* **2**, 2000, 338–353.

6. O. May, P. T. Nguyen, and F. H. Arnold, *Nat. Biotechnol.* **18**, 2000, 317–320.

7. F. C. Christians, L. Scapozza, A. Crameri, G. Folkers, and P. C. Stemmer, *Nat. Biotechnol.* **17**, 1999, 259–264.

8. S. R. Bolsover, J. S. Hyams, S. Jones, E. A. Shephard, and H. A. White, *From Genes to Cells*, Wiley-Liss, New York, 1997, pp. 167–178.

9. M. Fromant, S. Blanquet, and P. Plateau, *Anal. Biochem.* **224**, 1995, 347–353.

10. B. Steipe, in *Combinatorial Chemistry in Biology*, M. Famulok, E.-L. Winnacker, and C.-H. Wong, Eds., Springer-Verlag, Berlin, 1999, pp. 69–71.

11. J. E. Ness, M. Welch, L. Giver, M. Bueno, J. R. Cherry, T. V. Borchert, W. P. C. Stemmer, and J. Minshull. *Nat. Biotechnol.* **17**, 1999, 893–896.

12. C. Schmidt-Dannert, *Biochemistry* **40**, 2001, 13125–13136.

13. J-M. Zhou, S. Fujita, M. Warashina, T. Baba, and K. Taira, *JACS* **124**, 2002, 538–543.

14. K. FitzGerald, *Drug Discovery Today* **5**, 2000, 253–258.

15. K. Johnsson and L. Ge, in *Combinatorial Chemistry in Biology*, M. Famulok, E.-L. Winnacker, and C.-H. Wong, Eds., Springer-Verlag, Berlin, 1999, pp. 87–105.

16. C. Schaffitzel and A. Plückthun, *TRENDS Biochem Sci* **26**, 2001, 577–579.

17. A. D. Griffiths and D. S. Tawfik, *Curr. Opin. Biotechnol.* **11**, 2000, 338–353.

18. W. Kolanus, in *Combinatorial Chemistry in Biology*, M. Famulok, E.-L. Winnacker, and C.-H. Wong, Eds., Springer-Verlag, Berlin, 1999, pp. 37–54.

19. G. Georgiou, in *Evolutionary Protein Design*, F. H. Arnold Ed., Academic Press, San Diego, CA, 2001, pp. 293–311.

20. P. Forrer, S. Jung, and A. Plückthun, *Curr. Opin. Biotechnol.* **9**, 1999, 514–520.

21. K. D. Janda, Lee-C. Lo, Chih-H. Lo, Mui-M Sim, R. Wang, Chi-H. Wong, and R. Lerner, *Science* **275**, 1997, 945–948.

22. S. Atwell and J. A. Wells, *Proc. Natl. Acad. Sci. U.S.A.*, **96**, 1999, 9497–9502.

23. L. Sun, I. P. Petrounia, M. Yagasaki, G. Bandara, and F. H. Arnold, *Protein Eng.* **14**, 2001, 699–704.

PROBLEMS

1. Is the following statement true or false? Why? "Biological evolution has optimized the fitness of life over the last 4 billion years; therefore, there are no further possibilities for improvement of an enzyme's performance".

2. What is the main criterion used to decide between a directed evolution or a rational design protocol? Propose other criteria that could be used to support a decision.

3. When error-prone PCR is used to generate diversity, the error rate is usually tuned to between 2 and 15 errors for every 1000 base pairs. What would be the result of an error-prone PCR experiment if the error rate were tuned to 2 errors for every 25,000 base pairs? What if it is tuned to 2 errors for every 10 base pairs?

4. Why are PCR procedures limited to 20–30 cycles? Complete the following table (Table 8.3) to check your answer. A 5 ng sample of template (cDNA) is used, with a MW = 448,500. Assume that the entire template is copied in each cycle.

TABLE 8.3

Number of Cycles	Number of DNA Copies (For each template molecule)	DNA Mass (g)
5		
15		
30		
60		
500		

CHAPTER 9

Principles of Electrophoresis

9.1. INTRODUCTION

Electrophoresis is a bioanalytical tool used in fundamental research and diagnostic settings for the isolation and identification of high molecular weight biomolecules. The separation is based upon the mobility of charged macromolecules under the influence of an electric field. Mobility is a fundamental property of a macromolecule, and its value depends on the magnitude of its charge, its molecular weight, and its tertiary or quaternary structure (i.e., its shape). Because most biopolymers, such as proteins and nucleic acids, are charged, they can be separated and quantitated by electrophoretic methods.

An electrophoretic separation occurs in an intervening medium that separates two electrodes. At one end of the medium is the positively charged *anode*, and at the other is the negatively charged *cathode*, as shown in Figure 9.1. The intervening (support) medium may be as short as 10 cm or as long as 1 m. Throughout this medium, positively charged species will migrate toward the cathode and negatively charged species will move toward the anode.

The intervening medium consists of a liquid, usually a buffer, that is supported by an inert solid material such as paper or a semisolid gel. The liquid allows the movement of ions, while the solid support provides frictional drag. When a voltage is applied across the electrodes, a current is generated from the movement of ions in the electric field. The electric field strength, E, determines the rates of migration of the species in the support and can be varied experimentally.

$$E = (\text{applied voltage})/(\text{length of support medium}) \qquad (9.1)$$

Low-voltage electrophoresis is typically carried out at electric field strengths of 20 V/cm, while high-voltage techniques use field strengths of up to 200 V/cm.[1]

At such high-field strengths and applied voltages, it is reasonable to ask why redox processes are not important in electrophoresis, since electrochemical reactions are known to occur at applied voltages of <5 V. Figure 9.2 shows that the

Bianalytical Chemistry, by Susan R. Mikkelsen and Eduardo Cortón
ISBN 0-471-54447-7   Copyright © 2004 John Wiley & Sons, Inc.

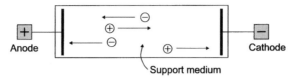

Figure 9.1. Movement of charged species in electrophoresis.

fundamental difference between electrolysis and electrophoresis may be represented by a resistor in a model circuit. Electrolysis occurs when the medium separating the two electrodes possesses a high intrinsic conductivity, so that the electrode–solution interfaces behave like capacitors, and these are connected by (ideally) zero resistance. The applied potential difference occurs across these capacitors, and the enormous field strengths generated in the few nanometers at the electrode–solution interfaces generate the electrochemical reaction that, in the circuit model, discharge the capacitors.

On the other hand, the intervening media used in electrophoresis have much lower conductivity, and an equivalent circuit for an electrophoresis cell includes a resistor between the capacitors at the electrode–solution interfaces. Across the support medium, potential is now (usually) a linear function of distance, and the electric field thus generated is responsible for driving the electrophoretic separation. Electrophoresis occurs at the electrodes used in electrophoresis, to maintain the

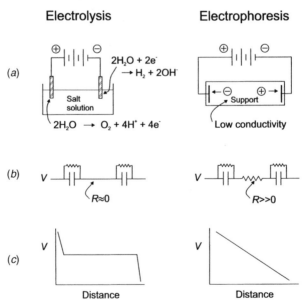

Figure 9.2. Experimental setup (*a*), equivalent electrical circuit (*b*) and dependence of potential on distance between electrodes (*c*) for electrolysis and electrophoresis.

current through the support; however, the voltage required to generate this current (<5 V) is insignificant relative to the large applied voltage.

The theoretical description of electrophoretic migration[2] begins with a consideration of a particle of charge q suspended in an insulating medium, and exposed to an electric field E. Fundamental laws of physics state that the electric force exerted on the particle will be equal to its viscous drag, as shown in Eq. 9.2:

$$qE = fv \qquad (9.2)$$

where f is the frictional coefficient, and v is the velocity of the particle. The product fv is the viscous drag exerted by the medium on the particle. In a vacuum, the frictional coefficent is zero, and velocity goes to infinity. In an electrophoretic support medium, the frictional coefficient is significant, and a characteristic velocity is achieved by charged particles as they migrate through the support.

To compensate for experimental variations in E, the electrophoretic mobility of a particle has been defined as a fundamental characteristic of a molecule under given conditions of support medium, pH, ionic strength, and temperature. The electrophoretic mobility, μ, usually reported in units of velocity per unit field strength, is given by Eq. 9.3:

$$\mu = v/E = q/f \qquad (9.3)$$

The value of μ is a characteristic of a given macromolecule, and has been shown to depend on its molecular weight and molecular shape (through the frictional coefficient f), and on its net charge q. This simple equation correctly predicts that mobility increases with q, decreases with increasing f, and is equal to zero for uncharged particles.

Attempts to rigorously apply Eq. 9.3 to experimental data have revealed a number of complications that are not accounted for by the simple theory. The support media used in electrophoresis are not strictly insulators, and usually contain buffer ions; charged species in such an electrolyte will attract ions, and this leads to shielding effects, in which the apparent charge of a macromolecule is somewhat lower in magnitude than its actual charge. In addition, the ionic atmosphere of the migrating particle is partially disrupted by the electric field and by the motion of the charged particles through the medium, so that the electric field is not strictly a constant throughout the support. For these reasons, mobility values reported for proteins and nucleic acids depend on experimental conditions.

Three basic forms of electrophoresis have been introduced. *Moving-boundary electrophoresis* is performed in a U-shaped cell, as shown in Figure 9.3. The analyte solution is initially located at the bottom of the U, and has a high concentration of a nonionizing component such as sucrose, glycerol, or ethylene glycol that increases its density and minimizes initial mixing at the boundaries of the dilute buffers present in the anodic and cathodic compartments. When the electric field is applied, migration of the components occurs in both directions, according to their charges, and the boundaries between analyte solution and buffer solutions move towards the

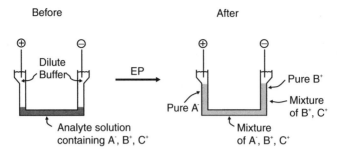

Figure 9.3. Moving-boundary electrophoresis.

electrodes as the separation proceeds. The migration of analyte species is normally monitored with refractive index detectors positioned near the upper ends of the cell. Moving-boundary electrophoresis was the first electrophoretic technique introduced, and is seldom used today except in situations where precise values of mobility are required.

Zone electrophoresis is used extensively in modern research and clinical laboratories. In this technique, the components of the analyte mixture separate completely to form discrete zones, or bands, that may be stabilized by a support medium or may exist as free zones. The analyte solution is applied to the medium as a spot or band, and the electric field causes the initial band to separate into component bands through migration, as shown in Figure 9.4.

Zone electrophoresis is used for the analysis of complex mixtures of biomolecules, for the determination of molecular weight and purity of isolated proteins and nucleic acids and for a variety of diagnostic tests. It is not used for the quantitative determination of mobility. The use of a support medium prevents mechanical disturbances and convection arising from temperature changes and high local concentrations of biopolymers from broadening the zones. Supports may act as adsorbents, as does paper, or as molecular sieves, as do agarose and polyacrylamide gels. Starch gels exhibit both properties, with analyte adsorption and molecular sieving both contributing to zone separation. The gels used in electrophoresis are continuous gels, rather than the beaded or particulate gels used in chromatography. In zone electrophoresis, analyte bands migrate at constant, characteristic velocities; for this reason, colored tracking species are added to samples for visualization of the progress of the separation.

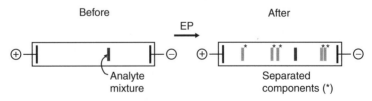

Figure 9.4. Zone electrophoresis.

Steady-state electrophoresis is distinct from moving-boundary and zone electrophoresis in that the ultimate positions of the zones are not time dependent. After electrophoresis has proceeded for a certain time, a steady state is achieved wherein the widths and positions of the zones are constant. In isoelectric focusing, for example, a stable pH gradient is created between the anodic and cathodic ends of the support; during a run, a biomolecule will migrate in this pH gradient until it reaches a pH at which its net charge is equal to zero. For proteins and polypeptides, this pH is the isoelectric point. Isoelectric focusing has very high resolving power, and is the topic of Chapter 11.

9.2. ELECTROPHORETIC SUPPORT MEDIA

9.2.1. Paper

Paper, or cellulose, electrophoresis was introduced in the early 1950s as a separation technique for proteins. Prior to the start of the electrophoretic run, the cellulose sheet is saturated with the running buffer and placed in a tank, as shown in Figure 9.5. The sample is then applied as a spot or a line, the tank is covered and the electric field is applied. After the separation has proceeded for a sufficient time, the paper is removed and dried. The separated components are then located by color, by fluorescence under UV light, or after staining with a dye. If the components have been radiolabeled, autoradiography of the entire paper will yield a separation pattern.

Components may be eluted from the paper following separation for further analysis. This is accomplished by simply cutting the zone of interest out of the paper, and either (a) soaking this paper in a small volume of elution buffer; (b) rinsing a triangular piece of the paper with an elution buffer, and collecting drops from the tip; or (c) electrophoresing the component of interest from the band into a small volume of buffer. Low-voltage paper electrophoresis is rarely used today, as it has been superceded by gel techniques. However, high-voltage paper electrophoresis is still useful for the separation of small molecules, such as amino acids and nucleotides; with these analytes, their low charge leads to low mobility, and their small size leads to significant diffusional spreading, or band broadening. The high-voltage technique allows separation to be accomplished rapidly, but because of the

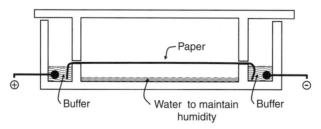

Figure 9.5. Experimental arrangement for paper electrophoresis.

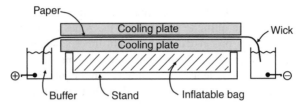

Figure 9.6. High-voltage paper electrophoresis with cooling.

high currents involved, heat is generated (as $i^2 R$), and cooling plates are necessary (Fig. 9.6) to maintain constant temperature. An inflatable bag is used under the cooling plates for support.

The complete resolution of mixtures of amino acids or nucleotides is not always possible after a single high-voltage run. For this reason, 2D techniques are sometimes used, where perpendicular electrophoresis runs or electrophoresis followed by perpendicular chromatography may be used. The careful selection of running buffers and chromatographic eluents allows the resolution of overlapped bands in the second dimension.

Cellulose acetate was introduced in 1957 as an alternative to cellulose (paper) as a support material. Cellulose possesses both electrophoretic and chromatographic separation properties, since many components will adsorb onto, or interact with, the hydroxyl groups on the support. Because of this, spots and bands exhibit tailing and resolution is low. Cellulose acetate minimizes the adsorption effect, since most of the hydroxyl groups have been acetylated. Separations using cellulose acetate are more rapid, with improved resolution and less tailing, leading to the facilitated detection of more concentrated spots or bands. In addition, the background staining of cellulose acetate is lower than paper, since stain adsorption is also minimized. Cellulose acetate has two other advantages over paper: it is transparent, facilitating optical detection of zones, and it dissolves easily in a variety of solvents, facilitating the elution and isolation of separated components.

9.2.2. Starch Gels

Starch was the first gel medium used for an electrophoretic separation, and was introduced in 1958. The gel is prepared by heating a paste of potato starch in the electrophoresis buffer until the starch grains burst and the heterogeneous mixture becomes transparent and homogeneous. The hot solution is then poured into a horizontal tank and cooled to ambient temperature, yielding a semisolid gel. As shown in Figure 9.7, a slot is cut into the gel, and sample is introduced as a slurry with starch grains. The surface of the gel is then sealed with wax or grease. Porous wicks are inserted directly into the ends of the gel, and function as electrical connections between the gel and the anodic and cathodic buffer compartments. Following the electrophoretic run, the semisolid gel is removed from the tank, and is cut into layers for staining. This preparative technique allows replicate layers to be stained with different reagents, and leaves macroscopic quantities of unstained

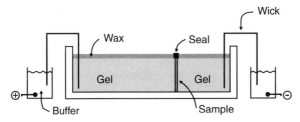

Figure 9.7. Starch gel electrophoresis.

material for isolation and further study. The positioning of the sample slot with respect to the anodic and cathodic ends of the gel is based upon an initial knowledge of the species of interest (its isoelectric pH or net charge) or is varied over multiple runs if analyte charge is unknown.

The migration of species in the starch gel depends on both the net charge and the molecular size. The gel has a sieving effect, so that small molecules tend to have greater mobilities. The sieving effect can be controlled to some extent through the pore size of the gel, which may be varied by varying the starch concentration; however, the porosity of starch gels may be controlled only over a limited range of concentration, and starch concentrations that are too high or too low yield gels that are excessively stiff or soft.

The starch polymer possesses negatively charged side chains (carboxylates) that can interact with proteins and hinder migration by an ion-exchange retention mechanism, so that starch gels, like cellulose, may possess both electrophoretic and chromatographic properties. The negative charge on the gel also leads to a phenomenon called *electroosmosis*, wherein positively charged counterions (H_3O^+) in the running buffer move toward the cathode. In effect, the buffer flows toward the cathode, and the overall transport of analyte species in the gel is the sum of transport due to electrophoretic migration and transport due to electroosmosis. The electroosmotic effect can be quantitated using blue dextran, an uncharged oligosaccharide that possesses zero electrophoretic mobility. Blue dextran may be added to a sample, and the migration of the blue dextran from the sample origin is measured after electrophoresis and subtracted from other measured distances to provide corrected migration distances. Starch has been superceded by polyacrylamide and agarose gels for most applications, but is still occasionally used for the separation of isoenzymes.

9.2.3. Polyacrylamide Gels

Polyacrylamide gel electrophoresis (PAGE) has replaced starch gel electrophoresis for the separation of proteins, small RNA fragments and very small DNA fragments. Polyacrylamide is more versatile than starch, because the molecular sieving effect can be controlled to a much greater extent, and because the adsorption of proteins to the gel is negligible. Polyacrylamide gels are prepared[3] by the reaction of acrylamide (monomer) with N,N'-methylenebis(acrylamide) (cross-linker) in the presence of a catalyst and initiator, as shown in Eq. 9.4. Initiators include ammonium persulfate and potassium persulfate, where the $S_2O_8^{2-}$ dianion decomposes into two $SO_4^{\bullet-}$ radicals, while the commonly used catalyst is tetramethylethylenediamine [TEMED, $(CH_3)_2N(CH_2)_2N(CH_3)_2$], which reacts with the sulfate radical anion to produce a longer lived radical species.

$$
\begin{array}{l}
\underset{\substack{|\\ C=O\\ |\\ NH_2}}{H_2C{=}CH} + \underset{\substack{|\\ C=O\\ |\\ NH\\ |\\ CH_2\\ |\\ NH\\ |\\ C=O\\ |\\ H_2C{=}CH}}{H_2C{=}CH} \xrightarrow[\substack{\text{Initiator}\\ 10-20 \text{ min}}]{\text{catalyst}} \\
\end{array}
\qquad (9.4)
$$

The concentration of TEMED used during gel casting determines the length of the polyacrylamide chains formed, and therefore the mechanical stability of the gel. Lower monomer concentrations require higher TEMED concentrations, but excess catalyst should be minimized, because it may interact with proteins or alter the pH of the running buffer. Because the polymerization that occurs during the casting of a polyacrylamide gel is a random process, a distribution of pore sizes will occur. It has been shown that the average pore size depends on the total amount of acrylamide used and the degree of crosslinking. The total monomer concentration is generally between 5 and 20% by weight.

Experimentally, PAGE may be employed in a column, for preparative separations, or in a slab, where several samples may be separated under identical conditions.[4] These configurations are shown in Figure 9.8. The preparation of a PAGE column begins with a tube of ~ 5–10 mm inner diameter, 10 cm or more in length. During gel polymerization, the gel mixture is gently overlayered with water to give a flat upper surface. Following casting, the water is removed, and the sample is applied. Samples are suspended in a concentrated sucrose solution (high density) and layered over the gel, to prevent mixing with the buffer in the upper (usually cathodic) electrode chamber. The run is begun at low current (1 mA for ~ 30 mins) to allow the entire sample to enter the gel. The current is then increased to 2–5 mA during the run. A tracking dye such as bromophenol blue (anionic) or methylene blue (cationic) is added to the sample to monitor the progress of the

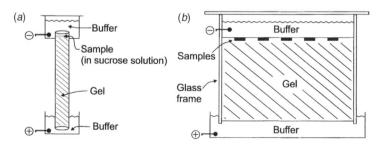

Figure 9.8. Experimental arrangements for (*a*) column and (*b*) slab PAGE.

separation. Buffers are often cooled and stirred to minimize i^2R heating and local pH changes that may occur near the gel–buffer interface.

When the tracking dye has migrated to the bottom of the column, the gel is removed using pressure, and proteins are fixed in place by precipitation [this occurs during a soaking step in a solution of trichloroacetic acid (TCA)] and are then stained for detection.

Slab PAGE allows simultaneous electrophoresis of a number of samples to be performed under identical conditions. Slab PAGE has a higher resolving power than column PAGE, and is often used for molecular weight and purity determinations. Because the slab is relatively thin, heat dissipation is more efficient in a slab than in a column. The slab can be used vertically or horizontally, but in practice, the horizontal slab method is used only when the monomer plus cross-linker concentrations are low and the gel is soft. A slab gel is prepared in the same way as a column gel, except that a comb, or slot-former, is inserted before polymerization. Removal of the comb after the gel has set leaves sample wells that are separated from each other by continuous strips of gel.

In zone electrophoresis, two components are considered separated when the difference in migration distances exceeds the width of the original sample band. Because of diffusional band broadening, the sharpness of bands diminishes with increasing run time. A method used to improve the resolving power of zone electrophoresis that is unique to gel media involves a stacking gel between the sample and the running gel, and is called "disc" electrophoresis,[5] due to the electric field strength *disc*ontinuity that yields a very narrow sample zone. Figure 9.9 shows the components of the gels and buffers used for this method.

The running gel in disc electrophoresis is cast in the presence of the running buffer, typically a high pH glycine buffer. The stacking gel is cast in the presence of a tris buffer at lower pH and ionic strength. Because the monomer concentration is much lower in the stacking gel, frictional drag is lower and analyte mobility is higher. The stacking gel improves resolution because of the pH discontinuity, and the buffering effect that occurs at the interface between the running and stacking gels. At pH 6.5, glycine exists as a zwitterion, with a mobility equal to zero because of zero net charge. At pH 8.7, glycine exists mainly as the anion, since the amino group has been deprotonated; its mobility is nonzero and it will migrate toward the anode. At the pH discontinuity, a deficiency of mobile ions exists

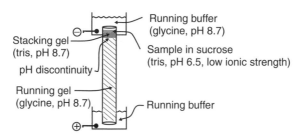

Figure 9.9. Disc column PAGE, with stacking gel of 2–4% monomer, and running gel of 5–20% monomer; typical values are 2.5 and 15%, respectively.

because of the shift of the glycine equilibrium toward the zwitterion. This causes a locally high electric field strength at the discontinuity, due to the high resistance of the uncharged interface. To maintain the flow of current in this region of the gel, anionic proteins migrate rapidly and stack in the dilute gel near the discontinuity. The proteins form narrow bands between the glycinate in the running gel and the chloride ions in the stacking gel. The narrow bands enter the running gel in this preconcentrated form, allowing improved resolution following the complete run. Run times as short as 20 min can be used with disc electrophoresis, because of its resolving power.

The pore size of a polyacrylamide gel controls the mobility and resolution of components because of the sieving effect of the pores on macromolecular species.[6] The pore size may be controlled by varying the total concentrations of monomer and cross-linker, and by varying their ratio. Gel compositions are defined by two parameters, their %T and %C values, that represent the total and crosslinker contents, respectively. These parameters are defined by Eqs. 9.5 and 9.6.

$$\%T = (\text{weight in grams of monomer plus cross-linker})/(100\,\text{mL}) \qquad (9.5)$$

$$\%C = 100 \times (\text{grams cross-linker})/(\text{grams monomer plus cross-linker}) \qquad (9.6)$$

The molecular weight of a protein and its net charge may be determined by performing electrophoresis on gels with constant %C, by varying %T from gel to gel. After separation, the migration of the protein is measured relative to a tracking dye, and the R_f value is calculated:

$$R_f = (\text{migration distance of protein})/(\text{migration distance of marker}) \qquad (9.7)$$

where the marker is not restricted by the gel. A Ferguson plot[7] is then constructed by plotting $\log(R_f)$ against %T. The antilog of the y intercept of this plot (%T = 0) yields the ratio of protein charge to marker charge, and the slope of the straight line is proportional to the protein's molecular weight. By running standard proteins of known molecular weight alongside the unknown, in a slab gel, a plot of Ferguson slope against molecular weight is used to determine the molecular weight of the unknown.

If the net charge on the unknown protein is not of interest, its molecular weight may be determined more quickly using the SDS–PAGE method.[8] In this method, a single gel is cast in the presence of $\sim 0.1\%$ SDS, which is also present in the running buffer. Protein samples (molecular weight markers and the unknown) are pretreated by heating in 1% SDS containing 0.1% mercaptoethanol. The mercaptoethanol reduces disulfide bonds that maintain quaternary and tertiary structure, to form random coils of the subunits, while the SDS combines with hydrophobic regions of the protein with a relatively constant binding of 1.4-g SDS/g of protein. This nonspecific binding interaction yields species with *constant charge per unit weight* (from the sulfate groups of SDS), since the contribution from charged protein side chains becomes negligible. Migration distances are then influenced only by sieving and are directly related to the molecular weight of the protein or subunit. A plot of distance against $\log(\text{MW})$ is linear with a negative slope, and can be used as a calibration curve for unknown determinations. Note that glycoproteins consistently bind SDS to a lesser extent than proteins,[9] so that an unknown glycoprotein should be calibrated using glycoprotein molecular weight standards. The SDS–PAGE can determine protein molecular weights to within 5% accuracy over the 10^4–10^6 Da range.

9.2.4. Agarose Gels

Agarose was used for the first time as an electrophoresis medium in the early 1970s. It is a linear polymer of D-galactose and 3,6-anhydrogalactose that is isolated from seaweed. Agarose contains $\sim 0.04\%$ sulfate, may be dissolved in boiling water and forms a continuous gel when cooled to $< 38\ °C$; the gel structure is maintained by hydrogen bonding.[10] As with polyacrylamide gels, the concentration of agarose in the gel determines the average pore size. Pore sizes are much larger with agarose than with polyacrylamide gels, so that nucleic acids and proteins too large to be separated on polyacrylamide gels can be separated and quantitated using agarose gels. Agarose slab gels are used in both vertical and horizontal modes. In the vertical mode, agarose concentrations as low as 0.8% can be made, allowing the separation of proteins and nucleic acids up to molecular weights of $\sim 5 \times 10^7$ Da. In the horizontal mode, agarose gels can be made from as little as 0.2% agarose, allowing molecular weights of $\sim 1.5 \times 10^8$ to be determined. Identical equipment is used for agarose and PAGE electrophoresis. Agarose gels tend to be relatively fragile (especially at very low agarose concentrations), and gels are either examined directly or carefully dried to a thin film prior to staining.

Agarose gels contain charged groups—mainly sulfate and some carboxylate groups. These charged groups interact with charged groups on proteins, and lead to ion-exchange effects; they may also lead to significant electroosmotic flow. The pretreatment of agarose in alkaline solution leads to the hydrolysis of these groups, and improves the sieving characteristics of the gels. The physical properties of agarose gels, especially the viscosity, are very sensitive to temperature fluctuations, so that strict control of temperature during electrophoresis is essential. Because of the large pores in agarose gels, their major area of application is in DNA separation and analysis.

9.2.5. Polyacrylamide–Agarose Gels

Composite gels containing both polyacrylamide and agarose have been used to achieve intermediate pore sizes in gels that possess good mechanical strength.[11] These gels may be used for nucleic acids, nucleoproteins, and other proteins too large to enter the pores of standard polyacrylamide gels.

9.3. EFFECT OF EXPERIMENTAL CONDITIONS ON ELECTROPHORETIC SEPARATIONS

Temperature is critical to the electrophoretic resolution of sample components. Joule heating produced by the electric current must be compensated by an efficient cooling system. In the absence of cooling, five factors contribute to decreased resolution. Convection currents occur because warmer solution near the center of the support has a lower density than cooler solution near the walls; water has a maximum density at 4 °C, and at this temperature, density variation with temperature is minimal; runs should therefore be performed at 4 °C where possible. Diffusion, which leads to the blurring or widening of zones, is more significant at higher temperatures, especially if the run requires several hours. A distortion of zones occurs when cooling is inadequate, since those parts of the zones in the warmer sections of the gel (the inside) move faster than those in the cooler (outside) portions; this unequal speed leads to curved bands, and sometimes to overlap with neighboring bands. In open vessels, evaporation may lead to surface dehydration of the support, and thus to local changes in ionic strength; for this reason covered vessels are used. Finally, viscosity varies with temperature; higher temperatures lead to softer gels and lower frictional coefficients, an effect particularly noticeable with agarose gels.

The pH of the running buffer can have dramatic effects on the separation achieved. This is not normally the case for nucleotides, polynucleotides, or nucleic acids—with these species, the net charge is always negative due to the acidic sugar-phosphate backbone, so that pH changes over the 4–10 range do not significantly affect separations. For proteins, however, the net charge is pH dependent. For pH values below the isoelectric point (pI), a protein will have a positive charge and will migrate toward the cathode. Conversely, if pH > pI, the protein will possess a net negative charge and will migrate toward the anode. It is possible for two or more proteins to give only one band after an electrophoresis run, if conditions are such that the proteins possess the same mobilities. For this reason, protein separations should be carried out at two or more pH values. Studies of the purity or homogeneity of a sample should always be carried out using multiple runs and different pH values.

The ionic strength of the running buffer also influences the quality of the separation achieved. Interactions between the surface groups of charged species and buffer ions result in an ionic atmosphere near charged macromolecules. Both the net charge and the mobility are affected by ionic strength, and protein analytes are significantly affected. The general rule for electrophoresis is that the ionic strength of the running buffer should be kept low in order to minimize counterion effects while maintaining adequate solubility of sample components.

9.4. ELECTRIC FIELD STRENGTH GRADIENTS

The variation of electric field strength with distance along the gel leads to the improved separation of slower moving components. Standard methods for electrophoresis gel preparation produce constant electric field strength across the gel, where the voltage change per unit distance is a constant over the entire gel; in these gels, the high-mobility species are well separated, but species near the origin are grouped together in a diffuse zone due to poor resolution. This resolution near the origin can be improved by generating an electric field strength gradient across

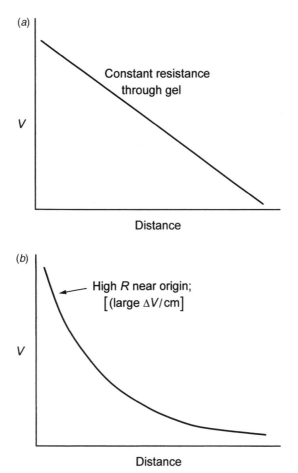

Figure 9.10. Voltage versus distance for (*a*) a linear electrophoresis gel, and (*b*) a gradient gel. From Ohm's law, with constant current, the observed voltage changes mirror the resistance properties of the gels.

the gel; the high field strength near the origin increases the velocity of the slower moving species. Figure 9.10 compares the profiles of voltage versus distance for linear and gradient gels.

Such gradients may be introduced by using either (a) wedge-shaped gels, or (b) ionic strength gradients. With wedge gels, the resistance increases with decreasing cross-sectional area, so that the thin end is near the origin, and the sample migrates toward the thicker end. These gels are cast in a slanted mold. Gels using ionic strength gradients are prepared from polyacrylamide cast with the ionic strength gradient. The dilute buffer end of the gel has higher resistance; this is the sample origin, so that the sample migrates toward the end that has lower resistance. A comparison[12] of the separations achieved for one of the DNA sequencing reaction

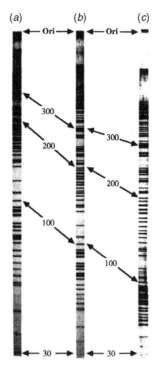

Figure 9.11. Comparison of the band separation of linear, wedge, and ionic strength gradient gels. The T reaction of a sequencing experiment carried out on M13 mp8 DNA was run on three different 40 cm 6% polyacrylamide gels, using (a) a standard linear gel, (b) a 0.35–1.05-mm wedge gel, and (c) a 0.05–0.50 M Tris buffer gradient. [Reprinted, with permission, from A. T. Bankier and B. G. Barrell, in "*Nucleic Acids Sequencing: A Practical Approach*", C. J. Howe and E. S. Ward, Eds., Oxford University Press, New York, 1989. © IRL Press at Oxford University Press 1989.]

mixtures (the T reaction) using a 6% polyacrylamide gel in the linear, wedge and ionic strength gradient configurations is shown in Figure 9.11.

9.5. DETECTION OF PROTEINS AND NUCLEIC ACIDS AFTER ELECTROPHORETIC SEPARATION

Following an electrophoretic run, the band from the tracking dye is often the only visible band. The detection of separated proteins and nucleic acids requires subsequent treatment of the separation pattern for visualization. This treatment may be performed directly on the gel, or may require a blotting step in which the entire separation pattern is transferred onto a thin membrane material. The choice of detection method depends on the concentrations of analytes in the separated zones and whether recovery of the purified sample is required.

9.5.1. Stains and Dyes[13]

The most common visualization method involves staining an entire gel with a species that interacts in a nonselective way with either proteins or nucleic acids. Protein gels are removed from the electrophoresis apparatus and "fixed" by immersion in 10% trichloroacetic acid, to precipitate proteins and prevent diffusional zone broadening and analyte loss into staining solutions. The gel is then rinsed, and immersed in a solution (0.1–0.2% w/v) of the stain, resulting in stain absorption throughout the gel. A destaining rinse step then leaves stain only where interactions with protein prevent its removal. Similar procedures are used for the detection of nucleic acids. Table 9.1 lists stains for proteins, glycoproteins, and nucleic acids.

A silver stain has been introduced that is 100 times more sensitive for proteins than Coomassie Blue, and 4 times more sensitive for nucleic acids than ethidium bromide.[14] This stain is based on the reduction of Ag^+ by thiol, tyrosine and amine functional groups in proteins and by the purine bases in DNA. The solid silver formed by reduction precipitates in the gel forming a permanent stain in the protein–nucleic acid zones, and is not reduced by buffer or gel materials, so that

TABLE 9.1. Electrophoresis Stains[a]

Stain	Absorption Max. (nm)	Use
Amido Black 10B	620	General protein stain
Coomassie Brilliant Blue R-250	590	General protein stain, 10 times more sensitive than Amido Black
Coomassie Brilliant Blue G-250	595	General protein stain
Alcian Blue	630	Glycoprotein
Uniblue A		Protein stain
Methylene Blue	665	RNA, RNase
Methyl Green	635	Native DNA, neutral or acidic tracking dyes
Fast Green FCF	610	Protein stain
Basic Fuschin	550	Glycoprotein, nucleic acids, sialic acid-rich glycoproteins
Pyronin Y	510	RNA, acidic tracking dye
Bromophenol Blue	595	Neutral and alkaline tracking dye
Bromocresol Green		Tracking dye for DNA agarose electrophoresis
Crocein Scarlet	505	Immunoelectrophoresis
Xylene Cyanole FF		Tracking dye for DNA sequencing
Toluidine Blue O	620	RNA, RNase, mucopolysaccharides
Ethidium Bromide		Fluorometric detection of DNA
Stains All		General protein stain

[a] See Ref. 13. [Reprinted, by permission, from P. G. Righetti, "*Isoelectric Focussing: Theory, Methodology and Applications*", Elsevier, New York, 1983. © 1983, Elsevier Science Publishers B.V.]

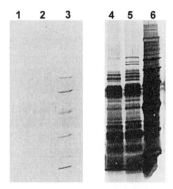

Figure 9.12. Comparison of identical polyacrylamide gels stained with Coomassie Blue (lanes 1–3) and Silver Stain (lanes 4–6). The initial sample concentrations are identical for lanes 1 and 4, 2 and 5, and 3 and 6. [Reprinted, with permission, from B. S. Dunbar, *"Two-Dimensional Electrophoresis and Immunological Techniques"*, Plenum Press, New York, 1987. © 1987 Plenum Press.]

background staining is minimal. Figure 9.12 shows a comparison of identical gels stained by coomassie blue and a silver stain.

After staining, gels may be photographed under UV or visible light, to yield qualitative information about the number and positions of the bands. Alternatively, quantitative information can be obtained from a scanning densitometer (Fig. 9.13) that measures light transmitted through the gel as a function of position in the gel.

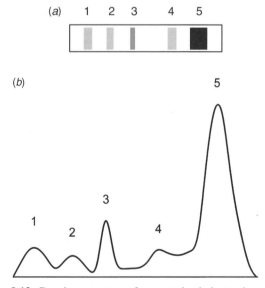

Figure 9.13. Densitometer trace from a stained electrophoresis gel.

The densitometer trace is used to measure peak positions (migration distance) and purity or relative concentrations of components by integration of the areas under the peaks. Modern densitometers perform this operation automatically. For quantitative work, overloading of the gels should be avoided, since this leads to very concentrated sample zones that may inhibit proportional staining and leads to underestimates of sample concentrations.

9.5.2. Detection of Enzymes by Substrate Staining[15]

Activity stains are of great importance during the isolation, purification, and characterization of enzymes, since a particular catalytic reaction is involved and the detection of this activity leads to the unequivocal identification of the zone of interest on the electrophoresis gel. Following separation, the gel is removed from the electrophoresis apparatus and is immersed in a minimal volume of a substrate solution. Detection relies on the formation of a colored product by enzyme in the zones containing the enzyme. Examples of activity stains are given in Table 9.2.

TABLE 9.2. Substrate Stains for Enzyme Detection[a]

Enzyme	Procedures
Red cell acid phosphatase	Phenolphthalein diphosphate followed by alkali
Phosphoglucomutase	Reduction of the tetrazolium salt of MTT
Placental alkaline phosphatase	β-Naphthyl phosphate and diazo salt of fast blue RR
β-Glucuronidase	8-Hydroxyquinoline plus blue RR salt
Cholinesterase	6-Bromo-2-naphthylcarbonaphthoxycholine iodide plus blue B salt
Glucose-6-phosphate dehydrogenase	Application of agar overlay containing a tetrazolium salt
6-Phosphogluconate dehydrogenase	Agar overlay with tetrazolium salt
Cytochrome oxidase	α-Naphthol plus dimethylpapraphenylenediamine
Esterases	α-Naphthylbutyrate plus diazo salt of fast blue RR
Lactic and malic dehydrogenases	Reduction of nitro blue tetrazolium salt
Acid phosphatase	Sodium α-naphthyl acid phosphate plus diazo salt of 5-chloro-o-toluidine
Phosphorylase	Reduction of silver phosphate to Ag by UV light
Succinate, β-hydroxybutyric and glutamate dehydrogenase	Reduction of nitro blue tetrazolium salt
Catalase	Inhibition of starch–iodide reaction (starch gel)
Caeruloplasmin	O-Dianisidine
Leucine aminopeptidase	Alanyl-β-naphthylamide plus diazotized O-aminoazotoluene
Hemoglobin (as peroxidase)	Benzidine plus hydrogen peroxide

[a] See Ref. 15. [Reprinted, with permission, from P. G. Righetti, "*Isoelectric Focussing: Theory, Methodology and Applications*", Elsevier, New York, 1983. © 1983, Elsevier Science Publishers B.V.]

The detection of enzymes by activity staining is important not only for the direct detection of separated enzymes, but also for the detection of noncatalytic proteins after staining with enzyme-labeled antibodies to the protein of interest. The ability to detect a particular catalytic activity greatly reduces background staining, and allows the unequivocal identification of a particular analyte species in very complex sample mixtures.

9.5.3. The Southern Blot[16]

This procedure was introduced by E.M. Southern in 1975, and is used to transfer DNA from agarose gels onto a nitrocellulose or nylon membrane for subsequent detection. Transfer to nitrocellulose is accomplished after the gel has been soaked in a denaturing solution containing $\sim 1.5\ M$ NaCl and $0.5\ M$ NaOH; this disrupts the base pairing of double-stranded DNA, so that only the single-stranded form is present in the gel. The gel is then neutralized to a pH of ~ 8, and placed in a transfer apparatus such as the one shown in Figure 9.14.

Once the gel has been placed in the transfer apparatus, buffer in the pan is drawn through the gel by the wicking action of the filter paper. The transfer proceeds over ~ 12 h. The upper filter papers are replaced when they are saturated with buffer. The nitrocellulose membrane is then removed, air-dried and heated to 80 °C for 2 h to ensure strong DNA binding.

Nylon membranes have largely replaced nitrocellulose for Southern transfers, because nylon has been shown to improve the transfer of small (< 200 base) DNA fragments, and it is more stable to pH extremes. Transfers to nylon follow the same procedure as outlined above, except that NaCl is not needed in the denaturing solution, and no baking step is required to enhance DNA binding to the membrane material. The DNA can be covalently bound to nylon membranes by a brief exposure of the blotted membrane to UV light. These UV-exposed membranes with covalently bound DNA allow multiple detection and rinsing cycles without diffusional zone broadening.

9.5.4. The Northern Blot[17]

The Northern blot is used to transfer RNA from agarose onto nitrocellulose membranes. Conditions for the transfer are similar to those used in the Southern blot, except for the following. Denaturation prior to transfer is done with methylmercuric

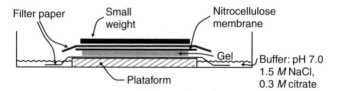

Figure 9.14. Apparatus for Southern blot.

hydroxide, glyoxal, or formaldehyde instead of sodium hydroxide, because NaOH would hydrolyze RNA at the $2'$-hydroxyl position. The RNA is not firmly bound to nitrocellulose until after the baking step at 80 °C, so that no rinsing step is used prior to this step. Gels with RNA are not prestained for photography prior to the transfer step, because the stain interferes with the transfer to nitrocellulose. Finally, special precautions are required to prevent the contamination of the samples with RNase, including the treatment of glassware and solutions, and the wearing of gloves during the entire procedure.

9.5.5. The Western Blot[18]

This procedure is similar in principle to Southern and Northern blots, but it is designed for the transfer of proteins from gels onto nitrocellulose membranes. An electrophoretic technique is often used to speed the transfer by $\sim$ 10-fold, and the apparatus used for such Western transfers is shown in Figure 9.15. The rapid electrophoretic process ensures that complete transfer occurs with minimal diffusional zone broadening.

9.5.6. Detection of DNA Fragments on Membranes with DNA Probes

DNA probes take advantage of the specificity of base pair hydrogen-bonding inter-actions over oligonucleotide lengths of 20 to >1000 base pairs. The DNA probe is an oligo- or polynucleotide to which a detectable label has been attached.[19] Hybri-dization of the probe to the target nucleic acid on the nitrocellulose or nylon mem-brane occurs through A-T and G-C hydrogen bonding. This interaction is relatively weak for individual base pairs, since only two (A-T) or three (G-C) hydrogen bonds exist. However, if hybridization occurs over a number of consecutive base pairs

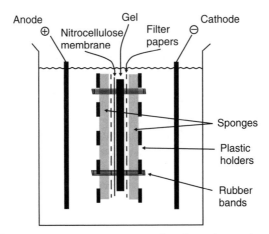

Figure 9.15. Apparatus for western transfer of proteins to nitrocellulose.

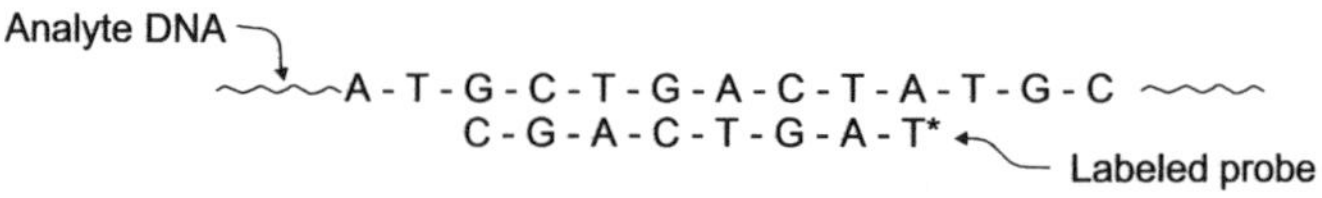

Analyte DNA ⌐

~~A - T - G - C - T - G - A - C - T - A - T - G - C ~~~~

C - G - A - C - T - G - A - T* ⌐

Labeled probe

Figure 9.16. Heteroduplex between target DNA and a labeled DNA probe.

(> 5), the process is exothermic at room temperatures. The oligonucleotide sequence and length of the DNA probe determines the selectivity of detection of a target species. For example, statistical considerations have shown that a 17-base probe may be used to locate a single, unique segment in a randomly ordered, six-billion base DNA sequence, such as the human genome.[20] Of course, DNA does not possess a statistically random sequence, and longer probes are generally used. Hybridization on the membrane is used to form a heteroduplex, as shown in Figure 9.16.

Heteroduplexes may be dissociated at low ionic strength or at high temperatures. The temperature at which the double-to-single-strand transition occurs is called the melting temperature, T_m. Probes with high $G + C$ content have higher T_m values, because a greater number of hydrogen bonds are present in the heteroduplex.

The DNA probes may be labeled with single atoms (i.e., radiolabeled), functional groups such as fluorophores, or side chains attached to enzymes, to allow the detection of bands in a separation pattern that contain the sequence of interest. The label attached to the DNA probe must not interfere with the hydrogen bonding of the probe to the target sequence. The hybridization step is carried out by immersing the nitrocellulose or nylon membrane in a solution containing excess DNA probe, and incubating at a temperature just below the T_m for the heteroduplex. Excess probe is then removed by rinsing, and the membrane is examined for the location of the label.

The DNA probes are often labeled by the nick-translation method.[21] This involves the combined activities of three enzymes: DNaseI, 5',3'-exonuclease and

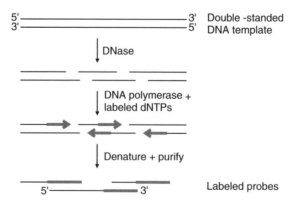

Figure 9.17. Labeling DNA probes by the nick-translation method.

5′,3′-polymerase. The latter two enzymes are present in the *E. coli* DNA polymerase complex. As shown in Figure 9.17, DNaseI randomly introduces single-stranded breaks, or nicks, by the hydrolysis of double-stranded DNA to create free 3′-hydroxyl groups. The exonuclease then removes one or more of the bases from the 5′-phosphoryl side of this nicked region, while the polymerase catalyzes the incorporation of labeled nucleotides onto the 3′-hydroxyl side, to fill the gap with labeled bases. The labeled nucleotides that are incorporated are complementary, that is, the resulting strand has the same sequence as the original strand. As the reaction proceeds, the nick shifts by one base in the 3′-direction; this process is allowed to continue until 30–60% of the total bases have been replaced by labeled nucleotides. Purification of the labeled probe species is unnecessary if the original DNA sample contained only the sequence of interest. By adjusting the DNaseI concentration, it is possible to produce short probes with many labels incorporated, or long probes with few labels.

The nick-translation method was originally used to incorporate radiolabeled nucleotides into DNA probes. More recent work has shown that nucleotides labeled with biotin can also be incorporated using this method, to yield probes of equivalent sensitivity. Uracil and adenine nucleotides that have been labeled with biotin are shown in Figure 9.18.

By itself, biotin is not a detectable label. It is used as a label because a subsequent irreversible reaction with the proteins avidin or streptavidin ($K > 10^{12} \, M^{-1}$) can be used for ultrasensitive detection, if these proteins have been labeled with an enzyme and an activity stain is used. Conjugates of these proteins with alkaline

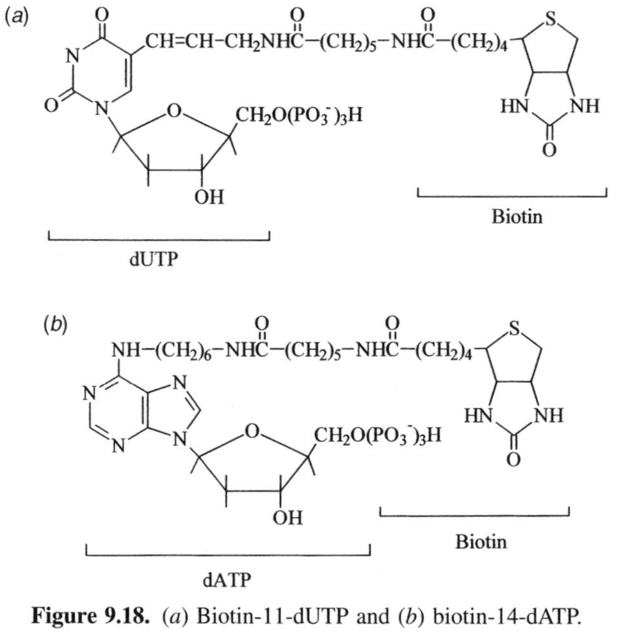

Figure 9.18. (*a*) Biotin-11-dUTP and (*b*) biotin-14-dATP.

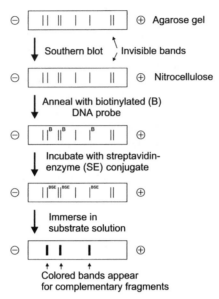

Figure 9.19. Detection of a DNA sequence in an electrophoretic separation pattern using a biotinylated DNA probe.

phosphatase, peroxidase, and other enzymes are commercially available, and are often used for this purpose. Figure 9.19 shows a summary of the sequence-specific detection of DNA, using electrophoresis, Southern blotting, hybridization with a biotin-labeled DNA probe, and detection using an enzyme–streptavidin conjugate and activity stain.

SUGGESTED REFERENCES

L. A. Lewis, *CRC Handbook of Electrophoresis*, CRC Press, Boca Raton, FL, 1980.

R. Westermeier, *Electrophoresis in Practice*, Wiley-VCH, New York, 2001.

G. J. Moody and J. G. R. Thomas, *Practical Electrophoresis*, Merrow, 1975.

D. F. Keren, *High Resolution Electrophoresis and Immunofixation*, Butterworths, London, 1987.

B. D. Hames and D. Rickwook, Eds., *Gel Electrophoresis of Proteins: A Practical Approach*, Oxford University Press, New York, 1981.

A. Chrambach, *The Practice of Quantitative Gel Electrophoresis*, VCH Publishers, Weinheim, Germany, 1985.

REFERENCES

1. H. Michl, in *Chromatography. A Laboratory Handbook of Chromatographic and Electrophoretic Methods*, 3rd ed., E. Heftmann Ed., Van Nostrand Reinhold Company, New York, 1975, pp. 282–311.

2. B. S. Dunbar, *Two-Dimensional Electrophoresis and Immunological Techniques*, Plenum Press, New York, 1987, pp. 4–5.

3. E. G. Richards and R. Lecanidou, in *Electrophoresis and Isoelectric Focussing in Polyacrylamide Gels*, R. C. Allen and H. R. Maurer, Eds., Walter de Gruyter, Berlin, 1974, pp. 16–22.

4. M. Melvin, *Electrophoresis*, John Wiley & Sons, Inc., New York, 1987, pp. 34–35.

5. U. K. Laemmli, *Nature (London)* **227**, 1970, 680–685.

6. A. Chrambach, *The Practice of Quantitative Gel Electrophoresis*, VCH Publishers, Weinheim, Germany, 1985, pp. 67–73.

7. A. Chrambach, *The Practice of Quantitative Gel Electrophoresis*, VCH Publishers, Weinheim, Germany, 1985, pp. 111–128.

8. B. D. Hames, in *Gel Electrophoresis of Proteins: A Practical Approach*, B. D. Hames and D. Rickwood, Eds., Oxford University Press, New York, 1981, pp. 1–86.

9. B. S. Dunbar, *Two-Dimensional Electrophoresis and Immunological Techniques*, Plenum Press, New York, 1987, p. 19.

10. P. Serwer, S. J. Hayes, and G. A. Griess, *Anal. Biochem.* **152**, 1986, 339–345.

11. P. M. Horowitz, J. C. Lee, G. A. Williams, R. F. Williams, and L. D. Barnes, *Anal. Biochem.* **143**, 1984, 333–340.

12. A. T. Bankier and B. G. Barrel, in *Nucleic Acids Sequencing: A Practical Approach*, C. J. Howe and E. S. Ward, Eds., Oxford University Press, New York, 1989, pp. 73–77.

13. P. G. Righetti, *Isoelectric Focussing: Theory, Methodology and Applications*, Elsevier, New York, 1983, p. 209.

14. C. R. Merril, *Nature (London)* **343**, 1990, 779–780.

15. C. R. Righetti, *Isoelectric Focussing: Theory, Methodology and Applications*, Elsevier, New York, 1983, pp. 230–231.

16. E. M. Southern, *J. Mol. Biol.* **98**, 1975, 503–517.

17. D. Grierson, in *Gel Electrophoresis of Nucleic Acids: A Practical Approach*, D. Rickwood and B. D. Hames, Eds., Oxford University Press, New York, 1982, pp. 29–31.

18. P. G. Righetti, *Immobilized pH Gradients: Theory and Methodology*, Elsevier, New York, 1990, pp. 158–160.

19. G. H. Keller and M. M. Manak, *DNA Probes*, Macmillan, New York, 1989, pp. 1–23.

20. M. Nei and W.-H. Li, *Proc. Natl. Acad. Sci. U.S.A.* **76**, 1979, 5269.

21. G. H. Keller and M. M. Manak, *DNA probes*, Macmillan, New York, 1989, pp. 76–79.

PROBLEMS

1. A mixture of cytochrome *c* and myoglobin are to be separated by polyacrylamide gel electrophoresis. Their isoelectric pH values (pI) are 9.6 and 7.2, and their molecular weights are 11.7 and 17.2 kDa, respectively. In which direction will each protein migrate (toward the anode or the cathode) at (a) pH 6.0 and (b) pH 8.5?

2. Two monomeric proteins, X and Y, of MW 16.5 and 35.4 kDa, respectively, were treated with SDS in order to separate them by SDS–PAGE.

(a) Should the sample mixture be applied to the anodic or the cathodic end of the gel?

(b) Which of the two proteins will migrate furthest from the origin?

(c) Migration distances of 9.2 and 2.6 cm were measured for the two standard proteins. An unknown protein migrated 5.2 cm through the same gel after SDS treatment. What is the molecular weight of the unknown protein?

3. A mixture of high molecular weight proteins and DNA is separated by agarose gel electrophoresis. Choose a combination of stains from Table 9.1 that would allow the protein bands to be distinguished from the DNA bands on the basis of color.

4. At what approximate pH should a Western blot be conducted, in order to transfer only the proteins in Problem 3 to a nitrocellulose membrane?

5. To detect the positions of electrophoretically separated DNA fragments, would ethidium bromide or a biotinylated DNA probe (used in conjunction with an avidin–enzyme conjugate and activity stain) be expected to provide positional data for all DNA fragments present in the sample? Which would provide selective data for DNA fragments containing a particular sequence?

CHAPTER 10

Applications of Zone Electrophoresis

10.1. INTRODUCTION

This chapter describes some applications of the principles and methods described in Chapter 9. Many variations of zone electrophoresis have been developed as specific tools for research and diagnostic tests. The examples given in this chapter, while not a comprehensive listing, are representative of the diversity of applications of zone electrophoresis. We will consider the determination of net charge, subunit composition and molecular weight of proteins, the determination of DNA molecular weight, the identification of isoenzymes, the diagnosis of genetic disease, DNA fingerprinting, DNA sequencing, and immunoelectrophoresis.

10.2. DETERMINATION OF PROTEIN NET CHARGE AND MOLECULAR WEIGHT USING PAGE[1]

Using the principles described in Chapter 9, multisubunit proteins can be separated according to molecular weight on polyacrylamide gels under nondenaturing conditions. A number of gels are prepared having constant %C and varying %T. The unknown protein and a number of standard proteins of known molecular weight are electrophoresed separately on these gels, with a tracking dye that is not restricted by the gel added to each sample. For each protein on each gel, an R_f value is calculated as the protein migration distance divided by the migration distance of the marker. The R_f values will always lie between zero and unity, provided that the tracking dye has the same charge *sign* as the protein. When this data has been tabulated, a Ferguson plot of $\log(R_f)$ versus %T is constructed for each protein. The net charge ratio (protein:marker) is obtained from the Ferguson plot, as the antilog of the y intercept (%T = 0). The negative slope of the Ferguson plot is called K_D, and is proportional to molecular weight. A calibration curve of K_D versus molecular weight is prepared from data obtained with the protein standards, and the

Bianalytical Chemistry, by Susan R. Mikkelsen and Eduardo Cortón
ISBN 0-471-54447-7   Copyright © 2004 John Wiley & Sons, Inc.

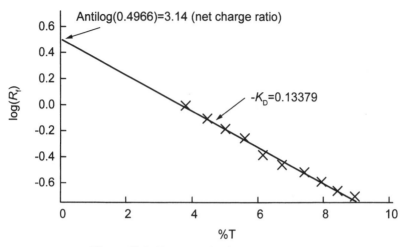

Figure 10.1. Ferguson plot for hemagglutinin.

K_D value of the unknown is used with this plot to obtain the molecular weight of the unknown.

This technique is illustrated with the protein hemagglutinin. A series of 11 gels were prepared using %C = 2, with %T varying from 4 to 9. The R_f values determined following electrophoresis and staining were used to construct the Ferguson plot shown in Figure 10.1. The y intercept of this plot yields a net charge for hemagglutinin of 3.13 and a K_D value of 0.13379.

Table 10.1 shows the K_D values obtained on the same gels for seven standards of known molecular weight. The calibration curve used to determine the molecular weight of hemagglutinin is then constructed, and the molecular weight of hemagglutinin is determined by interpolation of this curve, shown in Figure 10.2.

Note that this calibration curve exhibits a linear region up to molecular weights that result in very little migration in the polyacrylamide gels. Data obtained for very high molecular weight protein standards tend to lack precision, due to the difficulty

TABLE 10.1. Ferguson Slopes and Molecular Weights of Protein Standards

Protein	Molecular Weight (kDa)	K_D
Chymotrypsin	21.6	0.03868
Aconitase	66	0.07201
Plasminogen	81	0.08991
Lactoperoxidase	93	0.09336
Immunoglobulin G	160	0.12998
IgG dimer	320	0.25127
IgG trimer	480	0.31315
Hemagglutinin	Unknown	0.13379

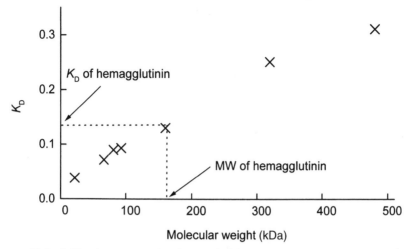

Figure 10.2. Calibration curve for hemagglutinin molecular weight determination. [Reprinted, by permission, from D. J. Holme and H. Peck, "*Analytical Biochemistry*", Longman, New York, 1998. © Addison Wesley Longman Limited 1998.]

in measuring very small migration distances. The molecular weight of hemagglutinin was thus determined to be 160 kDa, which is within the linear region of the calibration curve. If the unknown protein showed a molecular weight in excess of ∼ 300 kDa, the experiment would be repeated using a lower range of %T values and a series of standard proteins of higher molecular weight. This technique yields molecular weight values that are accurate to within 5%.

10.3. DETERMINATION OF PROTEIN SUBUNIT COMPOSITION AND SUBUNIT MOLECULAR WEIGHTS[2]

Tertiary and quaternary structure in multisubunit proteins is maintained in part by the presence of disulfide bonds between cysteine residues. When such a protein is exposed to a reducing agent such as mercaptoethanol or dithiothreitol, the disulfide bonds are converted to sulfhydryl groups, the subunits dissociate and the tertiary structure of the subunits is disrupted. In the presence of SDS, such protein subunits yield uniformly negative species. Proteins will bind SDS in a 1.4 g SDS to 1-g protein ratio, effectively obliterating any native charge on the protein, and this uniform negative charge/mass ratio results in electrophoretic migration from cathode to anode with distances that depend only on the size of the polypeptide chain. The SDS–polyacrylamide gel results in a simple molecular sieving action that allows small proteins to migrate rapidly, and restricts the migration of large proteins. Plots of log(relative molecular mass) (RMM, molecular weight) against relative mobility (migration distance) are linear, and allow the determination of unknown molecular weights. Figure 10.3 and Table 10.2 show data obtained for seven

TABLE 10.2. Protein Standards Used for SDS–PAGE[a]

Number	Protein	RMM	$\log_{10}$RMM
1	Cytochrome c (muscle)	11,700	4.068
2	Myoglobin (equine skeletal muscle)	17,200	4.236
3	IgG Light chain	23,500	4.371
4	Carbonic anhydrase (bovine)	29,000	4.462
5	Ovalbumin	43,000	4.634
6	Albumin (human)	68,000	4.832
7	Transferrin (human)	77,000	4.886

[a] See Ref. 3.

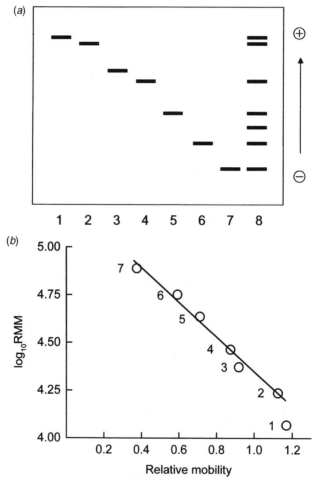

Figure 10.3. The SDS–PAGE (a) and MW calibration (b) for proteins in Table 10.2.[3] [Reprinted, with permission, from D. J. Holme and H. Peck, "*Analytical Biochemistry*", Longman, New York, 1998. © Addison Wesley Longman Limited 1998.]

proteins on a 5% polyacrylamide gel containing 0.1% SDS, after heating the samples in a mercaptoethanol–SDS solution.[3]

10.4. MOLECULAR WEIGHT OF DNA BY AGAROSE GEL ELECTROPHORESIS[4]

The large pore sizes of agarose gels allow the molecular sieving effect to be employed for the determination of DNA molecular weight. Due to the negatively charged phosphate groups on the DNA backbone, native DNA has a constant charge/mass ratio. The relationship between molecular weight and mobility is not as straightforward for DNA as it is for protein MW determinations by SDS–PAGE methods, however. Figure 10.4 shows plots of DNA length (in base pairs, n) against mobility, for linear and circular DNA. These plots show the general expected trend of decreasing mobility with increasing molecular weight, as well as decreasing mobility with increasing agarose content, or decreasing pore size. However, the semilogarithmic plot is nonlinear, especially at higher agarose concentrations, indicating that the simple sieving model does not strictly apply.

The mobility behavior of DNA has been modeled using polystyrenesulfonate polymers.[5] It has been shown that the mobility depends on the degree of *entanglement* of the polymer in the gel. Only weakly entangled polymers give the expected sieving behavior and linear calibration plots. This fundamental difference in mobility behavior between proteins and DNA is a result of the spherical, or random-coil, approximation that generally holds for proteins but does not hold for nucleic acid polymers. For DNA MW determinations, either the agarose concentration should be

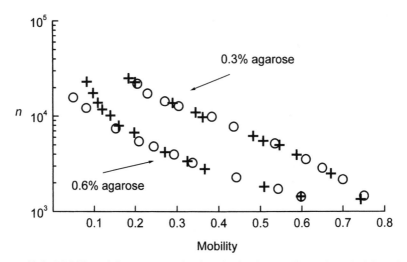

Figure 10.4. Mobility of linear (+) and relaxed, circular (o) DNA through 0.3 or 0.6% agarose gels, at 0.3 V/cm for 30 or 40 h, respectively, in 0.1 M Tris, 0.09 M boric acid, and 1 mM EDTA.

maintained at a minimal value, or a large number of bracketing MW standards should be employed to define the shape of the calibration curve.

10.5. IDENTIFICATION OF ISOENZYMES

Isoenzymes are enzymes from different species, or produced by different mechanisms within the same species, that catalyze identical reactions. Isoenzymes often differ from each other by only a few amino acid residues, and may have very similar molecular weights. In some cases, charged amino acid residues in one isoenzyme (e.g., lysine, arginine, aspartate, glutamate) may be substituted by uncharged amino acid residues in another, and this type of substitution will result in different net charges at a given pH, and different pI values for the different isoenzymes.

For example, human alcohol dehydrogenase exists in *more than twenty forms.*[6] These isoenzymes are all dimers, and all have molecular weights of ~ 80 kDa, or ~ 40 kDa per subunit. The different subunits have been called the α, β, $\gamma 1$, $\gamma 2$, π, and X subunits. All of the isoenzymes catalyze the conversion of ethanol to acetaldehyde according to Eq. 10.1:

$$CH_3CH_2OH + NAD^+ \xrightarrow{\text{alcohol dehydrogenase}} CH_3CHO + NADH + H^+ \quad (10.1)$$

Human alcohol dehydrogenase isoenzymes have been divided into three classes. Class I isoenzymes contain α, β, $\gamma 1$, and $\gamma 2$ subunit isoenzymes, are cationic at neutral pH, and are strongly inhibited by substituted pyrazole inhibitors. Class II isoenzymes contain the π isoenzyme subunit, are cationic at neutral pH, and are relatively insensitive to pyrazole inhibition. Class III isoenzymes contain the X subunit, are anionic at neutral pH and are insensitive to pyrazole inhibitors.

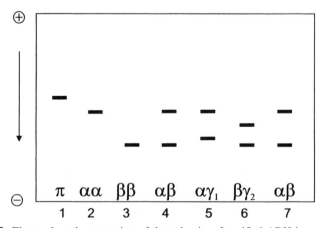

Figure 10.5. Electrophoretic separation of the subunits of purified ADH isoenzymes in the presence of 7 M urea. Lane 1 contains Class II ADH, with only π subunits, while the Class I isoenzymes in lanes 2–7 have subunit compositions as indicated in each lane.

Classes I and II have been studied using PAGE, with 7 M urea in the sample solution and gel, and with 10-mM dithiothreitol in the sample.[7] The urea interferes with hydrogen bonding and unfolds the protein, while DTT reduces the disulfide bonds. Because Class I and II isoenzymes are cationic at neutral pH, the samples are introduced at the anode and will migrate toward the cathode. While the masses of the different subunits are essentially identical, their net charges are significantly different, and this affects the charge/mass ratios that control electrophoretic mobilities. Figure 10.5 shows the stained gels obtained after a series of Class I and II alcohol dehydrogenase isoenzymes have been run.

10.6. DIAGNOSIS OF GENETIC (INHERITED) DISEASE

The DNA diagnostic techniques generally use analyte DNA that has been isolated from white blood cells. This high molecular weight DNA may be digested with *restriction endonucleases* to yield shorter, double-stranded fragments. The fragments are separated on agarose gels, transferred to nylon or nitrocellulose membranes, and examined for the sequence of interest using labeled DNA probes.

Restriction enzymes recognize certain regions, or sequences, of the DNA. These are relatively short, and occur at multiple sites over the entire native genome. Some examples of restriction enzymes and their recognition sites are listed in Table 10.3. Cleavage of the double-stranded DNA occurs where the slashes (/) are shown in the recognition sequence. The names of the restriction enzymes are derived from the species from which they are isolated (e.g., Eco is *E. coli*) and contain further classification numbers.

The size of the recognition site of a restriction endonuclease may vary between 4 and 9 (or more) base pairs. A small site leads to multiple cleavages and many fragments, since this small site is statistically likely to occur more often, while larger recognition sites lead to fewer fragments. Sickle-cell anemia is a recessive genetic disorder that results from a single-base substitution, or a point mutation, in the gene that codes for the β-globin sequence of tetrameric human hemoglobin. This point mutation results in the substitution of one amino acid in the entire β-globin

TABLE 10.3. Examples of Restriction Endonucleases and Their Recognition Sites

Enzyme	Recognition Site
Sca I	5′-AGT/ACT-3′
	3′-TCA/TGA-5′
Hind III	5′-A/AGCTT-3′
	3′-TTCGA/A-5′
Eco RI	5′-G/AATTC-3′
	3′-CTTAA/G-5′
Alu I	5′-AG/CT-3′
	3′-TC/GA-5′

polypeptide, and leads to low solubility in the deoxygenated form of hemoglobin. The DNA isolated from the white blood cells of patients is tested using the Mst II restriction endonuclease for digestion, followed by agarose gel electrophoresis and hybridization to a ^{32}P-labeled DNA probe, where the probe sequence contains the entire human β-globin coding sequence.[8] Individuals with sickle-cell anemia lack one of the Mst II cleavage sites, so that the DNA of these individuals will produce fewer fragments, and some normal fragment lengths will be missing. Figure 10.6 shows the autoradiogram of the separation patterns obtained by this method. Note that only the fragments that contain the β-globin coding sequence are visualized.

In Figure 10.6, it can be seen that differences in the restriction fragment lengths occur in the 1.15–1.35 kilobase region of the separation pattern. After quantitative

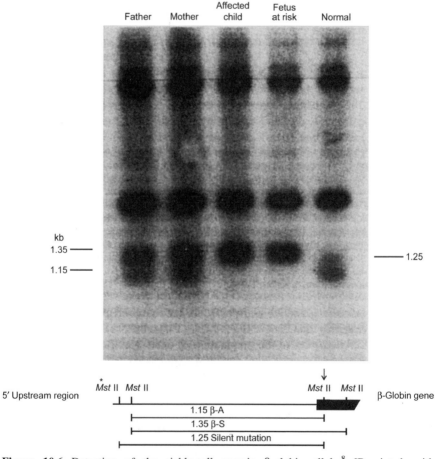

Figure 10.6. Detection of the sickle-cell anemia β-globin allele.[8] [Reprinted, with permission, from G. H. Keller and M. M. Manak, *"DNA Probes"*, Macmillan (Stockton Press), New York, 1993. © Macmillan Publishers Ltd. 1993.]

digestion of the DNA by the Mst II restriction enzyme, the normal (beta-A) DNA yields a 1.15 kilobase fragment. Because one of the restriction sites is absent on the sickle-cell gene (beta-S), this region of the DNA is not cleaved, and the resulting fragment is longer, at 1.35 kilobases. The patterns obtained for DNA from the mother and father both show the normal and the sickle-cell bands, while an affected child exhibits only the sickle-cell band at 1.35 kilobases. The results of this test clearly show that the fetus will have sickle-cell disease.

10.7. DNA FINGERPRINTING AND RESTRICTION FRAGMENT LENGTH POLYMORPHISM[9]

DNA fingerprinting methods are used in forensic labs for the identification of individuals, and in clinical labs for the diagnosis of genetic disease. These methods employ restriction enzymes that have been carefully selected by screening methods for their particular cleavage sites. Restriction enzymes cocktails are used in forensic methods to produce many fragments, so that sequence differences that normally occur between individuals produce many differences in fragment lengths. Diagnostic methods employ restriction enzymes that recognize a critical sequence where mutations of this sequence are known to be associated with the disease state.

The term *restriction fragment length polymorphism*, abbreviated RFLP, refers to the variety of fragment lengths from DNA digestion with restriction endonucleases. The various lengths can result in such distinct electrophoretic separation patterns that they are like fingerprints because they are unique to an individual. Simple polymorphisms, such as that seen in Figure 10.6 with the Mst II site lacking in sickle-cell patients, can be employed for diagnostics. Figure 10.7 illustrates the principles of RFLP methods.

If the total number of fragments resulting from quantitative digestion is relatively small, the detection step may involve simple ethidium staining directly on the gel, with no Southern blot. However, if many fragments are produced, Southern blotting and detection via hybridization with labeled DNA probes are necessary. The probes that are chosen hybridize with many fragments, so that a significant

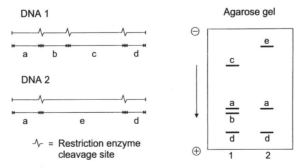

Figure 10.7. Restriction fragment length polymorphism for DNA identification.

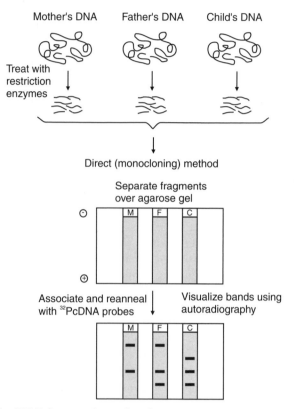

Figure 10.8. The RFLP for paternity testing. Step 1 (top) is the isolation and cleavage of native DNA into restriction fragments; Step 2 (middle) is the agarose electrophoretic separation, which yields smears of indistinguishable bands; and Step 3 (bottom) shows the pattern obtained after Southern blotting, hybridization with ^{32}P labeled probes, and autoradiography.

portion of the separation pattern is visualized, and individual DNA differences are readily observed. Figure 10.8 illustrates DNA fingerprinting for paternity testing. The separation pattern resulting from the child's DNA can be seen to contain bands present in both the mother's and the father's DNA, while a new band in the child's DNA suggests a possible new mutation. The critical reagent in RFLP methods is the restriction enzyme (or cocktail of enzymes), which must produce very distinct fragment lengths for individual DNA samples.

Diagnostic RFLP methods employ DNA that is readily available from the white cells of patient blood samples. Forensic methods, on the other hand, use DNA samples of unpredictable (usually low) quantity; furthermore, if the DNA sample is old and has been exposed to sunlight (UV radiation), the DNA may be present as cleaved fragments rather than intact genes. In such cases, the available DNA is amplified in concentration using the polymerase chain reaction (PCR).[10] PCR technology has revolutionized DNA analysis, and has made such general forensic tests feasible.

PCR amplification of DNA uses repetitive thermal cycling between three temperatures to (a) dissociate analyte DNA into single strands, (b) anneal primers, and (c) biosynthesize new DNA with the enzyme DNA polymerase. A solution containing the double-stranded DNA of interest, an excess of two unique primer oligonucleotides, a heat-stable DNA polymerase (Taq DNA polymerase is used; it is isolated from the thermophilic bacterium *Thermus aquaticus*), and an excess of deoxynucleotide triphosphates, is thermally cycled as follows. At 93–94 °C, the analyte DNA is denatured into single strands. The temperature is rapidly shifted to an annealing temperature between 37 and 72 °C (the best temperature is determined experimentally for a particular set of primers), and at this temperature, the single strands anneal, or form heteroduplexes, with the primers. The primers are single-stranded, 20–26 bases in length and 40–60% GC content. Two unique primers are needed for PCR amplification, with one annealing to each strand, near the 3'-end of the region of interest. Following the annealing step, the temperature is rapidly changed to 70–72 °C, where the DNA polymerase sequentially extends the primers one nucleotide at a time, moving toward the 5'-end of the region of interest. The nucleotides that are incorporated into the new double strands are complementary in sequence to the analyte strands. After this cycle, one copy of each analyte DNA molecule has been produced.

Repeated thermal cycling through the denaturation, annealing and extension temperatures yields copies of the copies, and an exponential amplification of the DNA concentration. In principle, the number of copies produced is equal to $2^n - 1$, where n is the number of complete thermal cycles used. Most PCR protocols specify over 25 cycles. In practice, the amplification is somewhat less than predicted, due to incomplete extension through the second primer sites. The usual size of the amplified region is 100–400 bases in length. Figure 10.9 shows a schematic representation of PCR amplification.

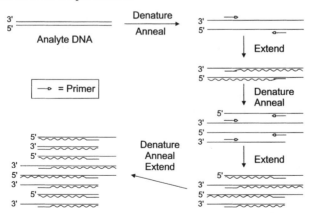

Figure 10.9. Schematic representation of the first three cycles of PCR amplification. The template DNA can be any piece of double-stranded DNA containing the sequence of interest, ranging in size from the PCR product itself to a human chromosome. The wavy lines indicate newly synthesized DNA; note that the primer is incorporated into this new strand and defines the 5'-end of the product. The straight lines indicate the template strands that are copied.

10.8. DNA SEQUENCING WITH THE MAXAM–GILBERT METHOD[11]

Determining the sequence of an unknown DNA sample may be accomplished using either enzymatic (the dideoxy method) or chemical (the Maxam–Gilbert method) reactions that yield a series of detectable fragments that differ in length by a single base. The Maxam–Gilbert method uses four chemical reactions that selectively cleaves DNA at dG, dC, dG+dA and dC+dT residues, and the products of these partial cleavage reactions are subjected to polyacrylamide gel electrophoresis. The initial step in the sequencing protocol involves labelling single-stranded DNA with ^{32}P at either the 5′- or the 3′-end (but not both). The labeled DNA is then divided into four portions, and subjected to four separate cleavage reactions. Each reaction results in partial cleavage of the fragment at particular bases, and, following PAGE, the sequence of the DNA is read from the pattern of bands on the autoradiogram.

For example, given an initial fragment, labeled at the 5′-end, that has the sequence ^{32}P-GCTGCTAGGTGCCGAGC, partial cleavage at (and removal of) the G residues will yield the following *detectable* fragments:

$$^{32}\text{P-GCTGCTAGGTGCCGAGC}$$
$$^{32}\text{P}$$
$$^{32}\text{P-GCT}$$
$$^{32}\text{P-GCTGCTA}$$
$$^{32}\text{P-GCTGCTAG} \tag{10.2}$$
$$^{32}\text{P-GCTGCTAGGT}$$
$$^{32}\text{P-GCTGCTAGGTGCC}$$
$$^{32}\text{P-GCTGCTAGGTGCCGA}$$

Many other fragments will also be produced, but will not be detectable, since they lack the ^{32}P label. Electrophoresis of this product mixture on a polyacrylamide gel will therefore yield eight bands—one for the initial fragment and seven for the reaction products. The initial fragment will be closest to the origin (the cathodic end of the gel), while the shortest fragment will migrate furthest from the cathode. It is very important that the cleavage reactions are not allowed to go to completion. If this occurs, only the shortest fragments will be present in each of the four reaction products. We will now consider the four degradation reactions.

1. *The G Reaction*: This reaction uses dimethyl sulfoxide (DMSO), hydroxide, and piperidine for the selective cleavage of DNA at deoxyguanosine residues, as shown in Reaction 10.3. DMSO methylates at the N7 position of guanine, and makes it susceptible to nucleophilic attack at the C8 position by hydroxide, which opens the ring. These two reactions also occur with adenine, and it is the next step that is selective for guanine. In the presence of piperidine, guanine (but not adenine) is displaced, and the deoxyribose unit is removed from the strand. This yields two DNA fragments, both of which will have phosphate groups present where they were attached to the removed sugar residue.

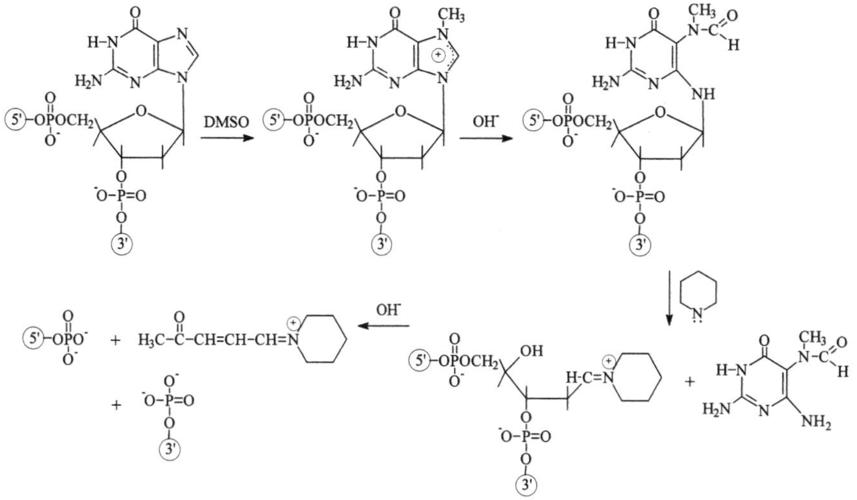

(10.3)

2. *The C Reaction*: The C reaction uses hydrazine, piperazine, and hydroxide to selectively remove dC residues from single-stranded DNA, as shown in Reaction 10.4. Hydrazine attacks cytosine at the C4 and C6 positions, opening the pyrimidine ring. This product then cyclizes into a new five-membered ring. The hydrazine will not attack thymidine residues if 1 M NaCl is present. Further reaction with hydrazine releases the ring, leaving the sugar residue in the DNA backbone as a hydrazone. Piperidine will react with this hydrazone, and in the presence of hydroxide, the sugar residue is removed, leaving two fragments that are phosphorylated as in the G reaction.

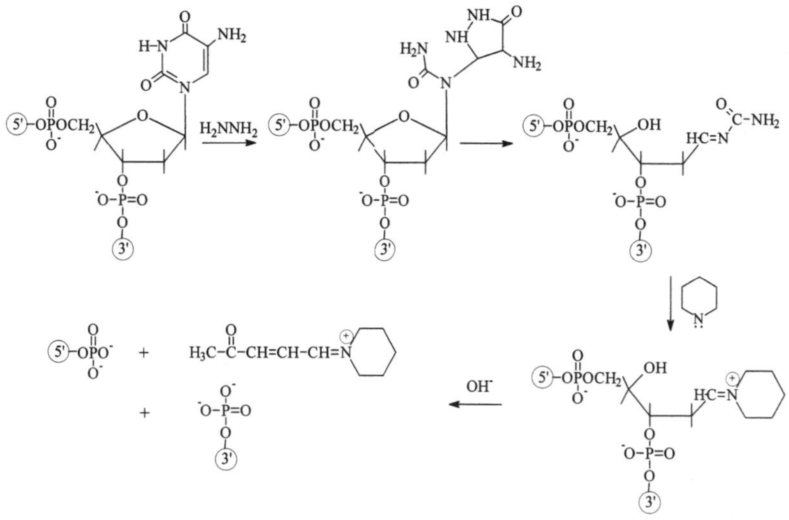

(10.4)

TABLE 10.4. Labeled Fragments Expected from the ^{32}P-ACTGTAGC Cleavage Using the Maxam–Gilbert Sequencing Method

G	G+A	C+T	C
^{32}P-ACT	^{32}P	^{32}P-A	^{32}P-A
^{32}P-ACTGTA	^{32}P-ACT	^{32}P-AC	^{32}P-ACTGTAG
	^{32}P-ACTGT	^{32}P-ACTG	
	^{32}P-ACTGTA	^{32}P-ACTGTAG	

Both the G and C reactions are specific for these bases. The two other reactions needed for sequencing result in cleavage at both G and A residues, and at both C and T residues, respectively.

3. *The G + A Reaction*: In this reaction, acid is used to weaken the glycosidic bond by protonating (rather than methylating as in the G reaction) at the N7 position of both A and G. This protonated form is susceptible to piperidine displacement, which then occurs as shown in Reaction 10.3.

4. *The C + T Reaction*: The C + T reaction is performed under identical conditions to the C reaction, except that the 1 *M* NaCl used in the C reaction is absent. Under these conditions, cleavage will occur at both C and T residues. For example, the fragments expected for the cleavage of the 5′-labelled fragment ^{32}P-ACTGTAGC in each reaction are shown in Table 10.4.

It can be seen from this table that all fragments resulting from the G reaction will also be present in the G + A reaction product, and, similarly, that all fragments resulting from the C reaction are also present in the C + T reaction product. These product mixtures are electrophoresed in separate lanes on a polyacrylamide

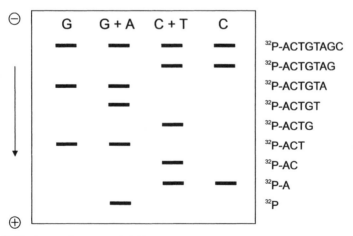

Figure 10.10. Autoradiogram of DNA sequencing gel obtained after Maxam–Gilbert sequencing reactions of ^{32}P-ACTGTAGC.

slab, are separated according to size, and autoradiography then reveals the pattern shown in Figure 10.10.

To read the sequence from the autoradiogram, we recall that the smallest fragments migrate the furthest from the cathode towards the anode. Furthermore, all of the lanes must have an identical band closest to the origin that results from the unreacted strand—if this is absent, then the reactions have been allowed to proceed

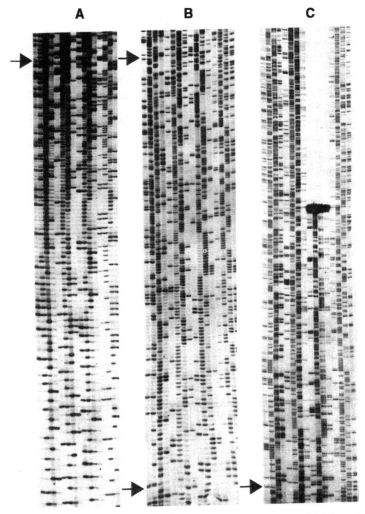

Figure 10.11. Autoradiograms of three 1-m sequencing gels, A = 16%T, B = 6%T, and C = 4%T. Each gel has 16 lanes, containing the four reaction products, in order (left to right) G, G+A, C+T, C, for each if four DNA samples. The arrows indicate crossover points from one gel to the next. [Reprinted, with permission, from R. F Barker, in *"Nucleic Acids Sequencing: A Practical Approach"*, C. J. Howe and E. S. Ward, Eds., Oxford University Press, New York, 1989. © IRL Press at Oxford University Press 1989.]

too far toward completion. Since the smallest fragment is known to be the ^{32}P label, and since this band occurs in the G + A lane, we know that the base closest to the label must be A. The second furthest migrating band is observed in both the C + T and the C lanes, so that the second base is a C. The third band occurs only in the C + T lane, and is therefore a T residue. We have now sequenced the first three bases, as ^{32}P-ACT. Continuing in this manner, the entire sequence of the fragment is easily read from the sequencing gel.

Polyacrylamide gels as long as 1 m have been used to enable the sequence of a 600-base DNA fragment to be read from a single gel.[12] More often, however, several gels are employed. As shown in Figure 10.11, for example, a gel with up to 20%T is used for bases 0–150, a 6%T gel for bases 140–300, and a 4%T gel is used to sequence bases 290–600+. It is necessary to include a region of overlap from one gel to the next, to ensure that no fragments remain undetected.

10.9. IMMUNOELECTROPHORESIS[13]

Antigen–antibody binding interactions can be induced after an electrophoretic separation pattern is blotted from the gel onto a membrane, in a technique called immunoblotting, and will also occur directly in the gels used for electrophoresis. Immunoelectrophoretic methods are used for the identification and quantitation of antigens.

The immunoblotting technique follows after the Western blotting of the proteins onto a membrane. The membrane with bound proteins is incubated in a solution containing a polyclonal antiserum or a monoclonal antibody, and is then rinsed. This primary binding step identifies the antigen of interest through the high selectivity of the binding reaction. Subsequent labeling steps, for example, with an enzyme-labeled anti-IgG followed by an activity stain, allow the detection of the antibody selectively bound to the antigen on the membrane. For the primary step, monoclonal antibodies are favored over polyclonals, because of the single epitope selectivity which reduces interferences that result in the detection of non-analyte proteins. If polyclonal antiserum is used, all antibodies present must be considered. A "good" antiserum contains 6–7 mg/mL immunoglobulins, and, of these, 20–30% are so-called specific antibodies directed against the immunogen. The remaining 70–80% of the antibodies are irrelevant to the antigen of interest, and are directed against environmental immunogens such as microorganisms and foodstuffs as well as against impurities in the injected immunogen. The affinity or avidity of the primary binding interaction is often significantly lower in immunoelectrophoretic methods than in typical immunoassays such as ELISA. This result is of antigen denaturation in the gel, and is particularly noticeable with SDS–PAGE separations.

Figure 10.12 shows examples of protein detection with polyclonal and monoclonal antibodies.[14] Note the large number of visible bands in the polyclonal immunoblot, and the fact that the monoclonal appears to interact with a different antigen epitope than most of the antibodies in the polyclonal mixture, since the one band

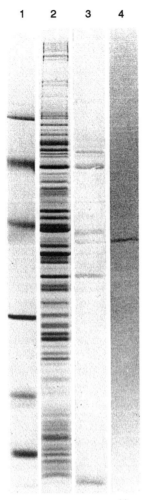

Figure 10.12. Immunoblots of *M. hyorhinis* antigens.[14] Lane 1: marker proteins; lane 2: Coomassie blue stain; lane 3: polyclonal antiserum immunoblot; lane 4: monoclonal antibody immunoblot. [Reprinted, with permission, from K. E. Johansson, in *"Handbook of Immunoblotting of Proteins"*, O. J. Bjerrum and N. H. H. Heegaard, Eds., CRC Press, Boca Raton, FL, 1988. © 1988 by CRC Press, Inc.]

visualized in the monoclonal immunoblot is barely visible in the polyclonal immunoblot.

Figure 10.13 shows a comparison of different labeling techniques, after an identical mouse monoclonal antibody was used in the primary immunoblotting step.[15] Lane A shows the indirect immunoperoxidase method, whereby the primary monoclonal is detected using a peroxidase-conjugated goat anti-mouse IgG antibody, followed by a peroxidase activity stain. Lane B shows the peroxidase anti-peroxidase

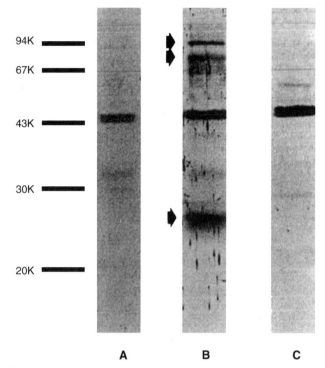

A   B   C

Figure 10.13. Comparison of immunoperoxidase (Lane A), peroxidase anti-peroxidase (Lane B) and biotin–streptavidin (Lane C) immunoblotting detection methods.[15] [Reprinted, with permission, from K. E. Johansson, in *"Handbook of Immunoblotting of Proteins"*, O. J. Bjerrum and N. H. H. Heegaard, Eds., CRC Press, Boca Raton, FL, 1988. © 1988 by CRC Press, Inc.]

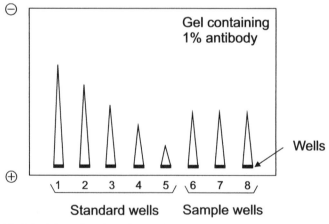

Figure 10.14. Principles of rocket electrophoresis. Rocket height is measured from the cathodic end of the well to the rocket tip, and is directly proportional to antigen concentration.

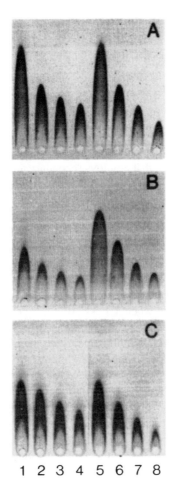

1 2 3 4 5 6 7 8

Figure 10.15. Dilution series of purified proteins and crude cell extracts from Chlamydomonas.[17] A: Antibodies (51 µg IgG/cm²) were raised against the large subunit of the protein RuBPCase; wells loaded with cell extract from 31.2, 15.6, 10.4, or 7.8×10^4 cells (wells 1–4) or with 600, 300, 150, or 75 ng of RuBPCase large subunit (wells 5–8); run performed at 14 V/cm for 24 h. B: Antibodies (74 µg/cm²) were raised against the small subunit of RuBPCase; wells 1–4 same as in A, wells 5–8 contain 300, 150, 75, or 40 ng RuBPCase small subunit controls; run at 8 V/cm for 24 h. C: Antibodies (20 µg/cm²) were raised to thylakoid membrane polypeptide; wells 1–4 contain extracts of 7.8, 6.2, 4.2, or 3.1×10^4 cells, and wells 5–8 contain purified thylakoid membranes corresponding to 20, 10, 5, or 2.5 µg of chlorophyll; run conditions as in B. [Reprinted, with permission, from F. G. Plumley and G. W. Schmidt, *Analytical Biochemistry* **134**, 1983, 86–95. Copyright © 1983 by Academic Press, Inc.]

method, in which rabbit anti-mouse IgG antibody is bound to the mouse monoclonal, and an anti-rabbit IgG antibody, with its two binding sites, is sandwiched between the rabbit anti-mouse IgG and a 2:1 peroxidase: rabbit anti-peroxidase complex. Finally, Lane C shows detection using a biotinylated anti-mouse IgG antibody followed by a streptavidin–peroxidase conjugate, which is stained for activity. Lanes A and C clearly indicate better immunoblot selectivity, and this results, in part, from limited use of polyclonal antisera in the immunological reactions.

A final example of immunoelectrophoresis methods is the quantitative method called rocket electrophoresis, that employs secondary antibody–antigen interactions for detection.[16] The rocket method uses thin agarose gels, typically 1–2%, that contain $\sim$1 % antibody; the antigen is a protein (i.e., univalent and multideterminate) and the antibody is a polyclonal antiserum. Wells are cut into the gel prior to the electrophoretic run, usually at the anodic end, in the same manner used for the single radial immunodiffusion method (cf. Chapter 5). Standards and unknown solutions are then placed in the wells. The samples are electrophoresed at low field strength (<20 V/cm) for $\sim$24 h using an acidic buffer, so that proteins are positively charged and migrate towards the cathode. Migration causes the dilution of the sample solutions, and when 1:1 Ag:Ab equivalence is reached, a precipitin line forms. Higher antigen concentrations lead to larger migration distances before the formation of the precipitin line. The lines formed are rocket shaped, and the height of the rocket, from the cathodic end of the well to the rocket tip, is directly proportional to antigen concentration. It has been shown that the best results with this method occur with long (>24 h) runs at low (5 V/cm) field strengths; under these conditions, rocket electrophoresis may be used to determine nanogram quantities of proteins. A protein stain such as Coomassie Blue or silver may be used to facilitate visual detection of the precipitin lines, provided that a prior rinse is used to remove unbound antibodies from the gel. Figure 10.14 illustrates the principles of rocket electrophoresis, while Figure 10.15 shows the quantitation of proteins present in cell extracts of a species of green algae.

SUGGESTED REFERENCES

B. S. Dunbar, *Two-Dimensional Electrophoresis and Immunological Techniques*, Plenum Press, New York, 1987.

L. A. Lewis, *CRC Handbook of Electrophoresis*, CRC Press, Boca Raton, FL, 1980.

C. J. Howe and E. S. Ward, *Nucleic Acids Sequencing: A Practical Approach*, Oxford University Press, New York, 1989.

J. Breborowicz and A. Mackiewicz, Eds., *Affinity Electrophoresis: Principles and Applications*, CRC Press, Boca Raton, FL, 1991.

G. H. Keller and M. M. Manak, *DNA Probes*, Macmillan (Stockton Press), New York, 1989.

REFERENCES

1. D. Rodbard and A. Chrambach, *Proc. Natl. Acad. Sci. U.S.A.* **65**, 1970, 970–977.

2. A. Chrambach, *The practice of Quantitative Gel Electrophoresis*, VCH Publishers, Deerfield Beach, FL, 1985, p. 177.

3. D. J. Holme and H. Peck, *Analytical Biochemistry*, 3rd ed., Longman, New York, 1998, pp. 400–402.

4. N. C. Stellwagen, *Biochemistry* **22**, 1983, 6180–6185.

5. D. L. Smisek and D. A. Hoagland, *Science* **248**, 1990, 1221–1223.

6. W. F. Bosron, T.-K. Li, and B. L. Vallee, *Biochem. Biophys. Res. Commun.* **91**, 1979, 1549–1555.

7. W. M. Keung, C. C. Ditlow, and B. L. Vallee, *Anal. Biochem.* **151**, 1985, 92–96.

8. G. H. Keller and M. M. Manak, *DNA Probes*, Macmillan (Stockton Press), New York, 1993, p. 16.

9. L. T. Kirby, *DNA Fingerprinting: An Introduction*, Macmillan (Stockton Press), New York, 1990.

10. W. C. Timmer and J. M. Villalobos, *J. Chem. Educ.* **70**, 1993, 273–280.

11. A. M. Maxam and W. Gilbert, *Methods Enzymol.* **65**, 1980, 449–560.

12. R. F. Barker, in *Nucleic Acids Sequencing: A Practical Approach*, C. J. Howe and E. S. Ward, Eds., Oxford University Press, New York, 1989, p. 133.

13. J. Renart, J. Reiser, and G. R. Stark, *Proc. Natl. Acad. Sci. U.S.A.* **76**, 1979, 3116–3120.

14. K. E. Johansson, in *Handbook of Immunoblotting of Proteins*, Vol. 1, O. J. Bjerrum and N. H. H. Heegard, Eds., CRC Press, Boca Raton, FL, 1988, pp. 159–165.

15. K. Ogata, in *Handbook of Immunoblotting of Proteins*, Vol. 1, O. J. Bjerrum and N. H. H. Heegard, Eds., CRC Press, Boca Raton, FL, 1988, pp. 167–176.

16. C. B. Laurell, *Anal. Biochem.* **15**, 1966, 45–52.

17. F. G. Plumley and G. W. Schmidt, *Anal. Biochem.* **134**, 1983, 86–95.

PROBLEMS

1. The standard and unknown proteins listed below were electrophoresed using SDS–PAGE. Estimate the molecular weight of the unknown protein.

Protein	MW (kDa)	Migration Distance (cm)
Aldolase	158	3.56
Catalase	210	3.23
Ferritin	440	2.86
Thyroglobulin	669	2.51
Unknown	?	3.03

2. A protein known to be highly purified is subjected to SDS–PAGE. Two bands are observed, and these correspond to molecular weights of 50 and 75 kDa. After staining and densitometry, the 75-kDa band is seen to possess twice the area of the 50-kDa band. What conclusions can be reached regarding the original protein?

3. Double-stranded DNA standards and an unknown were electrophoresed in a 0.6% agarose gel, stained with ethidium bromide and photographed under UV light. The negative of the photographed separation pattern was then scanned with a densitometer. The following data were obtained. Estimate the length of the unknown DNA, in base pairs.

Length (BP)	Migration Distance (cm)
9162	5.56
8144	7.84
7126	10.12
6108	13.50
5090	17.93
4072	23.79
3054	32.06
?	15.27

4. An oligodeoxynucleotide 10 bases in length was subjected to the Maxam–Gilbert procedure for sequence determination. The decamer was initially labeled with ^{32}P (as phosphate) at the 5'-end. The four individual sequencing reactions were carried out, and the mixtures were applied individually to four lanes of a 20%T polyacrylamide gel. Following electrophoresis, the following autoradiographic pattern (Fig. 10.16) was obtained:

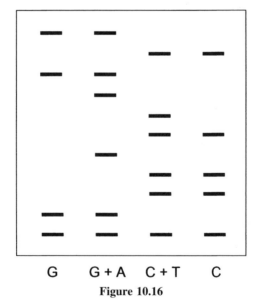

Figure 10.16

(a) Identify the anodic and cathodic ends of the gel as either the top or bottom of the diagram. At which end were the samples applied?

(b) Derive the base sequence of the oligodeoxynucleotide, starting from the 5'-end.

CHAPTER 11

Isoelectric Focusing

11.1. INTRODUCTION

Early experiments in the development of isoelectric focusing, a high-resolution steady-state electrophoresis method, occurred in 1912, with an electrolytic cell that was used to isolate glutamic acid from a mixture of its salts.[1] A simple U-shaped cell, such as that used for moving-boundary electrophoresis (Chapter 9), with two ion-permeable membranes equidistant from the center, created a central compartment that separated anodic and cathodic chambers, as shown in Figure 11.1. Redox reactions occurring in the anodic (Eq. 11.1) and cathodic (Eq. 11.2) electrolyte compartments generated H^+ and OH^- ions in the respective chambers:

$$2\,H_2O \rightleftharpoons O_2 + 4\,H^+ + 4\,e^- \quad \text{(Anode)} \tag{11.1}$$

$$2\,H_2O + 2\,e^- \rightleftharpoons 2\,OH^- + H_2 \quad \text{(Cathode)} \tag{11.2}$$

With this cell, glutamic acid placed in the central chamber can freely migrate across the ion-permeable membranes. As it crosses into the anodic chamber, the low pH causes protonation of the two carboxylates and the amine, so that a net charge of $+1$ is created on the molecule. This positively charged species then migrates toward the cathode, recrossing the membrane into the central chamber. When glutamic acid crosses the other membrane into the cathodic electrolyte compartment, the carboxylates and the amine group are deprotonated, and the molecule attains a net negative charge. This species migrates toward the anode, recrossing the membrane into the central compartment. The result of these migrations is that the glutamic acid becomes concentrated within the central compartment, at a pH close to its isoelectric pH, where it exists in the zwitterionic form with a net charge (and therefore mobility) of zero. The high concentration of glutamic acid at its pI value was the first recorded demonstration of the isoelectric focusing (IEF) effect.

A significant advance over this simple IEF experiment occurred in 1929,[2] when the number of compartments separating the anodic and cathodic electrolytes was increased to 12. This work clearly demonstrated the concept of establishing a stepwise

Bianalytical Chemistry, by Susan R. Mikkelsen and Eduardo Cortón
ISBN 0-471-54447-7   Copyright © 2004 John Wiley & Sons, Inc.

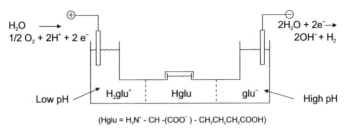

Figure 11.1. The first isoelectric focusing experiment, used to separate glutamic acid from a mixture of its salts.

variation in pH from one electrode chamber to the other. This stepwise gradient results in a steady-state pH distribution and an accumulation of *ampholytes* at their isoelectric pH values.

Ampholytes are chemical species capable of carrying net positive, zero, or negative charge, depending on the pH of the medium. Amino acids are common examples of ampholytes. At its pI, the net charge on an ampholyte is zero, so it will not migrate in an electric field. Ampholyte species will therefore migrate in the field until they reach a compartment in which pH = pI; they then focus in that compartment.

IEF was first successfully applied to proteins in 1938, when it was used to separate the protein hormones vasopressin and oxytocin from tissue extracts.[3] Twenty years later, ampholytes were first focused in a *continuous* pH gradient, stabilized by a dense sucrose medium, as an alternative to the multicompartment method. The continuous pH gradient in the sucrose medium was established by allowing acid and base to diffuse into opposite ends of the sucrose medium, held in a U-cell, from their respective electrode chambers. The stabilization of this continuous pH gradient with carrier ampholyte species led to modern IEF methods.

11.2. CARRIER AMPHOLYTES

The concept of carrier ampholytes in IEF was introduced in 1961.[4] Carrier ampholytes are species that are *amphoteric*, so that they reach an equilibrium position along the separation medium, and are good *electrolytes*, possessing both ionic conductivity, to carry current, and buffering capacity, to carry pH. They are used to generate stable pH gradients in the presence of the electric field, and are prefocused at their pI values before the sample is introduced. Ideally, a carrier ampholyte mixture consists of species that have identical diffusion coefficients and electrical mobilities, and differ in their pI values by only 0.05 pH unit, so that a linear pH gradient is generated. In practice, carrier ampholyte mixtures generate approximately stepwise changes in pH with distance along the separation medium.

Figure 11.2 shows the pH gradients calculated for mixtures of eight carrier ampholyte species possessing pI differences of 0.05 and 0.10 pH units. The individual species focus at their pI values, where their buffering capacity is low. A linear

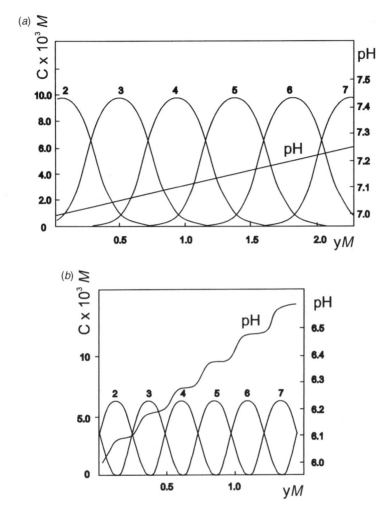

Figure 11.2. Concentration and pH as a function of distance calculated for mixtures of eight carrier ampholytes with pI differences of (*a*) 0.05 and (*b*) 0.10.[5] [Reprinted, with permission, from M. Almgren, *Chem. Scr.* **1** (No. 2), 1971, 69–75. Copyright © by The Royal Swedish Academy of Sciences.]

pH gradient is generated only for very small pI differences, and it can be seen that at 0.10 pH unit differences in the pI values of the carrier ampholytes, a stepwise dependence of pH on distance is found.

The first studies of carrier ampholytes were conducted using amino acids and dipeptides, but these species did not work well because their pK values for the amino and carboxylate groups are too far removed from their pI values. After they were prefocused, these species had very low buffering capacity. Good

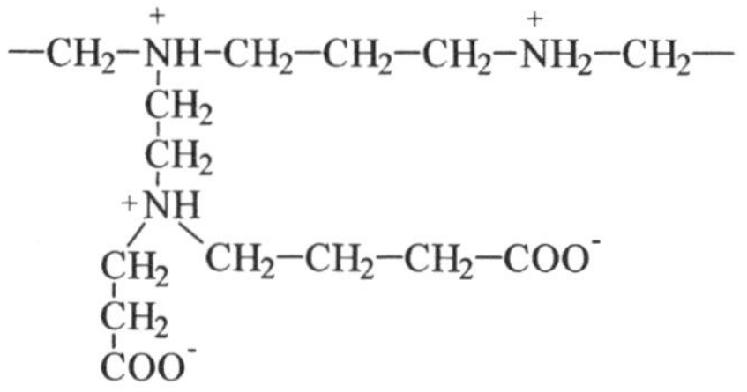

$$-CH_2-\overset{+}{N}H-CH_2-CH_2-CH_2-\overset{+}{N}H_2-CH_2-$$

$$\underset{\overset{|}{CH_2}}{\overset{|}{\underset{+NH}{\overset{|}{CH_2}}}}$$

$$\underset{\overset{|}{CH_2}}{\overset{\diagup}{CH_2}}\overset{\diagdown}{CH_2}-CH_2-CH_2-COO^-$$

$$\underset{COO^-}{\overset{|}{CH_2}}$$

Figure 11.3. Typical Ampholine component.[6] The pI values are varied by varying the number and type of the R_4N^+ and $RCOO^-$ groups.

buffering capacity is achieved with ampholytes for which pK values differ from the pI by 1.5 pH units or less.

In 1969, a patented mixture of synthetic carrier ampholytes was introduced as "Ampholine". These species are polyprotic amino-carboxylic acids that do not possess peptide bonds; each molecule in the mixture contains at least four protolytic groups, with at least one being an amine and at least one a carboxylic acid. Because all monovalent carboxylic acids possess pK values in the pH 4–5 range, and all monovalent aliphatic amines are protonated < pH 9–10, it was reasoned that chemical species containing multiple protolytic sites would provide a range of pK values. In fact, it was known that oligo- and polyamines of linear or branched structure, that have amino groups separated by ethylene bridges, have pK values spaced at graded intervals; tetraethylenepentamine, for example, has pK values of 9.9, 9.1, 7.9, 4.3, and 2.7 at 25 °C. If such polyamines are conjugated to an acidic group, for example, acrylic acid, the requisite criteria for carrier ampholytes are met. The structure[6] of a representative component of the Ampholine mixture is shown in Figure 11.3.

11.3. MODERN IEF WITH CARRIER AMPHOLYTES

IEF is commonly carried out on polyacrylamide gels. The preparation and polymerization of the gel are carried out in the presence of the ampholine mixture, in a vertical column or slab. The upper (anodic) reservoir is then filled with an acidic solution of, for example, phosphoric acid, and the lower (cathodic) reservoir is filled with a basic solution, for example, aqueous ethanolamine. These solutions are allowed to diffuse toward the center of the gel, to generate a pH gradient from the acid and base reservoir solutions. This step is followed by a prefocusing step before sample introduction, to allow the migration of the carrier ampholytes to their respective pI values. The carrier ampholytes thus focus at their pIs and stabilize the initial pH gradient. The sample is then layered at the anodic end in a sucrose solution, and the electrofocusing step is begun; this step may take several hours to reach steady state, with components migrating according to their mobilities until they focus at their pI values. Because IEF is a steady-state technique, the components are not subject to the diffusional broadening observed in zone electrophoresis, and very high resolutions can be achieved. Components with pI values that differ by as little as 0.02 pH units can be separated using the carrier ampholyte system, making it useful for the separation and quantitation of isoenzymes.

Isoelectric focusing has been used extensively for the identification of hemoglobin variants.[7] More than 300 variants of human hemoglobin are known to exist. The so-called "normal", adult hemoglobin is called HbA, and contains four polypeptide subunits called the α-, β-, γ-, and δ-globin polypeptides, and most of the known variants result from amino acid substitutions in the α- and β-globin chains. Some variants that are still considered HbA result in disorders that cause the loss of the heme groups, while variants called HbM cause either cyanosis (where Fe^{2+} is too easily oxidized to Fe^{3+}) or result in an abnormally high affinity of iron for O_2 (chronic hemolysis). Sickle-cell anemia is caused by one type of HbS in which valine is substituted for glutamate at position 6 of the β-globin polypeptide, and this is the most common human hemoglobin abnormality.

Hemoglobin and its variants have isoelectric pH values that vary over ~ 1 pH unit, in the 6.6–7.5 range, and IEF has become a standard diagnostic tool. Figure 11.4 shows an IEF map of 79 human hemoglobin variants that exhibit nearly 50 distinct pI positions.

A comparison of IEF and zone electrophoresis for the resolution of Hb variants has been carried out using a mixture of HbA and four known variants.[8] The stained gels are shown in Figure 11.5. It can be seen that the IEF technique allows all five bands to be distinguished, while the standard zone electrophoresis method yields only two broad bands, and would not be useful for diagnostic purposes.

IEF is thus a very powerful steady-state technique that works well for the separation of good ampholytes that possess steep titration curves near their pI values, such as proteins. As analyte moves through the pH gradient, its surface charge changes, and decreases according to the acid–base titration curve of the molecule. At the pI, the mobility is zero, and the species focuses. IEF is useful for proteins because of their sharp titration curves, but short peptides cannot be focused unless they possess at least one acidic or basic group in addition to the terminal amino and carboxylate groups, since without these groups the peptides will have a zero net charge over the entire pH 4–8 range. The highest molecular weight species that have been examined using IEF on polyacrylamide gels are ~ 750 kDa, because above this, the pore size of the gel severely limits mobility. The focusing of high molecular weight species requires long run times, because of low mobilities. IEF is a steady-state technique, and the resulting patterns, in principle, are not time dependent, and are not as sensitive to the technical skill of the operator as are those of zone electrophoresis methods.

Carrier ampholyte-based IEF methods are commonly used in situations where very high resolution of proteins according to their pI values is not required. Several problems exist with the use of carrier ampholytes that limit their resolving power. These include the low and uneven ionic strength that results in smearing of the most abundant proteins in the sample, the uneven buffering capacity and conductivity, the unknown chemical environment, a low sample loading capacity, and a

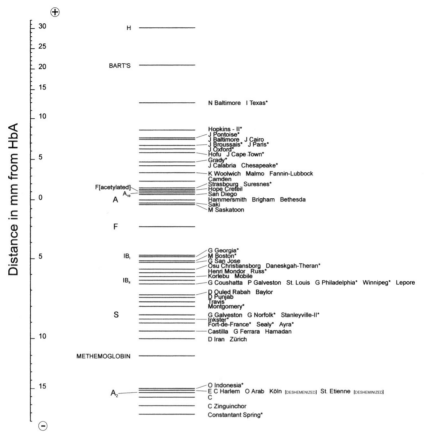

Figure 11.4. IEF map of human hemoglobin variants. The ferrous and ferric forms of normal HbA are indicated by IB$_I$ and IB$_{II}$, and the asterisks (*) indicate variants resulting from α-chain mutations.[7] [Reprinted, by permission, from P. Basset, Y. Beuzard, M. C. Garel, and J. Rosa, *Blood* **51** No. 5 (May), 1978, 971–982. © 1978 by American Society of Hematology.]

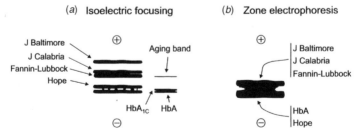

Figure 11.5 Migration patterns of a mixture of five Hb species (*a*) by IEF over the pH 6–9 range, and (*b*) by electrophoresis on cellulose acetate strips at alkaline pH.[8] [Reprinted, with permission, from P. Basset, F. Braconnier and J. Rosa, *J. Chromatogr.* **227**, 1982, 267–304. "An Update on Electrophoretic and Chromatographic Methods in the Diagnosis of Hemoglobinopathies". © 1982 Elsevier Publishing Company.]

long-term instability of the pH gradient that results from cathodic drift of the carrier ampholytes. Even with these known disadvantages, carrier ampholyte-based IEF is capable of resolving proteins that differ in their pI values by only 0.02 pH unit. The highest resolving power obtainable with IEF methods is achieved when immobilized pH gradients are used.

11.4. IMMOBILIZED pH GRADIENTS (IPGs)[9]

Gradients of pH can be immobilized directly into a polyacrylamide gel backbone by casting a gradient from one type of monomer to another, where the monomer species employed have protolytic functional groups. The buffering capacity is then held rigidly in place, and the pH gradient is generated during the gel casting process. Figure 11.6 shows an apparatus that is used for casting gradient gels. The diffusion of acid and base into the gel from opposite ends converts the acidic and basic groups to their correct forms (salts are used during casting). A prefocusing step is not required with IPGs as it is with carrier ampholyte IEF. Because of the control that is achievable during the casting of the gradient, linear pH gradients are readily constructed over very narrow pH ranges, allowing the resolution of proteins that differ in their pI values by only 0.001 pH unit.

Examples of the monomeric species used to generate the IPG are shown in Table 11.1. These compounds are all acrylamide derivatives, have pK values well distributed over the pH 3–10 interval, and are mixed with the acrylamide monomer solutions prior to gel casting. Note that *these species are not amphoteric*. The R group attached to the amide nitrogen contains either a carboxylate or an amine group, while pH extremes can be generated by the incorporation of a sulfonic acid group or a quaternary amine into the R group. The distance between the amide bond and the ionizable or buffering group is great enough that the incorporation of the monomer into the gel matrix does not significantly affect the pK of the buffering group. Once cast, these gels can be prepared in quantity, dried, and stored, so that a rehydration step just prior to use yields gels with reproducible gradients.

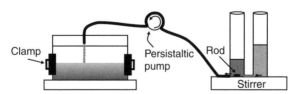

Figure 11.6. Experimental arrangement for casting an immobilized pH gradient gel. A linear pH gradient is generated by mixing two solutions with initially equal volumes, where one solution is dense and the other is, and they are titrated to the extremes of the desired pH interval. The compensating rod in the reservoir is used as a stirrer after the addition of catalysts and for hydrostatically equilibrating the two solutions.

TABLE 11.1. Monomers Used for the Preparation of IPG gels[a]

pK	Name	R[b]	MW
1.2	2-Acrylamido-2-methylpropanesulfonic acid	$-C(CH_3)_2(CH_2SO_3H)$	207
3.1	2-Acrylamidoglycolic acid	$-CH(OH)COOH$	145
3.6	N-Acryloylglycine	$-CH_2COOH$	129
4.4	3-Acrylamidopropanoic acid	$-(CH_2)_2COOH$	143
4.6	4-Acrylamidobutyric acid	$-(CH_2)_3COOH$	157
6.2	2-Morpholinoethylacrylamide	$-(CH_2)_2-N\bigcirc O$	184
7.0	3-Morpholinopropylacrylamide	$-(CH_2)_3-N\bigcirc O$	199
8.5	N,N-Dimethylaminoethylacrylamide	$-(CH_2)_2N(CH_3)_2$	142
9.3	N,N-Dimethylaminopropylacrylamide	$-(CH_2)_3N(CH_3)_2$	156
10.3	N,N-Diethylaminopropylacrylamide	$-(CH_2)_3N(CH_2CH_3)_2$	184
>12	QAE-Acrylamide	$-(CH_2)_2N(CH_2CH_3)_3$	198

[a] See. Ref. 9.
[b] General formula: $CH_2=CH-CO-NH-R$

With the monomers listed in Table 11.1, it is possible to create very narrow pH gradients, using one buffering species and one titrating species. For example, 3-morpholinopropylacrylamide (pK 7.0) used as a buffer, and 2-acrylamido-2-methylpropanesulfonic acid used as a titrant can be cast so that the final pH gradient covers the pH 6.5–7.5 range. The final pH range will be linear between pK -0.5 and pK $+0.5$ for the buffering species. IPGs with such narrow pH ranges are used for very high pI resolution.

Isoelectric focusing on such narrow pH range IPG gels have been used for the diagnosis of hemoglobinopathies, such as β-thalassemia, which is a recessive genetic disease. Figure 11.7 shows a comparison of IPG and carrier ampholyte gels used to separate hemoglobins present in 18-week-old fetal cord blood of normal and thalassemic fetuses.[10] A normal fetus will possess three types of hemoglobin at 18 weeks: fetal hemoglobin (HbF), acetylated fetal hemoglobin (HbF$_{ac}$), and adult hemoglobin (HbA). If the fetus is homozygous for β-thalassemia, no HbA is present at 18 weeks. The electropherograms show that the IPG system allows the clear resolution of the HbA and HbF$_{ac}$ bands, while the carrier ampholyte system yields a single band containing both HbA and HbF$_{ac}$. Thus the introduction of high-resolution IPGs has allowed the early diagnosis of the disease using routinely obtainable blood samples.

A second example of the improvements in isoelectric focusing generated by the introduction of IPG gels is shown in Figure 11.8, where protein samples obtained from bean seeds are subjected to IEF for up to 12 h, using carrier ampholytes and immobilized pH gradients.[11] In principle, proteins focused at their isoelectric pH values are not subject to changes in position with time, since IEF is a steady

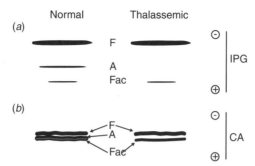

Figure 11.7. Separations of HbF, HbA, and HbF$_{ac}$ from fetal cord blood obtained at week 18 of pregnancy: (*a*) IPG gel covering pH 6.8–7.6 range; (*b*) carrier ampholyte gel covering the pH 6–8 Ampholine range; both gels were stained with Coomassie Brilliant Blue R-250 in Cu^{2+} solution.[10] [Reprinted, with permission, from M. Manca, G. Cossu, G. Angioni, B. Gigliotti, A. Bianchi-Bosisio, E. Gianazza, and P. G. Righetti, *American Journal of Hematology* **22** (No. 3, July), 1986, 285–293. "Antenatal Diagnosis of β-Thalassemia by Isoelectric Focusing in Immobilized pH Gradiens". © 1986 by Alan R. Liss, Inc.]

state rather than a time-dependent technique. However, carrier ampholyte-based IEF systems are known to exhibit cathodic drift that results from electroosmosis, with the slow drift of buffer, and carrier ampholytes, toward the cathodic end of the gel. Figure 11.8 shows that this drift is significant with carrier ampholytes after only 3 h of focusing, while gels with immobilized pH gradients exhibit superior time stability > 12-h runs.

A variety of precast IPG gels is commercially available from electrophoresis suppliers, and this represents another significant advantage over carrier ampholyte

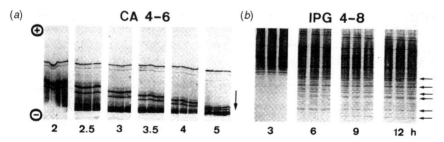

Figure 11.8. Comparison of (*a*) carrier ampholyte (pH 4–6 range) and (*b*) IPG (pH 4–8 range) isoelectric focusing pattern stability over time, for the separation of a mixture of bean seed proteins.[11] [Reprinted, with permission, from "Methodology of two-dimensional electrophoresis with immobilized pH gradients for the analysis of cell lysates and tissue proteins", A. Görg, W. Postel, A. Domscheit, and S. Günther, in *"Two-Dimensional Electrophoresis"*, Proceedings of the International Two-Dimensional Electrophoresis Conference, Vienna, November 1988. A. T. Endler and S. Hanash, Eds., VCH Publishers, Weinheim, Germany, 1989. © 1988 by VCH Verlagsgesellschaft mbH.]

systems, since gel-to-gel reproducibility is high when gels are initially prepared in large quantities.

The high-resolution and separation stability afforded by isoelectric focusing on immobilized pH gradient gels has also been exploited in 2D electrophoresis methods, where IEF in the first dimension is followed by a second separation method performed in a perpendicular direction.

11.5. TWO-DIMENSIONAL ELECTROPHORESIS[12]

Two-dimensional separation techniques are those in which a sample is subjected to two displacement processes that occur orthogonally to each other. The two dimensions should be based upon two completely different separation mechanisms, since identical mechanisms merely yield diagonal separation patterns. The resolving power of 2D techniques can be expressed by Eq. 11.3:

$$n_2 = n_1^2 \qquad (11.3)$$

where n_2 is the number of components resolvable by a 2D method and n_1 is the number resolvable by a one-dimensional technique.

To date, the most powerful combination of electrophoresis techniques employs IEF in the first dimension, resolving on the basis of pI, and SDS–PAGE in the second dimension to separate on the basis of molecular weight. The order is crucial, since the isoelectric point of a protein effectively disappears upon treatment with SDS, which yields a uniformly negative charge/mass ratio.

Two-dimensional electrophoresis was first reported in 1975, when it was used for the qualitative analysis of a very complex mixture of *E. coli* proteins for which 1D techniques are inadequate. Carrier ampholytes were used for column IEF in the first dimension, using pH 3–10 Ampholine, urea, a nonionic surfactant, acrylamide and cross-linker, TEMED, and persulfate during the gel casting process. This was followed by diffusion of 0.02 M NaOH and 0.01 M H_3PO_4 from cathodic and anodic reservoirs, a prefocusing step, and the 13-h IEF run. The IEF voltage was selected so that the total voltage × time product was between 5000 and 10,000 V-h. The gel was then extruded from the IEF column using a syringe and tubing, and equilibrated for 30 min in a solution of glycerol, mercaptoethanol, SDS, and buffer. The second dimension employed discontinuous SDS–PAGE on a slab of the same width as the length of the initial column used for IEF, as shown in Figure 11.9.

Three gels were used for the second dimension: (a) a loading gel of 1% agarose that surrounded the IEF gel to minimize protein loss into the top buffer, (b) a stacking gel of 5% polyacrylamide with 0.1% SDS at pH 7, and (c) a running gel, possessing a 9–14% polyacrylamide gradient with 0.1% SDS at pH 9. A running buffer of pH 9, containing 2% SDS was used during the 5-h run, and the entire slab was then removed, fixed and stained with Coomassie Blue. The resulting separation pattern is shown in Figure 11.10.

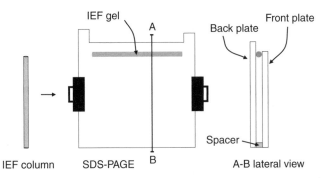

Figure 11.9. Electrophoresis chamber used for the second dimension (SDS–PAGE).

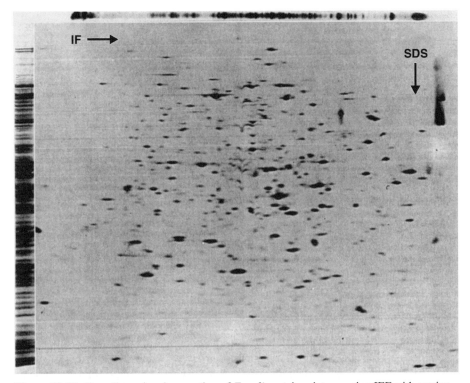

Figure 11.10. Two-dimensional separation of *E. coli* protein mixture, using IEF with carrier ampholytes, pH range 3–10, in the first (horizontal) direction, and SDS–PAGE was run from top to bottom in the second dimension on a 9–14% polyacrylamide gradient running gel cast with 0.1% SDS.[12] [Reprinted, with permission, from P. H. O'Farrell, *The Journal of Biological Chemistry* **250** (No. 10 May 25), 1975, 4007–4021. "High Resolution Two-Dimensional Electrophoresis of Proteins". Copyright © 1975 by the American Society for Biochemistry and Molecular Biology, Inc.]

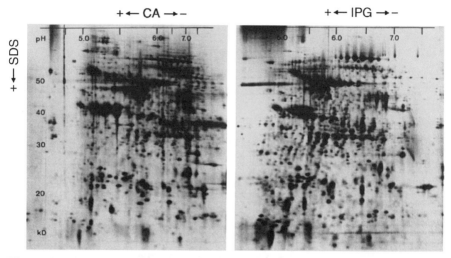

Figure 11.11. Comparison of 2D patterns obtained for bean seed proteins after first-dimension IEF with (*a*) carrier ampholytes and (*b*) an immobilized pH gradient. IEF proceeded in the horizontal direction, with pH as indicated across the gels, and SDS–PAGE occurred from top to bottom, with molecular weights indicated in (*a*).[13] [Reprinted, with permission, from A. Görg, W. Postel and S. Günther, *Electrophoresis* **9** (1988) 531–546. "The Current State of Two-Dimensional Electrophoresis with Immobilized pH Gradients". © 1988 by VCH Verlagsgesellschaft mbH, D-6940 Weinheim.]

Two-dimensional electrophoresis is now commonly used with immobilized pH gradients in the first dimension, and the term IPG–DALT has been coined for this 2D technique. Commercially available IPG gels may be used for the first dimension, and thin strips of this gel containing the focused proteins are equilibrated in an SDS–mercaptoethanol solution and then applied to the cathodic end of the SDS–PAGE slab. Figure 11.11 shows a comparison of 2D electrophoresis conducted using a nonlinear carrier ampholyte pH gradient and a linear IPG for isoelectric focusing in the first dimension.[13] Note especially the lack of smearing in the horizontal (IEF) dimension, as well as the improved resolution between pH 6 and 7 that occurs on the gel employing the first-dimension IPG.

IPG–DALT can be used for the separation of up to 5000 components in a protein mixture, and has now been incorporated as a qualitative tool for diagnostic analysis. It is not generally used for quantitation of proteins in a mixture, however, because loss of low molecular weight components often occurs during the SDS equilibration and transfer steps. An excellent overview of the capabilities and technical protocols of IPG–DALT is available in a review article.[14]

SUGGESTED REFERENCES

N. Catsimpoolas, *Isoelectric Focusing*, Academic Press, New York, 1976.

P. G. Righetti, *Isoelectric Focussing: Theory, Methodology and Applications*, Elsevier, New York, 1983.

P. G. Righetti, *Immobilized pH Gradients: Theory and Methodology*, Elsevier, New York, 1990.

N. Catsimpoolas and J. Drysdale, *Biological and Biomedical Applications of Isoelectric Focusing*, Plenum Press, New York, 1977.

Endler and Hanash, *Two-Dimensional Electrophoresis*, John Wiley & Sons, Inc., New York, 1988.

REFERENCES

1. K. Ikeda and S. Suzuki, *U.S: Pat.*, 1,015,891 (1912).

2. R. R. Williams and R. E. Waterman, *Proc. Exp. Biol. Med.* **27**, 1929, 56.

3. V. Du Vigneaud, G. W. Irwing, H. M. Dyer, and R. R. Sealock, *J. Biol. Chem.* **123**, 1938, 45–55.

4. H. Svensson, *Acta Chem. Scand.* **15**, 1961, 325–341.

5. M. Almgren, *Chem. Scr.* **1**, 1971, 69–75.

6. P. G. Righetti, *Isoelectric Focussing: Theory, Methodology and Applications*, Elsevier, New York, 1983, pp. 34–40.

7. P. Basset, Y. Beuzard, M. C. Garel, and J. Rosa, *Blood* **51**, 1978, 971–982.

8. P. Basset, F. Braconnier, and J. Rosa, *J. Chromatogr.* **227**, 1982, 267–304.

9. B. Bjellqvist, K. Ek, P. G. Righetti, E. Gianazza, A. Görg, R. Westermeier, and W. Postel, *J. Biochem. Biophys. Methods* **6**, 1982, 317–339.

10. M. Manca, G. Cossu, G. Angioni, B. Gigliotti, A. Bianchi-Bosisio, E. Gianazza, and P. G. Righetti, *Am. J. Hematol.* **22**, 1986, 285–293.

11. A. Görg, W. Postel, A. Domscheit, and S. Günther, in *Two-Dimensional Electrophoresis*, A. T. Endler and S. Hanash, Eds., VCH Publishers, Weinheim, Germany, 1989, p. 274.

12. P. H. O'Farrell, *J. Biol. Chem.* **250**, 1975, 4007–4021.

13. A. Görg, W. Postel, A. Domscheit, and S. Günther, in *Two-Dimensional Electrophoresis*, A. T. Endler and S. Hanash, Eds., VCH Publishers, Weinheim, Germany, 1989, p. 275.

14. A. Görg, C. Obermaier, G. Boguth, A. Harder, B. Scheibe, R. Wildgruber, and W. Weiss, *Electrophoresis* **21**, 2000, 1037–1053.

PROBLEMS

1. Explain the fundamental difference between zone electrophoresis as discussed in Chapters 9 and 10, and the steady-state technique of isoelectric focusing. Why do some IEF gels exhibit time-dependent behavior?

2. A perfectly linear pH gradient was immobilized across a 10-cm polyacrylamide gel using pK 7.0 3-morpholinopropylacrylamide titrated with 2-acrylamido-2-methylpropanesulfonic acid. This gel was used to focus a mixture of three Hb variants, with pI values of 7.42 (Hb A_2), 7.21 (Hb S), and 7.05 (Hb F). Calculate the positions of the three focused bands as distances from the cathodic end of the gel.

3. Suggest a method by which an enzyme-labeled antibody could be used to identify one component of the mixture of bean seed proteins in Figure 11.8. Why

would this type of method not be expected to work after either of the 2D separations shown in Figure 11.11?

4. When using 2D electrophoresis to separate complex mixtures of proteins, why is it necessary to perform isoelectric focusing in the first dimension?

5. Why is isoelectric focusing useful for proteins and glycoproteins, but not for nucleic acids? What would happen to DNA or RNA present in a mixture of proteins, following a long isoelectric focussing run over the pH 6–8 range?

6. A purified, unknown protein was examined using the Ferguson method (Chapter 10), and its pI and molecular weight were determined to be 7.8 and 210 kDa, respectively. The protein was exposed to dithiothreitol, to reduce disulfide bonds that connect subunits, and an IPG–DALT experiment was performed over the pH 6–8.5 range. The resulting separation pattern is shown below (Fig. 11.12):

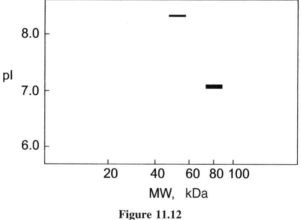

Figure 11.12

What new information is provided by these results regarding the unknown protein?

7. At what stage should molecular weight marker proteins be added in an IPG–DALT experiment? Why?

■■■■■■ **CHAPTER 12**

Capillary Electrophoresis

12.1. INTRODUCTION

Both zone and steady-state electrophoresis have been carried out inside capillaries that connect anodic and cathodic buffer reservoirs. Capillaries, filled with either buffers or gels, act as retainers of very small volumes of a support medium, and also control the electroosmotic flow rate. Electroosmosis, minimized for most high-resolution electrophoresis techniques, is of critical importance to capillary electrophoresis (CE), because it determines the rate of flow of the analyte solution through the capillary. Electroosmosis creates a flow of solvent through the capillary that is strong enough to elute all of the components (regardless of charge) at one end; it is thus readily automated, and much of the detection instrumentation developed for liquid chromatography has been adapted for use with capillary electrophoresis.

Capillaries offer several advantages over conventional columns and slabs for electrophoretic separations. There is enhanced dissipation of heat per unit volume through the capillary wall, which allows higher fields to be employed during runs— field strengths of 50 kV/m (500 V/cm) are not uncommon with capillary techniques. This allows rapid separations with minimal diffusional band broadening. Furthermore, the relatively small volume flow rates allow sampling from microenvironments, such as single cells.

The instrumental arrangement commonly employed in capillary electrophoresis is shown in Figure 12.1. With untreated silica capillaries, electroosmosis causes the buffer to flow from the anode to the cathode. Samples are introduced at the anodic end, and an on-column or post-column detector is placed at or near the cathodic end of the capillary. The high-voltage produced by the power supply and present in the anodic buffer reservoir is enclosed in a protective shield.

Many electrophoresis techniques have been adapted for use in capillaries. The most commonly used is capillary zone electrophoresis (CZE), where a flowing buffer fills the capillary, and separation is based on mobility differences. Capillaries have also been filled with polyacrylamide gels, and a great deal of effort has

Bianalytical Chemistry, by Susan R. Mikkelsen and Eduardo Cortón
ISBN 0-471-54447-7   Copyright © 2004 John Wiley & Sons, Inc.

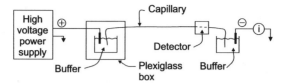

Figure 12.1. Instrumentation for CE.[1] [Reprinted, with permission, from A. G. Ewing, R. A. Wallingford, and T. M. Olefirowicz, *Anal. Chem.* **61**, 1989, 292A–301A. "Capillary Electrophoresis". Copyright © 1989, American Chemical Society.]

concerned the application of gel-filled capillaries to DNA sequencing, for the Human Genome Project. Isoelectric focusing has also been accomplished inside capillaries, and techniques have been developed to elute focused proteins from one end of the capillary for detection. These three main CE techniques will be the subject of this chapter.

The mathematical description of CZE separations assumes that heat dissipation is efficient (for air-cooled systems, this means thick-walled capillaries, while water-cooled capillaries are thin walled), and that the only significant cause of band broadening is the longitudinal diffusion of solute within the capillary.

We begin by initially ignoring electroosmosis, and considering only the migration of the charged analyte species in an electric field. The migration velocity, v (see Chapter 9), is related to the analyte's mobility, μ, and the electric field strength, E, by Eq. 12.1:

$$v = \mu E = \mu V / L \tag{12.1}$$

where V is the applied voltage, and L is the length of the capillary. From this expression, we can calculate the migration time of the analyte as

$$t = L/v = L^2/(\mu V) \tag{12.2}$$

From Eq. 12.2, we can see that the fast elution of sample components requires high voltages and/or short capillaries.

Through analogies to chromatographic methods, we can also calculate the number of theoretical plates possessed by a given capillary at a particular voltage toward a given analyte species. This number, N, is also called the separation efficiency; higher values generally yield better separations.

$$N = \mu V/(2\,D) \tag{12.3}$$

where D is the diffusion coefficient of the analyte. It should be noted that the number of theoretical plates, N, depends on the applied voltage, but does not depend on the length of the capillary.

12.2. ELECTROOSMOSIS

Electroosmosis results from the charge residing on the inner wall of the capillary. With untreated silica capillaries, this charge is negative, and positive counterions exist in a stagnant layer adjacent to the capillary walls. The cationic nature of this stagnant layer extends further out into solution that is mobile, covering a region that is roughly 3–300 nm from the capillary wall, for buffer concentrations of 10^{-3}–10^{-6} M, respectively. Figure 12.2 shows a schematic representation of the ionic environment at the silica–solution interface.

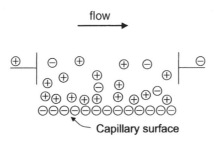

Figure 12.2. Ions at a silica–solution interface at the inner capillary wall in CE.

The potential that exists across this interfacial region is called the zeta potential, ζ, and may be described by Eq. 12.4:

$$\zeta = 4\pi\eta\mu_{EO}/\varepsilon \tag{12.4}$$

where η is the viscosity of the solution, μ_{EO} is the coefficient for electroosmotic flow, and ε is the dielectric constant of the solution. Using Eq. 12.4, the linear velocity of electroosmotic flow, U, is readily calculated as

$$U = [\varepsilon/(4\pi\eta)]E\zeta \tag{12.5}$$

Equation 12.5 indicates that the electroosmotic flow rate is large when ζ is large and when the interfacial region is small.

12.3. ELUTION OF SAMPLE COMPONENTS

Taking electroosmosis into account, the velocity and residence time of the analyte can be recalculated as

$$v = (\mu + \mu_{EO})V/L \tag{12.6}$$
$$t = L^2/[(\mu + \mu_{EO})V] \tag{12.7}$$

By using Eqs. 12.6 and 12.7, the separation efficiency, or number of theoretical plates, N, becomes

$$N = (\mu + \mu_{eo})V/2D \tag{12.8}$$

It should be apparent from Eq. 12.6 that μ, the analyte mobility, may be positive or negative (i.e., the analyte will migrate in either direction according to its charge), while μ_{EO}, the coefficient of electroosmotic flow, is a constant for given experimental conditions that will result in flow in one direction only. For untreated silica capillaries, if μ_{EO} is large enough, all ions will migrate toward the cathode (anions more slowly than cations), and uncharged species will be carried by the electroosmotic flow of the solution. The elution order of analytes will therefore be cations first, neutrals second and anions last.

Electroosmotic flow can be varied by increasing the viscosity at the capillary–solution interface using adsorbents, by modifying the silica walls to reduce the charge or change its sign, or by varying the pH of the buffer. The buffer concentration must be held at a significantly higher level than the analyte concentration, so that the electric field is constant throughout the capillary. However, the buffer concentration must also be low enough to avoid excessive i^2R heating; experiments are usually run at a variety of ionic strengths to determine optimal conditions.

While some techniques require the absence of electroosmotic flow during the separation itself (capillary gel electrophoresis and capillary isoelectric focusing), most common techniques exploit electroosmotic flow for sample introduction and detection. Electrophoresis in buffer-filled capillaries uses electroosmotic flow in an analogous manner to a chromatographic mobile phase: the flow is used to transport analyte from cathode to anode and separation occurs continuously between introduction and detection.

12.4. SAMPLE INTRODUCTION[2]

Two introduction methods are commonly employed in capillary electrophoresis. *Hydrodynamic* injection is based on siphoning, or gravity feeding the sample into the anodic end of the capillary. The anodic end is removed from the buffer reservoir and placed in the sample solution. The capillary end is then raised so that the liquid level in the sample vial is at a height Δh above the level of the cathodic buffer, and is held in this position for a fixed time t. This process has been automated for reproducibility, and the hydrodynamic flow rate has been shown to obey Eq. 12.9:

$$v_{hd} = \rho g r^2 \Delta h / 8 \eta L \qquad (12.9)$$

where v_{hd} is the volume flow rate, ρ is the density of the background electrolyte, g is the gravitational acceleration, r is the capillary inner radius, Δh is the difference in height between the two reservoir liquid levels, η is the background electrolyte viscosity, and L is the capillary length. This injection technique does not discriminate among ions, so that the relative concentrations of sample ions introduced onto the capillary will be the same as in the sample solution.

Electrokinetic injection involves drawing sample ions into the capillary interior with an applied potential. The anodic end of the capillary is removed from the buffer reservoir and placed into a sample vial along with the anode. An injection

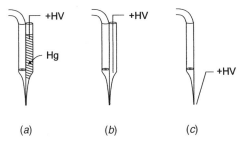

Figure 12.3. Three microinjectors used for the electrokinetic injection of intracellular fluids from single cells.[2]

voltage is applied for a brief time to cause migration and the electroosmotic introduction of sample. Because the applied potential induces a bulk electroosmotic flow of sample fluid into the capillary, the velocity of this bulk flow must be added to each ion's electromigration velocity. While both velocity components are proportional to the applied voltage (more precisely to the local electric field), the ion mobilities and thus the net injection velocities differ among the ions. This means that electrokinetic injection discriminates among the ions, with larger proportions of more mobile ions being introduced. Furthermore, if the conductivity of the sample is lower than that of the running buffer, the high sample resistance induces a locally high electric field that leads to larger, but less reproducible, injection quantities.

Both injection techniques have been shown to suffer from the effects of analyte diffusion, specially when the clean capillary is initially introduced into the sample solution. Diffusion occurs across the boundary area between analyte and buffer, which is defined by the cross-sectional area of the capillary. Both techniques also suffer from the effects of inadvertent hydrodynamic flow that results from the reservoir liquid levels being at slightly different levels. While these effects are significant for the buffer-filled capillaries used in capillary zone electrophoresis, they are both much less important when the capillary is filled with a gel.

Figure 12.3 shows three microinjector designs that have been used for electrokinetic injection of very small samples, from microenvironments such as single nerve cells. The capillary tip diameters in these injectors are 10 μm o.d. or less, and can penetrate and sample from single cells. The design shown in Figure 12.3(c) has been shown to be the most effective, because electrolysis occurs away from the capillary tip and therefore does not introduce bubbles that would insulate the capillary ends from each other.

12.5. DETECTORS FOR CAPILLARY ELECTROPHORESIS

Detectors used in the initial experiments with capillary electrophoresis were simple absorbance and fluorescence detectors that had been adapted from HPLC equipment. However, it soon became apparent that these instruments yielded poor

TABLE 12.1. Detectors Used in Capillary Zone Electrophoresis Instruments[a]

Classification	Property Measured	Detection Limit (mol)
Optical	Absorbance	10^{-15}–10^{-13}
	Fluorescence	
	(a) Precolumn derivatization	10^{-21}–10^{-17}
	(b) On-column derivatization	8×10^{-16}
	(c) Postcolumn derivatization	2×10^{-17}
	Indirect fluorescence	5×10^{-17}
	Thermal lensing	4×10^{-17}
	Raman	2×10^{-15}
Electrochemical	Conductivity	1×10^{-16}
	Current (amperometry)	7×10^{-19}
Radiochemical	^{32}P Emission	1×10^{-19}
Mass	Electrospray ionization MS	1×10^{-17}

[a] See Ref. 1. Reprinted, with permission, from A. G. Ewing, R. A. Wallingford, and T. M. Olefirowicz, *Anal. Chem.* **61**, 1989, 292A–301A. Copyright © 1989 American Chemical Society.

detection limits and sensitivity due to the nature of capillary techniques. Specific problems associated with detection in capillary electrophoresis include the small capillary dimensions, in particular, inner diameters of 100 μm or less (as low as 12.7 μm i.d.), as well as the small zone volumes—injection volumes range between 18 pL and 50 μL. This leads to zone lengths of 10 mm or less inside the capillary. Because off-column detectors (extending from the capillary exit) introduce significant zone broadening, on-column devices are generally preferred.

Table 12.1 shows the variety of capillary electrophoresis detection methods that have been tested to date, as well as their reported detection limits. While detection limits for instrumental methods are usually reported in concentration units, those reported for CE methods are generally given in moles because of zone broadening (the peak concentration at the detector is always less than the concentration injected) and the variety of injection volumes that are possible between instruments with different sized capillaries and different injection and operating potentials.

Detectors that have produced the lowest detection limits to date are based on fluorescence with precolumn derivatization, mass spectrometry, radiometry, and amperometry. Applications are described below for these detection systems employed with conventional CE instrumentation. For applications of these detection methods to CE analysis using microfabricated devices, the reader is referred to a review article.[3]

12.5.1. Laser-Induced Fluorescence Detection[4]

This detection method has been tested with a series of amino acids derivatized prior to injection with fluorescein isothiocyanate, to yield products that are excited at 488 nm and produce emission at 528 nm (Eq. 12.10):

$$
\text{S=C=N} \qquad + \qquad \text{OOC-CH-NH}_3^+ \qquad \longrightarrow \qquad \text{OOC-CH-NH-C-NH}
$$

(12.10)

The derivatized amino acids are separated in 5-mM carbonate buffer at pH 10, on a 50-μm i.d. fused silica capillary of 99 cm length, with a 25-kV separation potential. Electrokinetic injection is performed at the anodic end of the capillary, with a 2-kV injection potential applied for 10 s.

Figure 12.4 shows a schematic diagram of the laser-induced fluorescence detector. The fused silica capillary used for electrophoresis is placed $\sim$ 1 cm into the flow chamber of the sheath flow cuvette. The sheath stream surrounds the sample as it exits from the capillary, forming a thin stream in the center of the flow chamber. A focused laser beam (not shown) excites fluorescence. The fluorescence is collected at right angles, with a microscope objective, and passed through filters (cutting off radiation $<$ 495 nm and $>$ 560 nm), to reduce Raman and Rayleigh scattered light, and passed through an eyepiece fitted with a 200-μm radius pinhole. The pinhole restricts the field of view of the photomultiplier tube (PMT) detector to the illuminated sample stream. Precise alignment of the laser beam, objective, eyepiece pinhole, and PMT are required for the detection of maximum sample emission. The stainless steel body of the cuvette and the associated plumbing are held at ground potential.

Figure 12.5 shows an electropherogram recorded by this detector after the injection and separation of a mixture of 15 FITC-derivatized amino acids, and

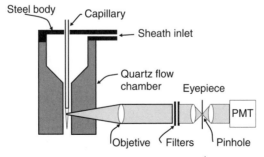

Figure 12.4. Laser-induced fluorescence detector for CZE.[4] [Reprinted, with permission, from Y.-F. Cheng and N. J. Dovichi, *Science* **242**, 1988, 562–564. "Subattomole Amino Acid Analysis by Capillary Zone Electrophoresis and Laser-Induced Fluorescence". Copyright © 1988 by AAAS.]

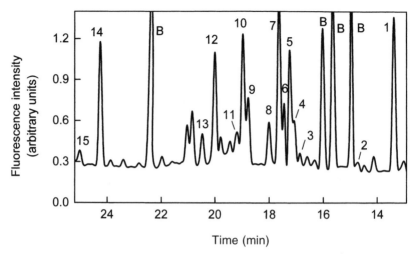

Figure 12.5. Data obtained with the LIF detector for 15 amino acids.[4] [Reprinted, with permission, from Y.-F. Cheng and N. J. Dovichi, *Science* **242**, 1988, 562–564. "Subattomole Amino Acid Analysis by Capillary Zone Electrophoresis and Laser-Induced Fluorescence". Copyright © 1988 by AAAS.]

Table 12.2 below indicates detection limits observed for each amino acid. While the detection of amino acids is not in itself of particular interest, peptides and proteins that are available only in very small quantities may be analyzed for their amino acid contents, following acid hydrolysis, using the instrumentation described here.

TABLE 12.2. Detection Limits Observed for Each Amino Acid[a]

Amino Acid	Concentration ($\times 10^{-11}$ M)	Mole ($\times 10^{-20}$)	Molecules
Ala	3.4	4.6	27,000
Arg	<0.5	<0.9	<5,700
Asp	6.6	6.8	41,000
Cys	2.4	3.3	20,000
Glu	2.6	2.8	17,000
Gly	2.8	3.7	22,000
Ile	1.7	2.5	15,000
Leu	7.0	10	61,000
Lys	8.6	15	90,000
Met	1.6	2.3	14,000
Ser	1.6	2.3	13,000
Thr	2.6	3.7	22,000
Trp	1.1	1.7	9,900
Tyr	1.1	1.4	8,400

[a] Reprinted, with permission, from J. B. Fenn, M. Mann, C. K. Meng, S. F. Wong, and C. M. Whitehouse, *Science* **246**, 1989, 64–71. "Electrospray Ionization for Mass Spectrometry of Large Biomolecules ". Copyright © 1989 by AAAS.

12.5.2. Mass Spectrometric Detection

Mass spectrometric detectors for capillary electrophoresis are necessarily post-column detectors and must be interfaced to the cathodic end of the capillary. These detectors consist of four main components: the *interface*, that joins the capillary to the ion source, the *ion source*, that generates ionic fragments from neutral analyte species, the *mass analyzer*, that distinguishes ions by their mass/charge (m/z) values, and the *ion detector*, that measures and amplifies the signal. The principles and instrumentation of bioanalytical MS are explained in Chapter 15.

Soft ionization methods produce few fragments under relatively mild conditions. The ionization method that has received the most attention in terms of its applicability to protein and DNA analysis is the *electrospray ionization* (ES) technique. This is a soft method that is capable of generating molecular ions from biological macromolecules present in solution. Table 12.3 gives examples of the charge and m/z ranges that have been observed with some biopolymer species in electrospray ionization mass spectrometers.

Because of the range of m/z values produced by species of constant mass, sophisticated software has been developed for data reduction with ES–MS, allowing the molecular weights of macromolecules to be determined with very high precision and accuracy. Because the molecular weights are so large, the natural abundances of the isotopes must be considered when analyzing m/z data. Signals for a single parent species are observed over a range of m/z values because of this isotopic distribution. These topics are considered in Chapter 15.

ES–MS has been interfaced to the cathodic end of capillaries used for capillary electrophoresis. Figure 12.6 shows a diagram of a typical instrumental apparatus, while Figure 12.7 shows a detailed schematic of one type of capillary–ionization source interface. The design of the interface is critical, and has been a very active area of research. With the device shown in Figure 12.7, the operational parameters are as follows. The separation potential, applied between the anodic and cathodic ends of the capillary, are between $+30$ and $+50$ kV, and the cathode is maintained at ~ 3–6 kV above ground. The focusing ring is maintained at $+300$ V, while the ion-sampling nozzle and skimmer are grounded. This electrical arrangement ensures ion formation and acceleration into the magnetic sector.

TABLE 12.3. Charges and m/z Values Obtained for Biological Macromolecules using ES–MS[a]

Compound	MW	Charge Range	m/z Range
Insulin	5,730	4–6	950–1450
Cytochrome c	12,400	12–21	550–1100
Alcohol dehydrogenase (subunit)	39,800	32–46	800–1300
Conalbumin	76,000	49–64	1200–>1500
Oligonucleotide (CATGCCATGGCATG)	4,260	6–11	350–710

[a] See Ref. 5.

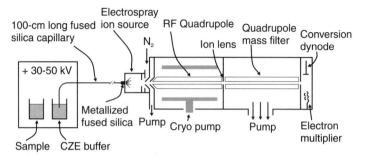

Figure 12.6. Typical CZE–ES–MS instrumentation.[6] [Reprinted, with permission, from R. D. Smith, J. A. Olivares, N. T. Nguyen, and H. R. Udseth, *Anal. Chem.* **60** (No. 5), 1988, 436–441. © 1988 by American Chemical Society.]

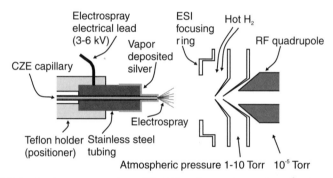

Figure 12.7. Ionization source for CZE–MS of protein analytes.[6] [Reprinted, with permission, from R. D. Smith, J. A. Olivares, N. T. Nguyen, and H. R. Udseth, *Anal. Chem.* **60** (No. 5), 1988, 436–441. © 1988 by American Chemical Society.]

Detection may proceed by monitoring a particular m/z value, which yields a so-called "single-ion electropherogram". These may be obtained at several different m/z values, for example, with one m/z value being monitored for each species of interest in the sample to give an additional dimension in selectivity. The data may then be combined to show the "reconstructed ion electropherogram" that shows the separation of all of the components. This is shown in Figure 12.8, for a separation of bovine insulin and sperm whale and horse heart myoglobin.

12.5.3. Amperometric Detection

Two developmental difficulties existed in the application of electrochemical methods to CE detection. The first involved the "cross-talk" or interference that resulted from the high voltages used in CE separations. These dc voltages may be as high as 50 kV, and the resulting electroosmotic currents can be up to six orders of magnitude greater than the faradaic currents measured at an amperometric detector poised at only a few hundred millivolts. Because of this, initial attempts resulted in the detector signal being overwhelmed by background signals. The second problem involved the small size of the capillary exit and the small volume flow rates, that are not amenable to the conventional amperometric detectors used for HPLC.

The first problem has been overcome through the use of an electrically insulating interface in the cathodic buffer reservoir. This interface is porous, and connects the separation capillary to a detection capillary. The porous contact allows the application of the ground potential in the cathodic reservoir, and electrically insulates the detection capillary from the applied separation voltage. The detection capillary channels the analyte solution to the two-electrode detector. A diagram of the instrumental setup is shown in Figure 12.9.

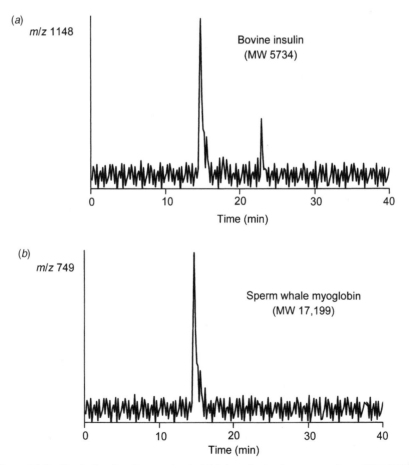

Figure 12.8. Single (*a–c*) and reconstructed (*d*) ion electropherograms from CZE–MS.[6]

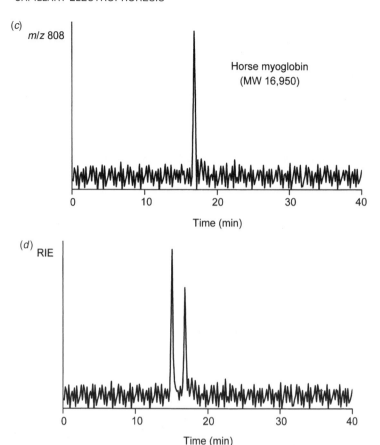

Figure 12.8   (*Continued*)

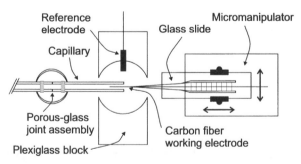

Figure 12.9. Amperometric CE detector with cathodic insulating assembly and ultramicroe-lectrode.[7] [Reprinted, with permission, from R. A. Wallingford and A. G. Ewing, *Anal. Chem.* **60** (No. 3), 1988, 258–263. "Amperometric Detection of Catechols in Capillary Zone Electro-phoresis with Normal and Micellar Solutions". © 1988 by American Chemical Society.]

The second problem has been overcome through the use of ultramicroelectrodes as working electrodes. These devices, initially made from carbon fibers, are 5–50 μm in diameter, and this small size allows their insertion into the capillary exit. Eluent flows around the working electrode, into an external buffer solution containing a reference electrode. The small size of the working electrode makes it invisible without magnification, and so a microscope and a micromanipulator are used for the exact positioning of the electrode in the detection capillary. A two-electrode system is used because the currents measured are very small, and do not result in the reference electrode drift that often occurs when large currents are measured, making an auxilliary electrode unnecessary.

The system shown in Figure 12.9 was used with a 26-μm i.d. capillary for the detection of catecholamines, and detection limits of 200–500 amol were obtained (for S/N = 2). When the capillary diameter was reduced to 12.7 μm, the same 5-μm carbon fiber working electrode yielded detection limits of 5–25 amol, almost two orders of magnitude lower. The improvement results from the collection efficiency of the working electrode. The smaller i.d. capillary allows less analyte to exit the capillary without being electrochemically converted.[7] This result contrasts directly with those obtained with optical detectors, where detection limits are worse for smaller capillaries because less detectable analyte is present in the light path.

Using a microinjector for electrokinetic microsampling, this instrument has been used for the detection of catecholamines present in the intracellular fluid of single nerve cells.[8] Figure 12.10 shows a diagram of the injection apparatus and an electropherogram generated with the working electrode poised at +750 mV versus SSCE, to oxidize catecholamines.

To date, amperometric detectors have only been applied to small, readily oxidized analyte species. A variety of alternate amperometric detection schemes are under development, however, including indirect amperometry. With this technique, the separation buffer contains a redox–active species that is continuously detected by the ultramicroelectrode. The analyte displaces the electroactive species, so that the electrochemical response is decreased during the elution of an analyte peak.

12.5.4. Radiochemical Detection[9]

Radiochemical CE detectors have been developed to test the application of CE to conventional DNA sequencing techniques, where the detection of ^{32}P provides the sequence information (see Section 10.8). The ^{32}P decomposes according to Eq. 12.11, emitting a β-particle (electron) that can be detected using scintillators.

$$^{32}P \longrightarrow {}^{32}S + \beta + \text{antineutrino} \qquad (12.11)$$

The half-life of ^{32}P is 14.3 days, and the average energy of the emitted electron has been measured as 0.57 MeV, with some electrons having energy in excess of 1 MeV. With this energy, the emitted electrons can travel up to 2 mm in aqueous solution, or 0.95 mm in fused silica. Thus, the emitted β-particles can penetrate capillary walls to enter a solid scintillator, where their energy is converted to light that is

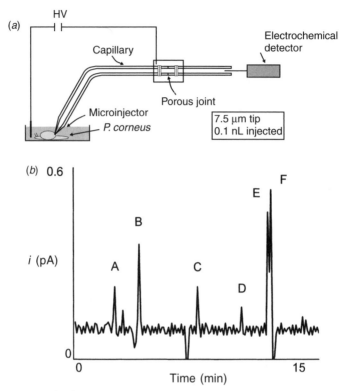

Figure 12.10. (*a*) Electrokinetic microinjector and (*b*) electropherogram generated from the removal and separation of cytoplasm samples from single nerve cells of *Planorbis corneus*, a pond snail.[8] [Reprinted, by permission, from R. A. Wallingford and A. G. Ewing, *Anal. Chem.* **60** (No. 18), 1988, 1972–1975. "Capillary Zone Electrophoresis with Electrochemical Detection in 12.7 μm Diameter Columns". © 1988 by American Chemical Society.]

detectable with conventional PMTs. A coincidence radiochemical detector is shown in Figure 12.11, where two PMTs are used, one on each side of the solid scintillator. This detector only registers a count if both PMTs respond simultaneously, and eliminates much of the background signal that occurs with single PMT detection. A CE separation of ^{32}P-labeled cytidine, adenosine, and guanosine triphosphate, obtained using the coincidence detector, is also shown.

12.6. CAPILLARY POLYACRYLAMIDE GEL ELECTROPHORESIS (C-PAGE)[10]

Polyacrylamide gels have been prepared inside capillaries, and CE separations performed with these capillaries show that electroosmosis has been eliminated, and

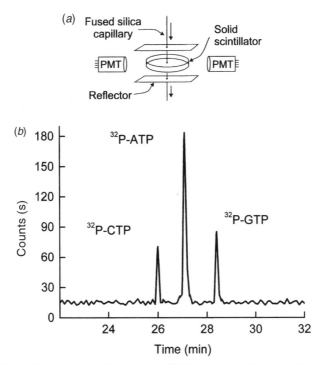

Figure 12.11. (*a*) Coincidence radiochemical CE Detector, and (*b*) results for the separation of ^{32}P-CTP, ^{32}P-ATP, and ^{32}P-GTP.[9] [Reprinted, with permission, from S. L. Pentoney, Jr., R. N. Zare, and J. F. Quint, in *"Analytical Biotechnology: Capillary Electrophoresis and Chromatography"* C. Horváth and J. F. Nikelly, Eds., ACS Symposium Series 434, American Chemical Society, Washington, DC, 1990. Copyright © 1990 by American Chemical Society.]

separations based on sieving may be performed. C-PAGE has been used for DNA sequencing, in a variety of configurations. We will consider one method in this section.

The dideoxy DNA sequencing method begins with the denaturation of double-stranded DNA (dsDNA) into single-stranded DNA. The ssDNA is then annealed with a fluorescent dye-labeled *primer*, which is an oligodeoxynucleotide ~ 20 bases long. The heteroduplex formed is then incubated in four separate reactions.

Each reaction mixture contains the heteroduplex, DNA polymerase, a mixture of the four dNTP species (N = G, A, T, or C), and a small amount of one of the four dideoxynucleotide triphosphates (ddNTP), where no 3'-OH group is present. The ratio of dNTP/ddNTP is high, about 1200:1. DNA polymerase sequentially extends the primer with bases complementary to those present on the opposite strand, but stops when one of the ddNTP species is incorporated. This yields a mixture of fragment lengths, all labeled, where polymerization has stopped at one

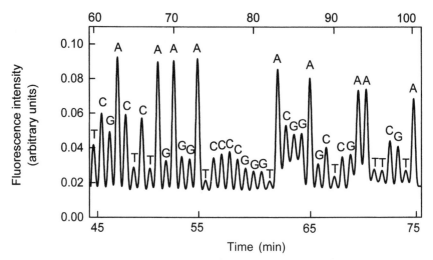

Figure 12.12. Portion of the C-PAGE electropherogram generated by dideoxy sequencing of a DNA sample, using a TAMRA-labeled 18-mer as a primer, and laser-induced fluorescence detection. The sequence is assigned based on peak intensities, which are 8:4:2:1 according to the relative volumes of the four reaction products combined in the sample.[10] [Reprinted, with permission, from H. Swerdlow, J. Z. Zhang, D. Y. Chen, H. R. Harke, S. Grey, S. Wu, N. J. Dovichi, and C. Fuller, *Anal. Chem.* **63**, 1991, 2835–2841. "Three DNA Sequencing Methods Using Capillary Gel Electrophoresis and Laser-Induced Flourescence". © 1991 by American Chemical Society.]

particular type base as a result of the incorporation of the particular ddNTP used in the reaction. The four individual products are then combined in volume ratios of 8:4:2:1, for A:C:G:T didioxy reaction products. Following separation by C-PAGE, the fluorescence label is detected via laser-induced fluorescence in a sheath-flow cuvette, and the sequence is read from the peak *intensities*.

The separation capillary is prepared by polymerizing 6%T, 5%C monomers with 30% formamide and 7 M urea inside a 37 cm long, 50 μm i.d. fused silica capillary. Electrokinetic injection at 200 V/cm for 30 s was used, since hydrodynamic injection does not work with gel-filled capillaries. Separation on the gel occurs by seiving, so that the shortest fragments elute first. Figure 12.12 shows the data obtained using a TAMRA[TM] dye label that is excited at 543.5 nm and emits at 590 nm. This method has a detection limit of 2 zmol (1 zmol = 10^{-21} mol) for each fragment.

12.7. CAPILLARY ISOELECTRIC FOCUSING (CIEF)[11]

Isoelectric focusing is a steady-state technique, and its application in capillaries requires the partial or total supression of electroosmotic flow. Capillaries used for CIEF can be derivatized with methylcellulose or linear polyacrylamide, to block the surface silanol groups, whose ionization is responsible for electroosmotic flow. A mixture containing the sample and the ampholytes are then loaded through-out the capillary, and the electric field is applied using acidic anode and basic cathode solutions. The current decreases with time, and reaches a minimum when the run is complete, at which time the focused pattern is static.

A technique called *salt mobilization* may be used to elute the focused protein bands. The field is turned off, and one of the reservoir solutions is changed to a solution containing acid (or base) plus a different cation (or anion) than H^+ (or OH^-). When the electric field is turned on again, electroneutrality causes a deficiency of either H^+ or OH^- to enter the capillary. The result is that the entire pH gradient moves toward the cathode if OH^- is deficient, or toward the anode if H^+ is deficient, and the focused pattern is eluted through a detector.

This method has been tested with a variety of covalently modified silica capillaries. It has been found that the derivatized capillaries tend to be unstable if pH values < 7.5 are used, and are especially unstable near the cathodic ends.

An alternative method, recently proposed, involves the use of modifiers in the running buffers, where the modifiers interact with the capillary walls to reduce, but not eliminate, electroosmotic flow. A continuous slow migration toward the detector thus occurs throughout the run. Using 0.1% methylcellulose as a modifier, species having pI values of 0.01 pH unit can be separated.

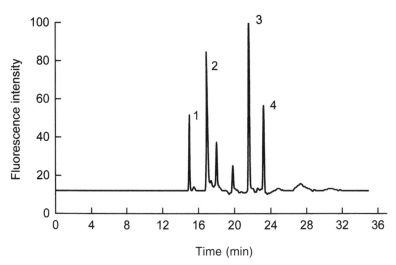

Time (min)

Figure 12.13. CIEF of cytochrome *c* (peak 1, pI 9.6), chymotrypsinogen A (peak 2, pI 9.0) and myoglobin (peaks 3 and 4, pI 7.2, and 6.8) on a 75-μm i. d. uncoated capillary, 60 cm long and 40 cm from anode to detector. Anolyte: 10-m*M* phosphoric acid; catholyte: 20-m*M* NaOH. Running buffer contains 0.1% methylcellulose, 1% TEMED, and 5% Pharmalyte 3–10 carrier ampholytes. The UV 280-nm detection.[11] [Reprinted, with permission, from J. R. Mazzeo and I. S. Krull, *Anal. Chem.* **63** (No. 24), 1991, 2852–2857. "Capillary Isoelectric Focusing of Proteins in Uncoated Fused-Silica Capillaries Using Polymeric Additives". © 1991 by American Chemical Society.]

Figure 12.13 shows such a separation, using a 40 cm, 75 μm uncoated fused silica capillary operated at 30 kV. A solution of ampholytes (5%) and sample (1 mg/mL of each protein) in the running buffer are loaded throughout the capillary. The application of the electric field results in slow, continuous separation and detection due to the slow electroosmotic flow. With this method, no salt mobilization is required.

SUGGESTED REFERENCES

C. Horvath and J. F. Nikelly, Eds., *Analytical Biotechnology: Capillary Electrophoresis and Chromatography*, American Chemical Society, Washington, DC, 1990.

O. W. Reif, R. Lausch, and R. Freitag, in *Advances in Chromatography*, Vol. 34, P. R. Brown and E. Grushka, Eds., Marcel Dekker, New York, 1994, pp. 1–56. (See also Chapter 4, by Z. El Rassi, pp. 177–250).

D. Schmalzing, S. Buonocore, and C. Piggee, *Capillary Electrophoresis-Based Immunoassays*, *Electrophoresis* **21**, 2000, 3919–3930.

REFERENCES

1. A. G. Ewing, R. A. Wallingford, and T. M. Olefirowicz, *Anal. Chem.* **61**, 1989, 300A–303A.

2. R. A. Wallingford and A. G. Ewing, in *Advances in Chromatography*, Vol. 29, J. C. Giddings, E. Grushka, and P. R. Brown, Eds., Marcel Dekker, New York, 1989, p. 22.

3. G. J. M. Bruin, *Electrophoresis* **21**, 2000, 3931–3951.

4. Y.-F. Cheng and N. J. Dovichi, *Science* **242**, 1988, 562–564.

5. J. B. Fenn, M. Mann, C. K. Meng, S. F. Wong, and C. M. Whitehouse, *Science* **246**, 1989, 64–71.

6. R. D. Smith, J. A. Olivares, N. T. Nguyen, and H. R. Udseth, *Anal. Chem.* **60**, 1988, 436–441.

7. R. A. Wallingford and A. G. Ewing, *Anal. Chem.* **60**, 1988, 258–263.

8. R. A. Wallingford and A. G. Ewing, *Anal. Chem.* **60**, 1988, 1972–1975.

9. S. L. Pentoney, Jr., R. N. Zare, and J. F. Quint, in *Analytical Biotechnology: Capillary Electrophoresis and Chromatography*, C. Horváth and J. F. Nikelly, Eds., American Chemical Society, Washington, DC, 1990.

10. H. Swerdlow, J. Z. Zhang, D. Y. Chen, H. R. Harke, R. Grey, S. Wu, N. J. Dovichi, and C. Fuller, *Anal. Chem.* **63**, 1991, 2835–2841.

11. J. R. Mazzeo and I. S. Krull, *Anal. Chem.* **63**, 1991, 2852–2857.

PROBLEMS

1. A CZE experiment was performed using an open 50-μm diameter silica capillary, 50 cm long. A 5-mM carbonate buffer at pH 10 was used, with a separation voltage of 25 kV, following a 15 s, 1-kV electrokinetic injection of an FITC-derivatized amino acid mixture. Detection using laser-induced fluorescence showed that the mixture contained three main components, with elution times of 4.5, 6.3, and 10.2 min. How would these elution times be expected to change if (a) The length of the capillary was increased to 1 m, (b) the separation voltage was increased to 50 kV, and (c) a length of 1 m *and* a voltage of 50 kV were used?

2. Untreated silica capillaries show a cathodic drift of buffer, that is a flow of buffer from the anodic to the cathodic end, allowing an on-column detector to be placed near the cathodic end of the capillary. At which end of the capillary should a detector be placed if the inner silanol surface of the capillary has been covalently modified with aminopropyltri(ethoxy) silane, "APTES" (cf. Chapter 4)?

3. Under acidic conditions, an analyte protein of unknown molecular weight is known to possess multiple positive charge, between +5 and +8. A mixture containing this protein and others were subjected to capillary electrophoresis in 10-mM trifluoroacetic acid, and the eluate from the capillary was fed directly into the ionization source of an electrospray mass spectrometer. As the protein eluted from the capillary, the m/z range of the mass spectrometric detector was scanned, and peaks were observed at m/z values of 938, 1071, 1250, and 1500. What is the molecular weight of the protein?

4. Detection limits obtainable with most CE detectors increase (worsen) as the diameter of the capillary decreases. However, ultramicroelectrode-based amperometric detectors behave in the opposite manner, with improved detection limits for smaller capillary diameters, allowing smaller quantities of sample to be used. Suggest a precolumn derivatization reaction that could be used to allow the electrochemical detection of amino acids following a CZE separation.

5. What is the main advantage of using a coincidence radiochemical CE detector over one that uses a single photomultiplier?

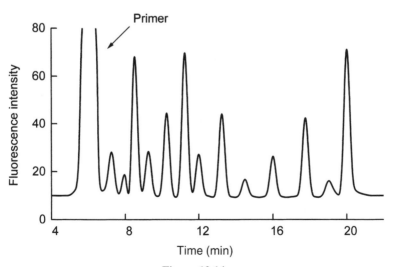

Figure 12.14

6. The C-PAGE electropherogram shown (Fig. 12.14) below was obtained following the separation of products of the dideoxy (Sanger) DNA sequencing reactions for a polydeoxynucleotide. A 12-base primer labeled with fluorescein isothiocyanate was used to initiate the four DNA polymerase reactions in the $3' \rightarrow 5'$ direction. The products of the four reactions that included ddGTP, ddATP, ddCTP, and ddTTP were, respectively, combined in the volume ratio 8:4:2:1. Using the electropherogram, determine the sequence of the first 13 bases that follow the primer. Remembering that the sequence of the original polydeoxynucleotide is complementary to the sequence of the dideoxy products detected, give the $5' \rightarrow 3'$ sequence of this section of the original polydeoxynucleotide.

7. Capillary isoelectric focusing was performed in a 75-μm i. d. capillary where the surface silanols had been derivatized with linear polyacrylamide. A mixture of proteins having pI values between 6 and 8 were focused using ampholytes covering the pH 3–10 range, using an anolyte of 20 mM HCl and a catholyte of 20 mM NaOH. A static separation pattern was indicated by a current minimum that occurred after 90 min at 15 kV. How can salt mobilization be used to elute the separation pattern at the anodic end of the capillary?

CHAPTER 13

Centrifugation Methods

13.1. INTRODUCTION

Gravitational sedimentation is a well-known separation method. It was empirically used for thousands of years, for example, to separate gold particles from sand. Particles that are in suspension in a fluid are influenced by different forces. If the particle has a higher density than the fluid in which it is immersed, it tends to migrate downward, following the gravity force direction. The frictional force acts to resist the movement (Fig. 13.1). The rate of this sedimentation may depend on the size as well the shape and the mass of the particles, and the viscosity and density of the fluid. Back diffusion, a force that relies on the concentration gradient, acts to counterbalance the tendency to settle.

The definition of "particle" in this context is very broad, and refers to every substance dissolved or in suspension in the medium, of either microscopic or macroscopic dimensions; this includes everything except the solvent or suspended fluid. The particles usually separated or studied in biochemistry are cells, subcellular organelles and large biomolecules, like DNA and proteins.[1]

Centrifugation methods are based on the effects of force fields stronger than the gravitational force, in order to accelerate processes that would otherwise occur over nonpractical time scales (months or years). Moreover, overcoming back-diffusion

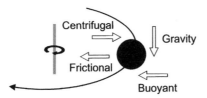

Figure 13.1. Forces acting on a particle during centrifugation. The gravitational force is not taken into account in practical applications, since its effect is very small in comparison with that of the centrifugal force, even with low-speed centrifuges.

Bianalytical Chemistry, by Susan R. Mikkelsen and Eduardo Cortón
ISBN 0-471-54447-7   Copyright © 2004 John Wiley & Sons, Inc.

forces allows the precipitation of small molecules that would not precipitate in our terrestrial gravitational field.

13.2. SEDIMENTATION AND RELATIVE CENTRIFUGAL *g* FORCE

By using Stokes' law, it is possible to calculate the sedimentation rate (v) for a particle.[2] The sedimentation rate is the result of two forces, the gravitational force (or centrifugal) that move the particle downward, and the frictional or drag force, resisting its motion through the solution. Stokes' law is valid for spherical particles that reach a constant velocity:

$$v = \frac{d^2(\rho_p - \rho_m)}{18\mu} \times g \tag{13.1}$$

where d is the diameter of the particle, ρ_p is the density of the particle, ρ_m is the density of the medium, μ is the viscosity of the medium, and g is the centrifugal force field. From this equation, it can be seen that the sedimentation rate of a particle is proportional to the square of the diameter of the particle. In addition, the sedimentation rate is proportional to the difference between the density of the particle and the density of the medium, and if this difference is zero, the sedimentation rate becomes zero. It can also be seen that sedimentation rate increases with increasing force field and decreasing viscosity.

The relative centrifugal force (RCF) is defined as the force field relative to the Earth's gravitational field; Equation 13.2 shows its definition:

$$RCF = \frac{F_{centrifugation}}{F_{gravity}} = \frac{m\omega^2 r}{mg} = \frac{\omega^2 r}{g} \tag{13.2}$$

where r is the distance between the center of rotation and the particle (cm); ω is the rotation speed in radians per second (rad/s), and m is the mass of the particle. The velocity in the centrifuge is usually expressed in revolutions per minute (rpm), and if $g = 980$ cm/s^2, we can write

$$\frac{\omega^2}{g} = \frac{[(2\pi)(rpm)/60]^2}{980} \tag{13.3}$$

$$= 1.119 \times 10^{-5}(rpm)^2 \tag{13.4}$$

Therefore,

$$RCF(g) = 1.119 \times 10^{-5}(rpm)^2 r \tag{13.5}$$

Since RCF is a ratio between two forces, it has no units; however, it is customary to follow the numerical value of RCF with the *g* symbol. In the centrifuge tube, the distance r increases from the tube top to the bottom, so that we can define an average g force (g_{av}) for a given rotor and speed, a minimum g force in the tube top and a maximum force in the tube bottom (g_{min} and g_{max}, respectively).

Biological particles. Stokes' law applies to spherical particles, which are large in comparison with the molecules that comprise the liquid medium, and are present at a concentration low enough to avoid modification of the liquid viscosity. Most biological particles are not spherical, and Strokes' law must be modified to take this into account. One approach to this problem is to consider that the biological particles' shapes could be approximated by "ellipsoids of revolution", or spheroids with one major and one minor axis. Calculations show that the frictional force over these ellipsoids is greater than that expected for spherical particles of the same volume.[3]

Centrifugal conditions are chosen according to the type of biological material. The chemical composition of the medium, as well as its temperature must be controlled to avoid denaturation. In addition, the hydrostatic pressure generated by the centrifugal force may be large enough to permeabilise cellular membranes[4] (affecting cell density) or split macromolecular associations, such as those that exist in multisubunit proteins. Centrifugal forces can also modify the molecular shape; the sedimentation rate is then nonlinearly dependent on the g force applied.

13.3. CENTRIFUGAL FORCES IN DIFFERENT ROTOR TYPES

There are four basic types of rotors: the swinging-bucket, fixed-angle, vertical, and zonal rotors. The vertical rotor could be classified as a fixed-angle rotor, in which the angle is extreme, but most authors consider that they are different enough to consider separately. A detailed description of these four types of rotors and their specifications can be found in Rickwood.[5]

The zonal rotors will not be described in detail here, since they are habitually chosen for large volume preparative protocols (typical capacities are 300–1700 mL), mainly using rate-zonal centrifugation. Basically, a zonal rotor consists of two half-cylinder sections, which screw onto one another. The sample is poured inside the rotor without the use of tubes.

13.3.1. Swinging-Bucket Rotors

The sample tube is inserted into a one-tube holder (bucket). During acceleration of the rotor, the bucket reorients from vertical (earth gravity force) to the horizontal position (centrifugal force). In Figure 13.2 the centrifugal g force direction is evident, showing that the particles in this rotor move first to the walls of the tube, then a bulk movement of particles and convection flow accelerates the precipitation process (called the "wall effect"). This rotor is generally used for rate-zonal separations. Note that the precipitation path is along the length of the centrifugation tube; therefore, this method is relatively inefficient for pelleting.

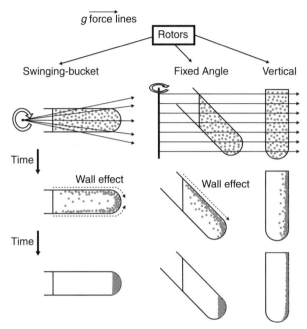

Figure 13.2. Centrifugal forces in different rotor types. The angle is measured with respect the rotation axis, these means than a fixed-angle rotor with an angle of 14° is related to a vertical rotor, and therefore its sedimentation pathlength is shorter than in a 40° rotor.

13.3.2. Fixed-Angle Rotors

In this rotor the sample tubes fit into holes in a solid rotor. The tube angle therefore remains constant during the loading, centrifugation, and unloading process. When the centrifugation force increases, the solution reorients inside the tube. For a given sample volume, the centrifugation path with a fixed-angle rotor is shorter than that with a swinging-bucket rotor. The particles reach the tube wall more rapidly, and then slide down the wall to form a pellet (Fig. 13.2). These rotors are more efficient for differential pelleting or other pellet-based procedures, and could be used for iso-pycnic experiments, which are equilibrium density gradient techniques, described later in this chapter. When the rotor angle is shallow, the sedimentation path is shorter, and the pelleting efficiency is increased. Fixed angle rotors are available with angles between 14° and 40°.

13.3.3. Vertical Rotors

In vertical rotors (Fig. 13.2) the tubes are fixed in a vertical position during centri-fugation, and when the centrifugal force increases the solution reorients to the g force. The reorientation process is explained later in this chapter. For now, this means that in this rotor the tube covers are important, and must be able to withstand a large hydrostatic pressure. The sedimentation path is the shortest of all kinds of rotors described here (the tube diameter), and the minimum *g* force, related to the minimal radius, is larger for this rotor, allowing faster separations at lower speeds. These rotors are the first choice for isopycnic centrifugation, can be used for rate-zonal centrifugation, and are never used for pellet-related techniques, because the pellet is formed over the entire tube wall, and usually detaches easily.

13.4. CLEARING FACTOR (*k*)

Pelleting capacity is related to the value of the clearing factor *k*, a variable that refers to the ability of a set of centrifugation conditions to separate or pellet particles. Smaller *k* values indicate greater efficiency. The parameter *k* is defined as follows:[6]

$$k = (t)(s)10^{13} \tag{13.6}$$

where *t* is the time required in hours (h), and *s* is the sedimentation coefficient, defined as

$$s = \frac{\ln(r_{max}/r_{min})}{\omega^2(t_2 - t_1)} \tag{13.7}$$

where ω is the angular velocity (rad/s) and the difference between the radius maximum (r_{max}) and minimum (r_{min}) determines the sedimentation path. When this path is shorter, *k* is smaller. A more useful expression, were the angular velocity is replaced by speed in rpm (and therefore the time units are minutes) is shown in Eq. 13.8:

$$k = \frac{2.53 \times 10^{11} \ln(r_{max}/r_{min})}{\text{rpm}^2} \tag{13.8}$$

The *k* value is given as a characteristic of a rotor, and applies to a particle in pure water at 20 °C, at the maximum speed allowed for the rotor. The corresponding k_{actual} values at lower speeds can be calculated using the following formula:

$$k_{actual} = k\left(\frac{\text{rpm}_{max}}{\text{rpm}}\right)^2 \tag{13.9}$$

If the sedimentation coefficients of the particles to be pelleted are known, an approximate prediction can be made of the time required for the separation, with the following formula:

$$t = \frac{k}{S} \tag{13.10}$$

were *t* is the time required in hours and S the sedimentation coefficient in Svedberg units.

Equation 13.10 is only an approximation. The k values are calculated on the basis that the centrifugation medium has the viscosity and density of water. An increase in either of these parameters will increase the value of k by changing the value of the constant in Eq. 13.8. Computer simulations can be used for more accurate predictions of precipitation/banding times using nongradient media. With gradient media, viscosity and density are functions of position, and a single k value does not apply; to predict the banding time in this situation, k' can be calculated for a determined gradient and a well-characterized particle.[6]

13.5. DENSITY GRADIENTS

The introduction of density gradients allowed several improvements in the analytical and preparative capabilities of centrifugal processes. First, it is possible to pour the sample in the top of a tube, in a more or less thin layer; differences in density between sample and separation medium avoids convection and delays the homogenization process. Second, it is possible to diminish convection fluid movements in the tube; convection can be produced by a difference in temperature between different parts of the centrifugal rotor. Third, it is possible to "focus" a particle when its reaches its own density, in a process analogous to isoelectric focusing.

13.5.1. Materials Used to Generate a Gradient

Density gradients may be generated in aqueous solutions from many materials. The desirable properties of an ideal material include high water solubility, high density, low viscosity, and low osmotic pressure of solutions, physiological and chemical inertness, and UV–vis light transparency.[7] We give here a short overview of water-soluble gradient materials, with their principal characteristics. Only in very specific situations are nonaqueous materials used (e.g., chloroform or benzene). More detailed information can be found in Price's excellent book.[8]

The most common materials used to generate density gradients are sucrose, Ficoll, and Cs salts. Sucrose solutions can be as concentrated as 65% w/w, with a maximum density of $1.32 \, \text{g/cm}^3$ at $4 \, °C$; however, concentrated solutions of sucrose have high osmotic strength and viscosity. Ficoll is a brand name for a synthetic polysaccharide with an average MW of 400,000 and a maximum density in aqueous solution of 1.23. It is very useful to separate osmotically sensitive particles, like mammalian cells.

Cesium salts (mainly CsCl and Cs_2SO_4) can be used to produce self-formed gradients by overnight centrifugation of a homogeneous salt solution. These gradients are formed *in situ*, and the steepness of the gradient depends on the rotor and g forces applied. The maximum density is very high: 1.9 and $2.1 \, \text{g/cm}^3$ for chloride and sulfate salts, respectively. These are typically used for the preparation and analysis of nucleic acids and for plasmid separation.

Colloidal particles of silica have also been used to generate density gradients. These particles are approximately 20 nm in diameter, and are often coated with an organic material. Percoll, for example, is a brand name for polyvinylpyrrolidone-coated inert silica particles. Percoll suspensions have the advantages of nontoxicity, low viscosity, and low osmotic pressure, and is the material of choice to separate membrane-bound particles, as well as cells and subcellular organelles. However, the maximum density of a Percoll suspension is only 1.2 g/cm^3.

Dense iodinated organic compounds (e.g., iodixanol) derived from X-ray contrast materials can be used in applications that require higher densities than Percoll (as high as 1.5 g/cm^3) as well as low ionic strength and low osmolarity.

13.5.2. Constructing Pre-Formed and Self-Generated Gradients

There are several ways to generate a gradient. Stepwise gradients can be formed by successively layering one solution over another solution of higher density. These stepwise gradients (4–6 layers) can be used as is, or can be transformed to a more or less linear gradient, by allowing diffusion to linearize the gradient or by freezing and thawing.

A more straightforward possibility is to use a device designed specifically to produce gradients, called "gradient generators". A simple gradient generator is depicted in Figure 13.3, and can be used to generate linear or exponential gradients. The two cylinders are connected at their base; one cylinder contains the lighter

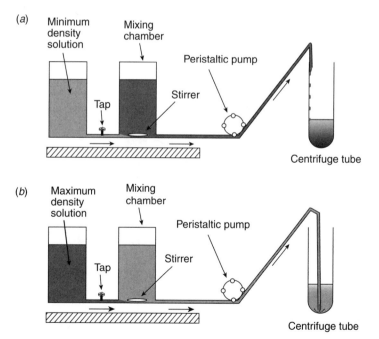

Figure 13.3. Device for generating density gradients. In (*a*), the more dense solution is in the mixing chamber; therefore, a gradient from dense to light will be formed, and the centrifuge tube is filled from the top. In (*b*), the positions of the solutions are inverted to generate a light to heavy gradient: in this case, the centrifuge tube is filled from the bottom.

solution (the solution with the lowest gradient-forming solute concentration) and the other contains the more concentrate or dense solution.

To generate a linear gradient the two chambers must have the same diameter, and the solution levels in the two chambers must remain approximately equal during gradient generation. Different gradients shapes can be produced using chambers with different diameters, or using motor driven pistons or syringe pumps. Automated systems that include gradient formers as well as a programmable gradient harvest capability, on-line absorbance (or other optical parameter) measurement and fraction collectors are also available.

Gradients can also be produced by centrifugation of a homogenous gradient-forming solution, such as aqueous cesium chloride. The self-generated gradients are very reproducible and predictable, but are not used frequently, because gradients generated in this way do not normally span a wide enough range of densities.

13.5.3. Redistribution of the Gradient in Fixed-Angle and Vertical Rotors

When the rotor begins to accelerate, the density gradient adjusts to the new g forces that act upon it. In swinging-bucket rotors, the centrifuge tube follows the direction of the g force, and the gradient undergoes no redistribution.

This is not the case with fixed-angle and vertical rotors, where the centrifuge tube angle is fixed and determined by the rotor design. Therefore, the gradient and sample redistribute in the tubes, as shown in Figure 13.4. This redistribution process takes place at low velocity (0–1000 rpm), to preserve the gradient, and

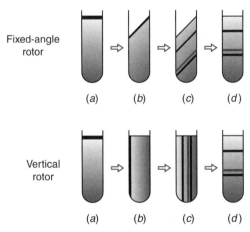

Figure 13.4. Redistribution of the sample layer and density gradients during centrifugations using fixed-angle and vertical rotors. In (a), the sample (with lower density than the gradient top) is layered, the rotor is at rest. In (b), the rotor begins to accelerate, and the sample begins reorientation. In (c), the sample separation proceeds. In (d) the rotor is at rest, and the particle bands are oriented by the Earth's gravitational force.

the acceleration must be controlled carefully. Once the reorientation process is finished, acceleration to the final velocity can proceed without limitation, and is followed by slow deceleration prior to sample harvest.

13.6. TYPES OF CENTRIFUGATION TECHNIQUES

It is possible to classify centrifugation techniques as preparative (when the main goal is to obtain a purified material) and analytical (for characterization of purified samples). Differential centrifugation is the simplest method, and does not require a density gradient medium; it is a typical preparative methodology. Techniques considered preparative but possessing analytical applications are isopycnic and rate-zonal centrifugation, both requiring density gradient media.

13.6.1. Differential Centrifugation[9]

Differential centrifugation is very useful for the separation of subcellular organelles and membranes. The starting material is usually a tissue homogenate or lysed cells. A standard procedure is shown in Figure 13.5, where consecutive pellets are separated as the supernatant is centrifuged to higher *g* forces for longer times.

Figure 13.6 illustrates the main disadvantage of differential centrifugation. The initial sample contains a homogeneous mixture of different particles. Smaller (or less dense) particles, which are near to the bottom, pellet together with large (or more dense) particles. The purity of the fractions maybe increased by resuspending

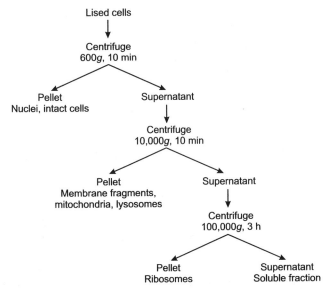

Figure 13.5. Typical differential centrifugation protocol. In order to obtain more purified fractions, the pellets could be resuspended in buffer and recentrifuged.

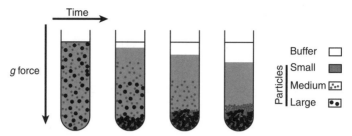

Figure 13.6. Sedimentation process in a multicomponent system (solvent with three sizes of particles). Note that the pellet contains a mixture of particle sizes.

the pellet and recentrifuging, but usually differential centrifugation is only the first step in a purification protocol.

13.6.2. Rate-Zonal Centrifugation

In rate-zonal centrifugation (or *velocity sedimentation*) the sample is loaded as a thin layer on the top of a density gradient medium. During centrifugation, the sample separates into bands, and the particles are separated on the basis of their different sedimentation coefficients (s). For biological particles, the coefficient s is mainly related to the size of the particles. This process is illustrated in Figure 13.7(a).

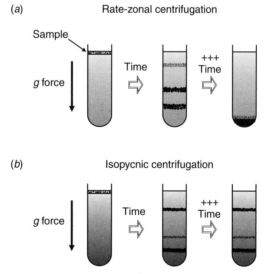

Figure 13.7. Comparison between (a) rate-zonal and (b) isopycnic centrifugation. Note that the gradient used in rate-zonal techniques is shallow, and if the centrifugation continues, the particles pellet. The diffusion process causes band broadening in rate-zonal centrifugation, but this is not evident when the particles are "focused" using isopycnic centrifugation.

The maximum density of the gradient usually does not exceed the density of the particles. Because of this, if the centrifugation continues indefinitely, the particles pellet at the tube bottom. The gradient is chosen to have the minimal density and viscosity possible to allow rapid separation and minimize diffusional band broadening.

In rate-zonal centrifugation the conditions for sensitive biological particles are gentle (low osmotic pressure, g forces, and viscosity) and typically the separation is faster than isopycnic separations. Applications include the separation of proteins and nucleoproteins, like ribosomes.

13.6.3. Isopycnic Centrifugation

In isopycnic centrifugation (or *equilibrium sedimentation*) the particles move through a density gradient until they arrive at their isopycnic points. This is the point where the density of the solution equals the density of the particle. At this point, the particle stops moving. With this technique, the time necessary for a given separation is not critical. The sample is separated into bands, and separation occurs on the basis of the buoyant density of the particles. Diffusional broadening is not significant, and the particles are "focused" in very narrow bands. In this technique the sample can be loaded at the top of the density gradient, at the bottom, or mixed with a gradient-forming solution; the particles move until they arrive at their equilibrium positions.

The isopycnic method has been used to dramatically demonstrate semiconservative DNA replication, using CsCl density gradients. Separation of DNA, RNA, protein, and carbohydrates can be performed in dense CsCl solutions, where the RNA pellets, the DNA forms bands, and protein and carbohydrates form a thin layer called a pellicle at the top of the gradient.

13.7. HARVESTING SAMPLES

Once the centrifugation is finished, there are two principal ways to collect fractions. One of these is to pierce the bottom of the plastic (disposable) centrifuge tube, and collect drops as different fractions; the collection is done from the higher density particles to the lower. Another possibility is the displacement of the gradient upward using a more dense solution; in this case, the fractions are collected from the top of the tube, from the lower to the higher density parts of the gradient. Each fraction can be examined separately (e.g., by UV spectrophotometry) and the data plotted as absorbance versus effluent volume or density. Modern instrumentation allows on-line measurement at the moment the centrifuged sample is harvested.

13.8. ANALYTICAL ULTRACENTRIFUGATION

Centrifugation techniques are used mainly in separation and purification protocols, or as a sample preparation step in analytical biochemistry. Historically, analytical ultracentrifugation was important for highly accurate absolute calculation of the molecular weight of biological materials and polymers. Information obtained from ultracentrifugation experiments applies to analytes present under native conditions, while many other techniques (e.g., size exclusion chromatography, gel electrophoresis, and MS) apply to denatured or fragmented biomolecules.

Analytical ultracentrifugation is carried out in low-volume centrifuge cells (1 mL or less), using high purity samples, in a specially designed centrifuge, equipped with an optical detection system to follow the sedimentation process continuously. In addition to the determination of molecular weight, it is possible to determinate molecular size and shape, sedimentation coefficients, density, and frictional coefficients. This technique has also been used to demonstrate the semiconservative replication of DNA, the conformational changes that arise when enzyme–substrate complexes form and for the study of macromolecular interactions. Like most other analytical techniques, samples are usually not recovered after the experiment. Typical samples are proteins, lipoproteins, nucleic acids, polysaccharides, subcellular particles, cells, and viruses.

13.8.1. Instrumentation

The first analytical ultracentrifuge was designed by Svedberg in 1923. Since then, the main improvements have dealt with data acquisition and processing, while the hardware design has not been significantly modified. A typical instrument is shown in Figure 13.8; it is basically composed of a high velocity centrifuge equipped with an optical detection system that is capable of monitoring the optical properties of the entire centrifuge cell.

The sample is poured into a cavity in a single or double sector centerpiece, made of an aluminum alloy or an epoxy resin material [Fig. 13.9(*b*)]. The centerpiece is fitted with transparent windows made of quartz, or sapphire if high *g* forces are required. A cell housing (with related parts including spacers, gaskets and screw rings) that holds the cell assembly fits into the ultracentrifuge rotor. A simplified

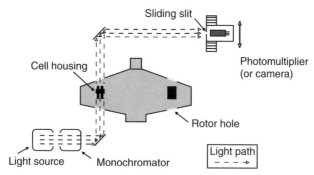

Figure 13.8. Analytical ultracentrifuge equipped with a UV–vis optical detection system. The "sliding slit" at the photomultiplier allows positional recording of the absorbance along the cell.

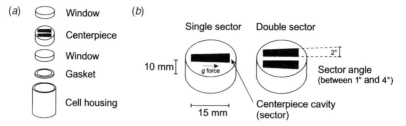

Figure 13.9. Basic components of typical single- and double-sector analytical ultracentrifuge cells. The sector angle is the angle subtended at the center of rotation.

representation of the assembly is shown in Figure 13.9(a). The double sector cells allow a reference to be run simultaneously and absorbance values subtracted from sample data, in order to calculate concentrations. More detailed descriptions of sample cells and analytical ultracentrifugation instrumentation can be found in Lavrenko et al.[11]

The horizontal orientation of the sample chamber allows positional detection of the analyte with the optical system. The force lines in these sectorial cells are radial (Fig. 13.10), and parallel to the cell walls, so that convection and "wall effects" are avoided.

Other optical techniques have also been used to monitor analytical ultracentrifugation experiments. These included refractive index (Schlieren optics) as well as interference patterns produced using Rayleigh- or Lebedev-type optics.[11] In both cases, data are acquired using photographic or video recording equipment. Examples of data obtained with these different optical systems are shown in Figure 13.11.

The light sources commonly used are high-power mercury arc lamps (100 W) or lasers; high power is necessary due the short exposure time of the centrifuge cell to the light path. In a typical rotor, the light passes through the cell for ~7 ms/s, and the rest of the time the light is directed onto the rotor.[12] High-quality optics are used to conduct the light from the source to the detector.

13.8.2. Sedimentation Velocity Analysis

In this technique, the particle moves toward an equilibrium position, but the experiment is terminated before equilibrium occurs. The change in velocity of the particle is interpreted using kinetic equations. Consider a simple two-component system,

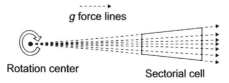

Figure 13.10. Force lines in a sectorial cell.

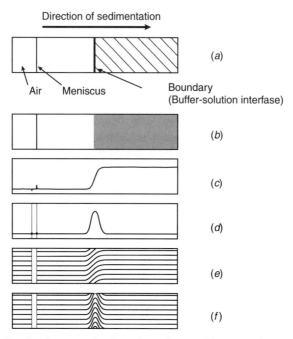

Direction of sedimentation

Air   Meniscus

Boundary
(Buffer-solution interfase)

(a)

(b)

(c)

(d)

(e)

(f)

Figure 13.11. Results for a moving-boundary ultracentrifuge experiment using different optical detection systems and a double-sector cell. Part (a) is a graphical representation, (b) is the result of an uv photograph, (c) is a plot of absorbance versus distance (from b), (d) is a photograph obtained with Schlieren optics, (e) is an interference diagram obtained using Rayleigh optics, and (f) is another interference diagram, obtained with Lebedev optics.

like a protein dissolved in buffer [e.g., 1 mg/mL BSA in 0.1 M phosphate buffer]. In a sectorial cell, it is possible to follow the movement of the boundary by optical techniques (Fig. 13.11). This technique is also called "moving boundary analysis", and it is used to determine the sedimentation coefficient s, defined as

$$s = \frac{dr/dt}{\omega^2 r} = \frac{d\ln r}{dt} \times \frac{1}{\omega^2} \qquad (13.11)$$

where ω is the angular velocity (in rad/s) and r is the distance of the boundary from the center of rotation, in cm. It is possible to integrate r in Eq. 13.11 with respect to time (between t_1 and t_2), to arrive at Eq. 13.12, were time is in seconds, and r in centimeters:[13]

$$s = \frac{2.1 \times 10^2 \log_{10}(r_2/r_1)}{(\text{rpm})^2 (t_2 - t_1)} \qquad (13.12)$$

A plot of $\ln r$ against t (Eq. 13.11) or two measurements of boundary position (Eq. 13.12) can be used to determine the value of s. Usually, the sedimentation coefficient is expressed in Svedberg units, where 1 S = 10^{-13} s.

The s value can be used to calculate the molecular weight of a macromolecule, using Eq. 13.13:

$$s = \frac{M(1 - \bar{v}\rho)}{Nf} \tag{13.13}$$

where M is the molecular weight, $\bar{v}$ is the partial specific volume, ρ is the solvent density, N is Avogadro's number, and f is the frictional coefficient of the particle. The f value can be calculated through the diffusion coefficient (D):

$$D = \frac{RT}{Nf} \tag{13.14}$$

By substituting Eq. 13.14 into Eq. 13.13, we obtain the Svedberg equation:

$$M = \frac{sRT}{D(1 - \bar{v}\rho)} \tag{13.15}$$

The diffusion coefficient indicates the extent of boundary broadening during the experiment, but is not easy to determine, limiting the use of this method.

In order to use Eq. 13.15 to obtain molecular weights, experimentally obtained s and D values must be corrected to ideal conditions. The observed s and D values are dependent on solution conditions as well as temperature. It is possible to convert an observed s value to a standard value for a solution with the density and viscosity of water at 20 °C using the following equation:[13]

$$s_{20,w} = (s_{T,m}) \frac{\eta_{T,m}(\rho_P - \rho_{20w})}{\eta_{T,m}(\rho_P - \rho_{Tm})} \tag{13.16}$$

where s_{Tm} is the experimentally determined sedimentation viscosity using medium m and temperature T, $\eta_{T,m}$ and $\eta_{20,w}$ are the viscosity of the medium at T and the viscosity of water at 20 °C, respectively, ρ_P is the density of the particle in study, $\rho_{T,m}$ and $\rho_{20,w}$ are the density of the medium at T and the density of water at 20 °C, respectively. The sedimentation coefficient has been calculated for a variety of biological particles, and Figure 13.12 shows some examples of these values.

The sample concentration affects the observed s values; for example, when protein concentration increases, lower s values are observed. A number of experiments with different protein concentrations are necessary to correct this concentration-dependent bias, to allow extrapolation of the s value to a protein concentration of zero. This new value extrapolated to zero concentration is known as $s_{20,w}^0$, and is used in the calculation of MW (M) using Eq. 13.15. But if is known that the protein or macromolecule has a simple globular shape, instead of calculating D it is possible to use the following empirical formula:

$$s_{20,w}^0 = 0.00248 \, M^{0.67} \tag{13.17}$$

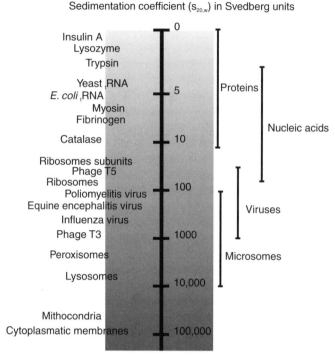

Figure 13.12. Sedimentation coefficients (*s*) for some biological particles, with values expressed in Svedberg units. Nuclear particles have values on the order of 10^6–10^7 S, while whole cells have values between 10^7 and 10^8 Svedberg units.

Other empirical formulae have been proposed for different biological materials, including single strand, double strand and circular DNA.[14]

13.8.3. Sedimentation Equilibrium Analysis

In this type of experiment, particles are allowed to move until they arrive at their equilibrium positions. This means that sedimentation proceeds until sedimentation forces are balanced by diffusion. There are several approaches to this technique. One of these is to scan the cell, to obtain the absorbance profile at different times; this allows calculation of concentration as a function of distance from the rotation axis (dc/dr). The following expression is then used to calculate the particle mass:

$$M = \frac{RT}{(1 - \bar{v}\rho)\omega^2} \frac{1}{rc} \frac{dc}{dr} \tag{13.18}$$

It is possible to use this technique to study more complex systems, including polydisperse samples in which interactions occur between several different components. Details regarding the mathematical methods can be found in Schachman.[15]

13.9. SELECTED EXAMPLES

13.9.1. Analytical Ultracentrifugation for Quaternary Structure Elucidation[16]

Analytical ultracentrifugation methods have been employed to study the serine acetylransferase (SAT) of E. coli, in order to elucidate how many protein subunits are present in the functional enzyme. In bacteria, L-serine is converted to cysteine in a two-step process; SAT catalyzes the first reaction in this metabolic pathway, described as "sulfur fixation". The SAT subunit has been sequenced, and its MW is 29.3 kDa.

Sedimentation equilibrium experiments were performed, using three SAT concentrations (0.096, 0.73, and 3.54 mg/mL) in buffer, at 8000 rpm and 5 °C until the concentration versus radial distance reached equilibrium, typically within 24 h. The cell was scanned every 3 h at 280 nm, and the absorbance values were used to calculate concentration profiles. The analyses of equilibrium solute (SAT) distributions were done using computer programs provided by the ultracentrifuge manufacturing company. The results obtained for three protein concentrations and for a control trimeric protein, chloramphenicol acetyltransferase (CAT), in which each monomer has a MW = 24,965, are shown in Table 13.1. These data, which have been corrected for concentration using the following equation:[17]

$$s = \frac{s^0}{(1 + k_s c)} \tag{13.19}$$

where k_s is the concentration dependence coefficient and c is the concentration corrected for radial dilution, reveal that the SAT protein is hexameric. Moreover, the slight correlation between the concentration and the experimental MW of the holoenzyme is consistent with a small degree of self-association.

Additional sedimentation velocity experiments were done using three SAT concentrations, at higher velocity (40,000 rpm), under otherwise similar experimental conditions. The cell was scanned every 30 min at 280 nm for 14 h. The data

TABLE 13.1. Equilibrium Sedimentation Derived Molecular Weights[a]

Enzyme	Concentration (mg/mL)	MW Enzyme	MW Subunit	Apparent Quaternary Structure
SAT	0.10	169,700 ± 1060	29,260	5.8
SAT	0.73	177,600 ± 1060	29,260	6.1
SAT	3.60	184,300 ± 1060	29,260	6.3
CAT	0.50	75,100 ± 740	24,965	3.0

[a] A control protein, CAT, is used to show data reliability. [Reprinted, with permission, from V. J. Hindson, P. C. E. Moody, A. J. Rowe, and W. V. Shaw, *The Journal of Biological Chemistry* **275**, No. 1, 2000, 461–466. "Serine Acetyltransferase from *Escherichia coli* Is a Dimer of Trimers". Copyright © 2000 by The American Society for Biochemistry and Molecular Biology, Inc.]

obtained at each protein concentration indicate that the protein does not self-associate. The sedimentation coefficient ($s_{20,w}^0$) was calculated as 7.08 ± 0.09 S, from which the frictional ratio of SAT was found to be 1.53; this value is high when compared with the expected value for a globular protein (~ 1.15), supporting the hypothesis that SAT subunits are not organized in a compact motif. This hypothesis was confirmed using other techniques. The authors conclude that the fundamental building block is probably a trimer, with SAT arranged as a pair of loosely staked trimers.

13.9.2. Isolation of Retroviruses by Self-Generated Gradients[18]

Retroviruses are viruses in which the genetic material is RNA. Several retroviruses are related with human diseases, like influenza or HIV. Some human retroviruses have glycoproteins as important surface structures; these are usually of crucial importance for infectivity, and therefore for virus characterization.

Human cells from patients suffering from multiple sclerosis were cultivated, and the supernatants were separated from cells and debris after centrifugation at 1000 g at 4 °C for 30 min. The supernatant was transferred to a 60-mL centrifuge tube, and underlaid with 4 mL of a concentrated solution of iodixanol, an iodinated, nonionic density gradient medium, in HEPES–NaOH (60 mM, pH 7.4, 0.8% NaCl). This dense underlay medium is called a "cushion", and prevents the pelleting of particles less dense. Particles then concentrate at the interface between the supernatant and the cushion. The tubes were centrifuged in a fixed angle rotor (22.5°) at 45,000 g (4 °C, 2 h) and the supernatant removed from the tubes by aspiration, leaving only of 4–5 mL in the proximity of the cushion. The cushion and the overlying supernatant were combined, and the volume measured for final regulation of the concentration of the gradient medium.

This mixture, adjusted to contain 20% of iodixanol, was transferred to 11-mL tubes and centrifuged in a vertical rotor at 364,000 g (4 °C, 3.5 h). Under these conditions, a gradient with a density from 1.02 to 1.25 was formed. Following centrifugation, the gradient was collected through a puncture in the tube bottom, in 0.5-mL fractions. Each fraction was assayed by PCR to determine retrovirus concentration.

The authors showed that the iodixanol gradient medium is better for this purpose than the previously used sucrose medium. Sucrose gradient media have high viscosity and osmolarity when compared with iodinated gradient media; these properties can alter the external (glycoprotein) features of this virus, reducing infectivity; moreover, the gentle separations in iodixanol media allow the concentration of small amounts of virus in comparison with sucrose gradients.

13.9.3. Isolation of Lipoproteins from Human Plasma

Lipoproteins transport lipids in blood; an early classification system was established using centrifugation methods, which are based on their different densities. Lipoproteins are classified as chylomicrons, largest and lowest in density

($\rho < 0.96$ g/mL), with low protein concentration (2%). Very low density lipoproteins (VLDL), low-density lipoproteins (LDL), and high-density lipoproteins (HDL) are progressively smaller, more dense and rich in protein, increasing the ρ and protein content up to 1.210 g/mL and 40–55% for HDL.

Figure 13.1 shows the forces acting during centrifugation. If the particle ρ is lower than that of the centrifugation medium, the buoyant force becomes important and the particle moves upward. Techniques in which the particles to be analyzed move upward, reaching the surface of the centrifugation medium, are called flotation or buoyant centrifugation separations.

Standard centrifugation procedures used for fractionation of lipoproteins in human blood plasma are made by *sequential flotation* using KBr, NaBr, NaCl, or a combination of these salts to produce a solution with determined density, in which a selected lipoprotein fraction moves upward.

In a typical procedure, the human plasma is centrifuged (usually in a fixed angle rotor) at 50,000 g for 30 min; under these conditions, chylomicrons float as a milky layer. The infranatant is removed, and centrifuged at 100,000 g for 20 h, when a floating layer of VLDL is obtained, as well as a pellet, which contains the plasma proteins, LDL and HDL plasma lipoprotein fractions. The pellet is resuspended and mixed with a concentrated salt solution, in order to obtain a final density of 1.063 g/mL, and centrifuged at 100,000 g for 28 h. A floating layer of LDL is obtained, and the pellet is then resuspended and adjusted to a density of 1.21 g/mL. Centrifugation at 100,000 g for 48 h produces a floating layer of HDL.

Using self-formed iodixanol gradients, a novel rapid method for the separation of plasma lipoprotein has recently been described.[19] The centrifugation time is reduced to ~4 h, but the gradient must be collected carefully, in 0.1–0.2-mL fractions.

SUGGESTED REFERENCES

P. Sheeler, *Centrifugation in Biology and Medical Science*, John Wiley & Sons, Inc., New York, 1981.

D. Rickwood Ed., *Centrifugation-A Practical Approach*, IRL Press, Oxford, UK, 1984.

J. Graham. *Biological Centrifugation*, BIOS Scientific Publishers Limited, 2001.

REFERENCES

1. H.-W. Hsu, "Separation by Centrifugal Phenomena", *Techniques of Chemistry*, Vol. XVI, John Wiley & Sons, Inc., New York, 1981, p. 111.

2. D. Rickwood and G. D. Birnie, in *Centrifugal Separations in Molecular and Cell Biology*. G. D. Birnie and D. Rickwood, Eds., Butterworths, London, 1978, pp. 1–3.

3. P. Sheeler, *Centrifugation in Biology and Medical Science*, John Wiley & Sons, Inc., New York, 1981, pp. 17–20.

4. V. Sitaramam and M. K. J. Sarma, *Proc. Natl. Acad. Sci. U.S.A.* **78**, 1981, 3441–3445.

5. D. Rickwood, Ed., *Centrifugation-A Practical Approach*, IRL Press, Oxford, UK, 1984, pp. 295–304.

6. P. Sheeler, *Centrifugation in Biology and Medical Science*, John Wiley & Sons, Inc., New York, 1981, pp 169–176.

7. C. A. Price. *Centrifugation in Density Gradients*, Academic Press, New York, 1982, p. 114.

8. C. A. Price. *Centrifugation in Density Gradients*, Academic Press, New York, 1982, pp. 114–126.

9. D. Rickwood, Ed., *Centrifugation-A Practical Approach*, IRL Press, Oxford, UK, 1984, pp. 168–189.

10. H.-W. Hsu, "Separation by Centrifugal Phenomena", *Techniques of Chemistry, Vol. XVI*, John Wiley & Sons, Inc., New York, 1981, pp. 220–228.

11. P. Lavrenko, V. Lavrenko, and V. Tsvetkov, in *Analitycal Ultracentrifugaton V*, H. Cölfen volume Ed., F. Kremer and G. Lagaly, Eds., *Progress in Colloid and Polymer Science*, Vol. 113, 1999, pp. 14–22.

12. R. Gauglitz, in *Analitycal Ultracentrifugaton*, J. Behlke guest Ed., F. Kremer and G. Lagaly, Eds., *Progress in Colloid and Polymer Science*, Vol. 99, 1995, pp.199–208.

13. P. Sheeler, *Centrifugation in Biology and Medical Science*. John Wiley & Sons, Inc., New York, 1981, pp. 23–25.

14. B. D. Young, in *Centrifugation-A Practical Approach*, D. Rickwood, Ed., IRL Press, Oxford, UK, 1984, p. 133.

15. H. K. Schachman, *Ultracentrifugation in Biochemisty*, Academic Press, New York, 1959, pp. 201–247.

16. V. J. Hindson, P. C. E. Moody, A. J. Rowe, and W. V. Shaw, *J. Biol. Chem.* **275**, 2000, 461–466.

17. H. K. Schachman, *Ultracentrifugation in Biochemisty*. Academic Press, New York, 1959, p. 91.

18. A. Møller-Larsen and T. Christensen, *J. Virol. Methods* **73**, 1998, 151–161.

19. J. M. Graham, J. A. Higgins, T. Gillott, T. Taylor, J. Wilkinson, T. Ford, and D. Billington, *Atherosclerosis* **124**, 1996, 125–135.

PROBLEMS

1. (a) Calculate the RCF maximum (RCF_{max}), minimum (RCF_{min}), and average (RCF_{av}), for a centrifugation tube in which the meniscus is 12 cm from the rotation axis and the bottom at 22 cm. The rotor is driven at 27,000 rpm.

 (b) If the rotor is driven at half-velocity (13,500 rpm) will the *g* force applied to the solution be approximately one-half?

2. (a) A titanium fixed-angle rotor can be run at a maximum speed of 52,000 rpm. It has $r_{max} = 10.8$ cm and $r_{min} = 3.2$ cm. Calculate the *k* (clearing) value.

 (b) Compare the obtained value with the ones presented in Table 13.2. Choose the best rotor for the separation of soluble proteins from a mammalian cell homogenate.

3. A vertical rotor data sheet specifies $k = 304$. The tubes for this rotor have an internal diameter of 2.4 cm, with $r_{min} = 13.6$ cm. What is the maximum speed for this rotor? (Note that in real situations, aging of the rotor and accumulation

TABLE 13.2. Values of k and Total Capacity for Ultraspeed Rotors[a]

Rotor Type	RCF_{max} (g)	RCF_{min} (g)	Total Capacity (mL)	k
Swinging-bucket	90,300	39,300	102	338
Swinging-bucket	285,000	119,000	84	137
Swinging-bucket	484,200	254,000	26.4	45
Fixed-angle (18°)	59,200	29,600	940	398
Fixed-angle (14°)	94,500	63,300	210	113
Fixed-angle (29°)	511,000	220,800	112	38
Vertical	70,000	50,400	312	123
Vertical	240,600	173,400	280	34
Vertical	510,000	416,600	40.8	8

[a] Superspeed rotor with RCF between 5000 and 50,000 g have k values between 400 and 5000 or higher.

of stress over the rotor structure must be considered, so the maximum velocity allowed for safe operation is gradually decreased.)

4. If the k factor is 300 for a swinging-bucket aluminum alloy rotor, with $r_{max} = 19.5$ cm and $r_{min} = 8$ cm, calculate k_{actual} for speeds of 10,000, 20,000, 30,000, and 40,000 rpm. The maximum speed for this rotor is 50,000 rpm.

5 How much time is necessary to pellet influenza virus particles ($s_{20,w} = 700$) in a rotor with $k = 180$? What is the pelleting time for eukaryotic ribosome subunits ($s_{20,w} = 40$ and 60 for small and large subunits, respectively)? Assume that the medium viscosity and density are similar to water at 20 °C.

6. A protein increases its distance from the rotation axis from 7.6 to 9.8 cm during a 4-h experiment, in which the rotor speed is 40,000 rpm. Calculate its sedimentation coefficient in Svedberg units.

CHAPTER 14

Chromatography of Biomolecules

14.1. INTRODUCTION

Chromatographic methods are used for the separation of components of a mixture present in a flowing *mobile phase*, based on their different strengths of interaction with a *stationary phase*. Stronger interaction with the stationary phase yields a longer retention time, which are the times between sample introduction and component elution. This chapter concerns liquid chromatography (LC), where the mobile phase is an aqueous solution, and the stationary phase is a particulate solid or a semisolid particulate gel. The separation mechanisms that will be addressed include size exclusion (gel filtration), affinity and ion-exchange chromatography. All of these methods are applicable to biological macromolecules present in aqueous solution, allow separation under mild conditions and may be used on either preparatory or analytical scales.

Classical LC employs a vertical column packed with a slurry of stationary-phase particles. The particles are allowed to settle into a densely packed bed, and excess mobile phase is removed. The sample is introduced at the top of the column, followed by clean mobile phase, and is gravity-fed through the column by the continuous removal of mobile phase from a small exit at the bottom of the column. Modern adaptations of this method include the addition of a postcolumn pump or the use of mild pressures at the top end of the column to speed component elution.

High-performance liquid chromatography employs very fine particles, and uses high-pressure pumps to force the mobile phase through the column. The mobile phase passes from a reservoir through the pump, an injector, the column and a flow-through detector to a waste collection reservoir. The injector allows the introduction of sample mixtures into the flowing mobile phase with little or no disturbance to the flow.

14.2. UNITS AND DEFINITIONS

Flow rates may be expressed in linear or volume units. Linear flow rates refer to the column length traveled by the mobile phase per unit time, and are usually expressed

Bianalytical Chemistry, by Susan R. Mikkelsen and Eduardo Cortón
ISBN 0-471-54447-7 Copyright © 2004 John Wiley & Sons, Inc.

in centimeter per minute (cm/min). Volume flow rates are commonly used, and are expressed in milliliters per minute (mL/min). The time between sample introduction and component elution is the retention time of the component and is symbolized t_r. The adjusted retention time, t_r' takes into account the time required for a nonretained solute to pass from injector to detector with no interactions:

$$t_r' = t_r - t_m \tag{14.1}$$

where t_m is the time required for the nonretained solute to reach the detector. An unretained solute has $t_r = t_m$, and thus has $t_r' = 0$. The relative retention of two components, 1 and 2, is called α and is defined in Eq. 14.2.

$$\alpha = t_{r2}'/t_{r1}' \tag{14.2}$$

where $t_{r2}' > t_{r1}'$, so that α values are always greater than unity. Under a given set of experimental conditions, an individual component has a characteristic capacity factor, k:

$$k = (t_r - t_m)/t_m = t_r'/t_m \tag{14.3}$$

A large value of k for a given component usually means that a good separation will be achieved; however, large k values imply long elution times. Values of k between 2 and 20 are generally considered useful. The capacity factor is related to the partition coefficient, K, of the solute (S) between the mobile and stationary phases, where $K = [S]_{stationary}/[S]_{mobile}$, and to the relative volumes of stationary and mobile phases at equilibrium, V_s and V_m, according to Eq. 14.4:

$$k = K(V_s/V_m) \tag{14.4}$$

The relative retention of two components, α, may also be expressed in terms of K:

$$\alpha = t_{r2}'/t_{r1}' = k_2/k_1 = K_2/K_1 \tag{14.5}$$

14.3. PLATE THEORY OF CHROMATOGRAPHY[1]

Consider a column that is divided into N segments of equal length, and that each segment is just long enough to allow complete equilibration of solute partitioning between the stationary phase and the mobile phase, according to its partition coefficient. Each of these segments is called a *theoretical plate*, and a column of length L will have a height equivalent to a theoretical plate (HETP) given by L/N. A "good" column will have large N values ($10^4 -> 10^5$), and a small HETP. Columns packed with a small-particle stationary phase have been shown to yield higher N values than those packed with larger particles.

Theoretical plates are determined experimentally for a given solute, by measuring the retention time and the peak width at the base:

$$N = (t_r/\sigma)^2 = 16(t_r/w)^2 \tag{14.6}$$

where σ is the standard deviation of the (Gaussian) solute elution profile, and w ($= 4\sigma$) is the width of the elution profile measured at the base of the peak, which is readily measured by triangulation of the peak. In practice, N values are determined periodically for a representative solute to check column performance for degradation.

The plate theory assumes that complete equilibration occurs in each of the N segments of the column. This assumption is not applicable to all solutes, and because of this, the rate theory was developed.

14.4. RATE THEORY OF CHROMATOGRAPHY[2]

The rate theory takes into account the finite rate at which the solute can equilibrate between the mobile and stationary phases. The shapes of the bands that are predicted by the rate theory depend on the rate of elution, the diffusion of the solute along the length of the column, and the availability of different paths for the solute molecules to follow. The value of the HETP now depends on v, the volume flow rate of the mobile phase, according to the van Deemter equation, Eq. 14.7:

$$\text{HETP} = A + B/v + Cv \tag{14.7}$$

where A, B, and C are constants for a given column/mobile phase/solute system. The van Deemter equation implies that HETP not only varies with flow rate, but also passes through a minimum at an optimum value of the flow rate.

Each of the terms in Eq. 14.7 can be conceptually associated with different phenomena present during a chromatographic separation. The first term, A, results from the different paths that the solute molecules in the mobile phase take in traveling around or through the stationary-phase particles; its value is independent of flow rate, and smaller particles yield smaller values of A and hence smaller HETP values. The second term, B/v, arises from longitudinal diffusion as a result of the flow profile, shown in Figure 14.1.

The fastest flow of solute occurs in the center of the column; closer to the column walls the flow is slower, and approaches zero in the stagnant layer contacting the walls. Solute in the center of the column at the front of the solute zone is

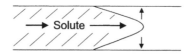

Figure 14.1. Flow profile for injected solute travelling through a packed stationary phase.

therefore present at higher concentration than solute near the walls at the same distance along the column. Because of this concentration gradient, solute diffuses toward the walls, and slows down. The result is that an injected plug of analyte develops more diffuse boundaries as it travels through the column. The final term in Eq. 14.7, Cv, results from the finite time that is required for equilibration of the solute between the mobile and stationary phases. If the solute zone moves rapidly but the solute cannot escape the stationary phase rapidly, then solute present in the mobile phase moves ahead, broadening the band.

In an efficient column, the values of A, B, and C are all small, resulting in a small (10^{-4}–$<10^{-5}$ m) HETP value. Very small stationary-phase particles, of the order of 3–10 μm, have been shown to yield the best efficiencies.

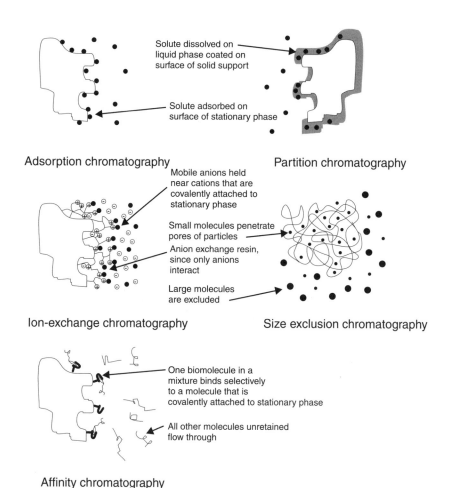

Figure 14.2. Retention mechanisms for the five major classes of chromatography.

The resolution of two components may be quantitated using Eq. 14.8:

$$\text{Resolution} = \Delta t_r / w_{av} = \frac{\text{Difference in retention times}}{\text{Average baseline peak width}} \tag{14.8}$$

Resolution values have been shown to increase with the square root of column length. Thus, doubling the length of a column will increase resolution by a factor of $(2)^{1/2}$.

Figure 14.2 shows the five major classes of chromatography in common use. Each is based on a unique retention mechanism, and three of these mechanisms are of particular interest for the separation and quantitation of biological macromolecules. Size-exclusion chromatography (SEC), also called gel filtration, separates species on the basis of molecular size, with small molecules being retained and large species eluting first. Affinity chromatography involves selective binding interactions such as antibody–antigen and enzyme–substrate interactions. Ion-exchange chromatography is used to separate species on the basis of molecular charge and the distribution of molecular charge; the stationary phase possesses either positive or negative charge, and interacts only with oppositely charged solutes. All three of these retention mechanisms may be applied to macromolecular solutes that are present in aqueous solutions, and function well under mild conditions of temperature and pH.

14.5. SIZE EXCLUSION (GEL FILTRATION) CHROMATOGRAPHY

Neutral, particulate gel media have been shown to give separations related to the size of the components of a mixture. This was discovered independently for polymers dissolved in organic solvents, and for biomolecules in aqueous media. Separation is nondestructive, and will occur under mild conditions. In addition, separations are usually independent of the composition of the mobile phase; thus, the mobile-phase composition may be selected to suit the species being separated, for the stability of these components. One disadvantage of this independence is that gradients with varying eluting power do not exist for gel filtration separations.

The stationary phases used for SEC are particulate gels, in which the solute can penetrate through the entire macroscopic volume of the packed particles. The separation conditions are chosen to avoid both specific and nonspecific adsorption, so that elution volumes depend only on the sizes of the solutes, as a result of molecular sieving. Elution volumes are never greater than the total bed volume of the stationary phase. Large solutes elute first, with smaller species eluting later, and it is not unusual to achieve 100% recovery of a purified solute. The first gel used for the separation and purification of proteins was Sephadex^{TM}, and this gel medium is still in routine use. It consists of particles of cross-linked dextran, and has been prepared with different pore sizes, to give different fractionation ranges.

The SEC separation is governed by Eq. 14.9.

$$V_e = V_o + K_{av} V_g \tag{14.9}$$

where V_e is the elution volume of the solute, V_o is the void volume of the column (the volume occupied by the mobile phase), V_g is the volume occupied by the gel, and K_{av} (av = available) is the partition coefficient for the solute between the gel phase and the surrounding liquid phase.[3] If only steric exclusion governs the separation,

$$0 \leq K_{av} \leq 1 \qquad (14.10)$$

Figure 14.3 shows a representative separation of species of high, intermediate, and low molecular weights.

High MW species are too large to enter the pores of the gel, and so elute with the void volume. Very low MW species enter all pores, and are the last to elute. Species of intermediate MW are separated based on their average abilities to enter and exit the pores of the gel, with larger molecules tending to be excluded more often. The retention coefficient is given by

$$R = V_o/V_e \qquad (14.11)$$

where R is small for highly retained (small) solutes, and approaches unity for high molecular weight solutes.

The shapes of solutes are also important in their retention behavior; it has been shown that DNA restriction fragments (rod-shaped) have K_{av} values that are more sensitive to molecular weight than those obtained with denatured (random-coil) proteins. In fact, the SEC parameter governing the retention is the hydrodynamic volume of the solute, which is related to its radius of gyration, R_g. The molecular weight of a solute is related to its radius of gyration by Eq. 14.12:

$$R_g = k(MW)^a \qquad (14.12)$$

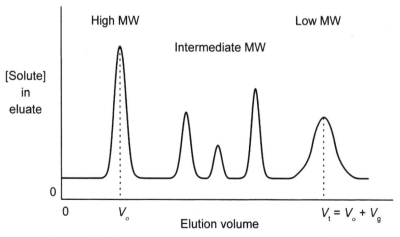

Figure 14.3. Representative chromatogram obtained with SEC.

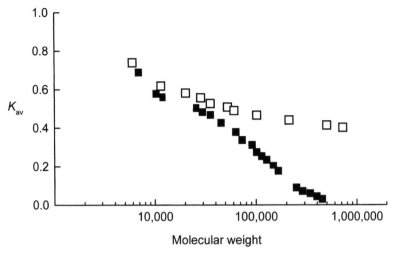

Figure 14.4. Selectivity curve from SEC of DNA (■) and protein (□) molecular weight standards on a 106 × 10 mm Superose 6 gel filtration column with 0.02 M Tris-HCl pH 7.6 containing 0.15 M NaCl as the eluent.[4] [Reprinted, with permission, from H. Ellegren and T. Låås, *Journal of Chromatography* **467**, 1989, 217–226. "Size - Exclusion Chromatography of DNA Restriction Fragments. Fragment Length Determinations and a Comparison with the Behaviour of Proteins in Size-Exclusion Chromatography". © 1989 Elsevier Science Publishers B.V.]

where $a = 1$ for rods, $a \approx 0.5$ for flexible coils, and $a = 0.3$ for spheres. From this equation, it is apparent that the radius of gyration increases more rapidly with molecular weight for rods (DNA) than for spheres (proteins). Figure 14.4 shows the dependence of K_{av} on log(MW) for a series of protein and DNA standards.[4]

For proteins, a mathematical model of retention has been developed that works well for Sephadex gels.[5] The solute is treated as a sphere of radius r_s, while the gel is a network represented by infinitely long, straight rods of radius r_x. The rods are randomly distributed, and have an average density of L units of rod length per unit volume of gel. The values of L and r_x may be calculated from known dimensions of dextran chains, and then K_{av} may be found from Eq. 14.13:

$$K_{av} = \exp\{-\pi L(r_x + r_s)^2\} \tag{14.13}$$

Equation 14.13 has been used to evaluate the hydrated radii of proteins by determining K_{av} experimentally on three Sephadex gels, G-75 (fractionation range 3–80 kDa), G-100 (fractionation range 4–150 kDa) and G-200 (fractionation range 5–600 kDa). The results of these experiments are shown in Table 14.1.

In addition to steric exclusion, other types of interactions between solute and gel are possible. In particular, electrostatic interactions may be very significant with some gels, especially if the gel has charged groups and can act as an ion exchanger. If the eluent has low ionic strength, these charged groups create a Donnan potential

TABLE 14.1. Determination of r_s for Proteins Using SEC on Sephadex[a]

| | K_{av} on Sephadex | | | |
Substance	G-75	G-100	G-200	r_s (Å)
Cytochrome c	0.43	0.59	0.72	16.4
Human serum albumin	0.04	0.19	0.41	36.1
Human immunoglobulin G	0.00	0.05	0.21	55.5

[a] See Ref. 5.

between the inside and outside of the gel particles. This can happen if, for example, a solute has a nonzero net charge but is too large to enter the pores, while its small counterions can enter the gel. This results in ion-exchange properties in localized areas of the gel. These types of electrostatic interactions may be suppressed by increasing the ionic strength of the eluent, which can be accomplished using volatile electrolytes that are readily removed from the separated components following the chromatographic run.

Adsorption of the solute onto the gel matrix can also occur, but is not normally encountered in SEC. It may be counteracted by changing the eluent. The ideal eluent is a good solvent for both the solute and the substance forming the gel matrix. In addition, hydrogen-bond-breaking agents such as urea or guanidinium, or chaotropic ions, or surfactants may be added to eluents to improve SEC characteristics.

To test for either adsorptive or electrostatic interactions, the SEC separation is performed at a variety of temperatures. If the separation occurs by size alone, the retention coefficient $R(= V_o/V_e)$ is independent of temperature; very small variations may be observed as a result of gel swelling or microstructural changes to the gel. The presence of a significant dependence of R on T indicates the presence of a mechanism other than size exclusion. While R should not vary with T, diffusion coefficients increase with T and so zone broadening occurs, leading to decreased resolution with increasing separation temperatures.

Normally, an approximately linear relationship is observed between K_{av} and log(MW) over the fractionation range of the gel. The expanded relationship is, in fact, sigmoidal, as shown in Figure 14.5. The calibration curve can be linearized using the logit transformation, where $logit(K_{av}) = \ln(K_{av}/(1 - K_{av}))$ is plotted against log(MW).

Resolution and separations in SEC are described by the van Deemter equation (Eq. 14.7). The main contributions to zone broadening, in order of importance, are as follows. The first factor involves the kinetics of partitioning between the mobile phase and the gel; there is a limited rate at which equilibrium can be established, and this rate depends on the solute's diffusion coefficient. Second, differences in the lengths of different stream paths in the packed bed of irregularly shaped particles result in "eddy diffusion"; with an ideally packed column, this is not a significant problem, but in practice, eddy diffusion may be significant. Finally, longitudinal

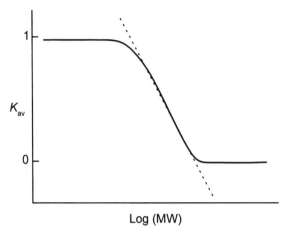

Figure 14.5. Calibration plot of K_{av} against log(MW).

diffusion within the column contributes to band broadening, but this is a relatively small effect.

The effect of flow rate has been examined as a factor that influences bandwidth, and therefore resolution. Figure 14.6 shows a typical plot of band width as a function of flow rate. At low flow rates, longitudinal diffusion is significant, and leads to larger bandwidths than at the optimum. At high flow rates, the kinetics of partitioning becomes the dominant factor that leads to increased bandwidths and decreased resolution. The contribution of eddy diffusion is independent of flow rate.

The size of the particles used in the stationary phase has also been shown to exert a significant effect on resolution. Small particles result in fast equilibration of solutes, as well as smaller contributions of eddy diffusion to band broadening. There is a practical limitation in the use of small particles, however. The smaller

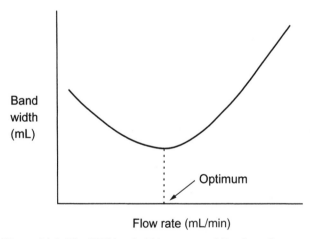

Figure 14.6. The SEC bandwidth versus mobile phase flow rate.

the particle, the higher is the resistance to the flow of mobile phase. Thus, small particles require higher pressures to elute solute at a given flow rate. The gels used in SEC stationary phases are usually soft, or elastic, so that higher pressures can result in compression and column blockage.

The separation temperature also affects resolution, in that diffusion coefficients increase with temperature, so that partitioning equilibria are faster at higher T. This equilibration time is the factor that usually limits resolution.

The length of the column is important, because the number of theoretical plates (N) increases with the square root of column length; however, practical limitations are imposed on column length. Components eluted from the column are diluted with respect to their initial concentrations, and reasonable separations yield dilution factors of $\sim$2, for standard column lengths (20 cm).

SEC mobile phase composition is, in principle, not an important factor. In practice, however, electrostatic effects may be important. Under physiological conditions (0.9% NaCl), Donnan effects are usually negligible. The eluent pH may be altered to change the net charge on macromolecules, to minimize Donnan effects. For example, the separation of nucleic acids is normally performed at low pH, so that some phosphate groups are protonated. In some cases, charged groups on the stationary phase may be chemically removed prior to column packing. For protein separations, denaturing additives, such as urea or guanidinium–HCl, are used to produce random coils of polypeptides for molecular weight determinations. SDS and disulfide bond reducing agents may also be added to the mobile phase.

Finally, the viscosity of the samples may be important if macromolecule concentrations are high enough to contribute to band broadening as a result of unstable flow.

14.6. GEL MATRICES FOR SIZE EXCLUSION CHROMATOGRAPHY[6]

Gels used for bioanalytical SEC must have hydrophilic properties. Polysaccharides, such as dextran and agarose, as well as synthetic polymers, such as polyacrylamide, have been used successfully for SEC. The monomers and low molecular weight polymer forms of these polymers are water soluble, and must be cross-linked to create insoluble gel beads. Some types of gels may be dried and rewetted reversibly, such as dextran. These are called aerogels, and are known to maintain their open structure when they are dehydrated. Xerogels, on the other hand, shrink when dried, and swell when they are rewetted, so that alterations in pore size may occur. Table 14.2 lists some of the commercially available SEC stationary-phase materials, along with their molecular weight fractionation ranges for globular proteins.

The trade names of the stationary phase media represent particular polymers formed under well-controlled conditions. Sephadex^{TM} consists of dextran, cross-linked with epichlorohydrin in alkaline solution. Sephacryl^{TM} is an allyl dextran, cross-linked with N,N'-methylenebis(acrylamide). Sepharose^{TM} is a beaded form of agarose gel, where aggregates of the polysaccharide chains are formed.

TABLE 14.2. SEC (Gel Filtration) Stationary Phase Materials

Name (^{TM})	MW Fractionation Range (kDa)
Sephadex G-10	−0.7
Sephadex G-15	−1.5
Sephadex G-25	1.0–5.0
Sephadex G-50	1.5–3.0
Sephadex G-75	3.0–80
Sephadex G-100	4.0–150
Sephadex G-150	5.0–300
Sephadex G-200	5.0–600
Sephacryl S-100	1.0–75
Sephacryl S-200	3.0–160
Sephacryl S-300	5.0–1000
Sephacryl S-400	20–20,000
Sephacryl S-500	20–3,000,000
Sepharose 2B	70–40,000
Sepharose 4B	60–20,000
Sepharose 6B	10–4000
Bio-Gel P-2	0.1–1.8
Bio-Gel P-4	0.8–4.0
Bio-Gel P-6	1.0–6.0
Bio-Gel P-10	1.5–20
Bio-Gel P-30	2.5–40
Bio-Gel P-60	3.0–60
Bio-Gel P-100	5.0–100
Bio-Gel P-150	15–150
Bio-Gel P-200	30–200
Bio-Gel P-300	60–400
Bio-Gel A-0.5m	<10–500
Bio-Gel A-1.5m	<10–1500
Bio-Gel A-5m	<10–5000
Bio-Gel A-15m	40–15,000
Bio-Gel A-50m	100–50,000
Bio-Gel A-150m	1000–150,000

Bio-Gel P^{TM} is a beaded polyacrylamide gel, and has separation properties similar to Sephadex^{TM}. Bio-Gel A^{TM} is a beaded agarose gel, like Sepharose; it has the largest pore sizes known, and is even capable of separating small virus particles.

14.7. AFFINITY CHROMATOGRAPHY

Chromatography may be considered to have two extremes. At one extreme, long, efficient columns of low-to-moderate selectivity are used for the separation of complicated sample mixtures into many component peaks or bands. This is the case for (a) normal-phase chromatography, where the stationary phase is hydrophilic (silica, alumina) and the mobile phase is hydrophobic; (b) reversed-phase chromatography, where aqueous or semiaqueous mobile phases are used with a hydrophobic stationary phase such as silica-bound C18; (c) ion-exchange chromatography; and (d) gel filtration.

The other extreme involves the use of short, inefficient columns of very high selectivity to separate a small number of components (often only one) from hundreds of unretained solutes, which is the basis for affinity chromatography. As the name implies, affinity chromatography relies on the specific binding interactions that occur between biochemical recognition agents and their ligands. The high selectivities afforded by these interactions can yield separations in a minute or less, using columns as short as 1 cm.

Affinity ligands are those species that are bound to a solid support to form the stationary phase. They may be "specific" in the sense that an antibody binds to only one solute or epitope, or they may be "general", such as nucleotide analogues or lectins, that bind to certain groups of solutes. Affinity ligands may have a low or high molecular weight: For example, an immobilized antibody is an affinity ligand for antigen purification and quantitation, while an immobilized antigen may be used as an affinity ligand for antibody purification. Table 14.3 shows examples of a number of general affinity ligands and their selectivities.

As an example of the steps involved in affinity chromatography, we will consider the purification of an enzyme present in a mixture of proteins, using an immobilized competitive inhibitor as an affinity ligand. The first step after column preparation is the application of the sample; during this step, the enzyme that binds to the inhibitor is retained, or "adsorbed" on the column. This is followed by a rinse step, whereby all nonbinding species are removed. The third step involves elution, and this step

TABLE 14.3. General Affinity Ligands and Their Binding Partners[a]

Ligand(s)	Selectivity
Cibachron blue dye, or derivatives of AMP, the NADH, or NADPH	Dehydrogenase enzymes, through binding at NAD(P)H binding site
Lectins: concanavalin A, lentil lectin, wheat germ lectin configurations	Polysaccharides, glycoproteins, glycolipids, and membrane proteins with certain sugar
Trypsin inhibitor, methyl esters of amino acids, or D-amino acids	Proteases
Phenylboronic acid	Species containing cis-diol groups, such as glycosylated hemoglobins, sugars, and nucleic acids
Protein A	Immunoglobulins, binding through F_c fragment
Nucleic acids, nucleosides or nucleotides	Nucleases, polymerases, nucleic acids

[a] See Ref. 7.

must involve a reversal of enzyme–inhibitor binding; this may be affected by a change in pH, or the use of an eluent containing free substrate or inhibitor. Finally, the column is regenerated by returning to the initial mobile-phase composition. Classical affinity chromatography employs a gravity feed of the mobile phase through a vertical column, while high-performance affinity chromatography (HPAC) uses a pump to force mobile phase through the column. HPAC allows the use of smaller stationary-phase particles, yielding columns with higher efficiencies, and the same basic equipment is used as in HPLC methods (pump, injector, column, detector, recorder).

Ideal support media for affinity chromatography are rigid, stable, have a high surface area, and do not adsorb any species nonspecifically. Classical AC employs organic gels, such as dextran, agarose, cellulose, or polyacrylamide for stationary-phase materials. For example, Sepharose 6B has 6% agarose, a 100-μm particle size, pores averaging 150 Å in diameter, and an operating pH range of 4–7. It can withstand only 1 psi pressure, but can be cross-linked with 2,3-dibromopropanol to improve its chemical and mechanical stability. The cross-linked form can be used over the pH 2–9 range, at up to 6 psi pressure.

The so-called "medium performance" AC supports can withstand higher pressures, and have particle sizes of ∼40 μm. Examples include hydrophylic polyacrylamides (pH 1–14, 100 psi), as well as the bonded-phase silica/controlled-pore glass supports (pH 2–8, >1000 psi).

High-performance affinity chromatography employs bonded-phase silica with particle sizes of 10 μm, having pore sizes of 300 Å, as well as nonporous particles of 1.5-μm diameter or less. With these support media, pressures in excess of 2500 psi may be used, and these are the pressures commonly encountered in HPLC systems.

14.7.1. Immobilization of Affinity Ligands

Affinity ligands are immobilized through covalent binding to the support material. A dense, stable coverage of the support is desired, and most methods for immobilization use two steps: support activation and coupling to the affinity ligand (see Section 4.2 regarding enzyme immobilization methods).

An effective immobilization method must consider the following factors. The functional groups on the ligand that are available for coupling may be one or more of amine, thiol, alcohol, and carboxylic acid. Spacer arms are often needed when small solutes are immobilized so that the immobilized ligand may access the binding site of a macromolecule; these spacers are often substituted hydrocarbon chains, and may alter the binding properties through hydrophobic or ionic interactions (neutral hydrophilic spacers such as polyethylene glycol have been very effective). The pH during immobilization must be controlled, since this determines the reactivity of the functional groups and may cause irreversible damage of the ligand or support at extreme values. Finally, the density of active groups on the support must be considered: multipoint attachment of large ligands to a support yields high stability, but may distort the binding site and decrease affinity; and improper

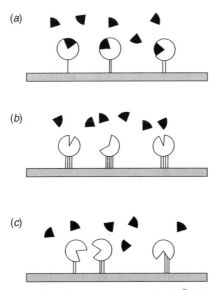

Figure 14.7. Effects of immobilization on a macromolecule.[8] (*a*) Ideal immobilization; (*b*) Altered three-dimensional (3D) structure; and (*c*) improper orientation and spacing.

orientation or spacing of the ligands may also lead to reduced affinity, so that immobilization conditions must be optimized. Figure 14.7 illustrates the effects of some of these factors.[8]

Many support materials are commercially available in a preactivated form, requiring only the coupling of the desired affinity ligand. These materials already have a spacer arm incorporated, and possess an active functionality for ligand immobilization. Table 14.4 lists a number of reactions of activated supports with ligand functional groups. For all of these illustrated reactions, the R group of the ligand may be a small molecule, such as a hapten or inhibitor, or a macromolecule, such as a protein, carbohydrate, or nucleic acid that has been modified, if necessary, to possess the indicated functional group.

14.7.2. Elution Methods

Elution methods are classified as either biospecific or nonspecific. Biospecific elution is accomplished by the addition of a low molecular weight inhibitor or competitor to the mobile phase, causing competitive binding. For example, the binding between immobilized glucosamine and a lectin may be reversed by the addition of glucose as a modifier to the mobile phase; glucose in the eluent interacts with the lectin binding sites, and a lectin–glucose complex is eluted. The immobilized glucosamine and the soluble glucose in the eluent compete for lectin (analyte) binding sites. On the other hand, if the lectin is immobilized for the purification of a glycoprotein, glucose may again be used as a mobile-phase modifier, but in this case the glucose will displace the glycoprotein from the column by competition with the

TABLE 14.4. Reactions of Activated Support Materials with Affinity Ligands

Activation Method	Reaction with Affinity Ligand
Cyanogen bromide	$-O-C{\equiv}N$ + RNH_2 → $-O-\underset{\underset{NH}{\|\|}}{C}-NHR$
Active ester (NHS)	$-\overset{O}{\overset{\|\|}{C}}-O-N$ (ring) + RNH_2 → $-\overset{O}{\overset{\|\|}{C}}-NHR$
Epoxide	$-\overset{O}{\overset{\diagup\diagdown}{CH-CH_2}}$ + RNH_2 (RSH, ROH) → $-\overset{OH}{\overset{\|}{CH}}-CH_2-NHR$
Tresyl chloride	$-CH_2-OSO_2-CH_2CF_3$ + RNH_2 (RSH) → $-CH_2-NHR$
Carbonyldiimidazole	$-O-\overset{O}{\overset{\|\|}{C}}-N{\diagup}N$ + RNH_2 → $-O-\overset{O}{\overset{\|\|}{C}}-NHR$
Thiol/disulfide exchange	$-S-S-$ (pyridyl) + RSH → $-S-S-R$
Diazonium	(phenyl)$-\overset{+}{N}{\equiv}N$ + $R-$(phenyl)$-OH$ → (phenyl)$-N{=}N-$(phenol)R

analyte (glycoprotein) for immobilized lectin binding sites. In this case, the free glycoprotein elutes.

Nonspecific elution methods are also common in affinity chromatography. These methods are used to induce binding reversal, and may consist of one or more of the following changes to the mobile phase: (a) pH change (e.g. from 7 to 3); (b) addition of denaturants such as urea or guanidinium; (c) addition of chaotropic ions such as thiocyanate or perchlorate; (d) addition of organic solvents; or (e) change of ionic strength. Elution conditions must be determined empirically, and must be mild enough so that the support, ligand, and analyte are not irreversibly damaged. Nonspecific elution methods are generally faster than specific methods, because the rate of dissociation of bound analyte from the support is increased. This results in less band broadening, or a more concentrated sample in the eluted zone.

Figure 14.8 illustrates an affinity chromatographic separation of the five isoenzymes of human lactate dehydrogenase, using immobilized AMP as an affinity ligand, and biospecific elution using a continuous gradient of NADH concentration.[9] AMP is a competitive inhibitor of lactate dehydrogenase, and binds at the NADH binding site. AMP was bound to a Sepharose stationary phase, to

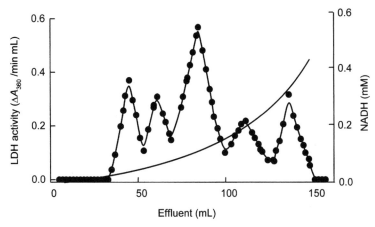

Figure 14.8. Bioelution of LDH isoenzymes from a Sepharose–AMP affinity column using a concave NADH gradient. [Reprinted, by permission, from P. Brodelius and K. Mosbach, *FEBS Lett.* **35**, 1973, 223–226. "Separation of the Isoenzymes of Lactate Dehydrogenase by Affinity Chromatography Using an Immobilized AMP-Analogue". The exclusive © for all languages and countries is vested in the Federation of European Biochemical Societies.]

generate an N^6-(6-aminohexyl)adenosine-5′-monophosphate linkage. This type of separation may be accomplished using biospecific elution methods, but would not be possible with nonspecific elution, where significant changes in the structures of all of the isoenzymes may be expected to occur during elution, yielding one major elution zone.

14.7.3. Determination of Association Constants by High-Performance Affinity Chromatography[10]

Association constants for the reactions of biomolecules with their binding partners can be evaluated using affinity chromatography. Basically, an immobilized affinity ligand competes with a soluble ligand for the binding sites of the biomolecule, and the retention volume of the biomolecule is measured as a function of the concentration of soluble ligand. Samples injected contain a fixed concentration of the biomolecule, and varying concentrations of soluble ligand.

The theoretical approach to relate retention volumes to association constants is based on a model that consists of a monomeric biomolecule that has a single binding site (e.g., it is applicable to the F'_{ab} fragments of antibodies, but not to intact immunoglobulins). The biomolecule interacts with an immobilized ligand and a soluble ligand, each of which also has a single binding site. The model is depicted in Figure 14.9. Because of the general applicability of the model, the biomolecule is represented as A, and the immobilized ligand and soluble ligand are shown by the symbols X and L, respectively. The species to consider in developing the relationships are therefore A, X, L, AX, and AL, and they are related through

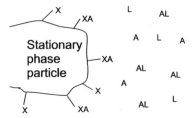

Figure 14.9. Model of species present during the biospecific elution of biomolecule A from a column having immobilized affinity ligand X, using ligand L.

the two equilibrium expressions:

$$A + L \rightleftharpoons AL, \qquad K_{AL} = [AL]/([A][L]) \qquad (14.14)$$

and

$$A + X \rightleftharpoons AX, \qquad K_{AX} = \{AX\}/([A]\{X\}) \qquad (14.15)$$

where molar concentrations are represented by square brackets (mol/L), and surface concentrations are shown with parentheses (mol/m^2 or mol/cm^2). The only species present on the surface of the stationary-phase particles that are represented using surface concentrations are X and AX.

The only assumptions needed are (a) that A, X, and L are univalent (this implies that identical epitopes are present on X and L for binding to A), (b) that the soluble ligand L does not interact with X or XA, and (c) that neither A nor L interact non-specifically with the stationary-phase particles.

From expressions for mass balance, the total concentration of A in the mobile phase, $[A]_t$, and the total quantity of A bound to the stationary phase, Q_A, can be represented by Eqs. 14.16 and 14.17:

$$[A]_t = [A] + [AL] \qquad (14.16)$$

$$Q_A = \{AX\} \times \text{(surface area)} \qquad (14.17)$$

where $[A]_t$ is a concentration represented in molar units and Q_A is a total quantity given in moles. Since the total quantity of immobilized ligand, Q_X, is represented in a form similar to Eq. 14.17 as

$$Q_X = (\{X\} + \{AX\}) \times \text{(surface area)} \qquad (14.18)$$

then the ratio of Q_A to Q_X is simply the surface concentration ratio:

$$Q_A/Q_X = \{AX\}/(\{X\} + \{AX\}) \qquad (14.19)$$

By using the expression for K_{AX} given in Eq. 14.15 to substitute for $\{AX\}$, and rearranging to cancel $\{X\}$ yields

$$Q_A/Q_X = [A]K_{AX}/(1 + [A]K_{AX}) \tag{14.20}$$

Now, Eq. 14.16 can be used to substitute $([A]_t - [AL])$ for $[A]$, and K_{AL} (Eq. 14.14) is introduced to substitute for $[AL]$, yielding Eq. 14.21:

$$Q_A = (Q_X[A]_t K_{AX})/(1 + [A]_t K_{AX} + [L]K_{AL}) \tag{14.21}$$

A general expression that relates retention volumes to Q_A for many forms of chromatography is shown in Eq. 14.22:

$$V_r - V_o = (\delta Q_A/\delta[A]_t)_{[A]_t=0} \tag{14.22}$$

where V_r is the uncorrected retention volume, V_o is the void volume of the column, and the derivative is evaluated at $[A]_t = 0$ so that the availability of binding sites on the stationary phase does not limit the reaction. Taking the derivative of Eq. 14.21 with respect to $[A]_t$, and evaluating it at $[A]_t = 0$ yields Eq. 14.23:

$$V_r - V_o = (Q_X K_{AX})/(1 + [L]K_{AL}) \tag{14.23}$$

which rearranges to the final form used in the experimental determination of K_{AL}:

$$1/(V_r - V_o) = 1/(Q_X K_{AX}) + [L]K_{AL}/(Q_X K_{AX}) \tag{14.24}$$

In Eq 14.24, it can be seen that V_o, Q_X, K_{AX}, and K_{AL} are all constants. Retention volumes are measured as a function of $[L]$, and a plot of $1/(V_r - V_o)$ against $[L]$ is constructed. This plot is linear, and the ratio of the slope to the y intercept yields the value of K_{AL}. Note that this method does not require knowledge of either Q_X or K_{AX} for the determination of K_{AL}. It should also be noted that $[L]$ is the free ligand concentration ($[L] = [L]_t - [AL]$), but in practice, $[L] \approx [L]_t$ because $[L]_t$ is usually much greater than $[A]_t$.

An example of this type of experiment is shown in Figure 14.10.[11] The information desired is the association constant for the binding of ribonuclease (RNase) and thymidine 5'-monophosphate (5'-TMP). A stationary phase was prepared using Sepharose, with TMP bound to yield thymidine 3'-(p-Sepharose-aminophenylphosphate)-5'-phosphate (i.e., 5'-TMP was immobilized through the 3'-hydroxyl group onto the stationary phase). Purified RNase was injected into mobile phases containing free 5'-TMP concentrations ranging from 7.5×10^{-5} to 5.0×10^{-4} M. In this experiment, constant-volume fractions of the eluent were collected, and the absorbances of the eluted fractions were measured at 260 nm. A more modern version of this experiment would simply use a UV detector with a flow-through cell, and monitor absorbance as a function of time at constant flow rate; the elution volumes

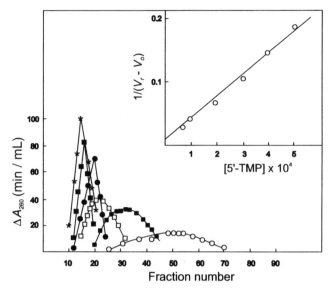

Figure 14.10. Elution profile of RNase as a function of free [5′-TMP], using an affinity column with immobilized 5′-TMP. The concentrations of 5′-TMP in the mobile phase were $5.0 \times 10^{-4}\,M, 4.0 \times 10^{-4}\,M, 3.0 \times 10^{-4}\,M, 2.0 \times 10^{-4}\,M, 1.0 \times 10^{-4}\,M$, and $7.5 \times 10^{-5}\,M$ for the earliest to the latest eluted fractions, respectively. The inset shows the plot according to Eq. 14.24 that was used to determine the association constant for the RNase-5′-TMP binding reaction.[11] [Reprinted, with permission, from B. M. Dunn and I. M. Chaiken, *Biochemistry* **14** (No. 11), 1975, 2343–2349. "Evaluation of Quantitative Affinity Chromatography by Comparison with Kinetic and Equilibrium Dialysis Methods for the Analysis of Nucleotide Binding to Staphylococcal Nuclease". © 1975 by American Chemical Society.]

of RNase would then be calculated by multiplying the flow rate by the retention times. It is readily apparent from Figure 14.10 that higher free 5′-TMP concentrations yield smaller retention volumes, since the RNase elutes in earlier fractions. The inset shows a plot of $1/(V_r - V_o)$ against [5′-TMP], according to Eq. 14.24, and the value of K_{AL}, determined by dividing the slope of this plot by the y intercept, is $3.8 \times 10^4\,M^{-1}$.

14.8. ION-EXCHANGE CHROMATOGRAPHY

Ion-exchange chromatography is based on the equilibration of solute species between the solvent, or mobile phase, and charged sites fixed on the stationary phase. Anion-exchange columns have cationic sites immobilized, and anionic

TABLE 14.5. Functional Groups on Representative Ion-Exchange Gels

Type	Classification	Name	Functional Group
Cation	Strong acid	Sulfopropyl (SP)	$-OCH_2CH_2CH_2SO_3^-H^+$
	Weak acid	Carboxymethyl (CM)	$-OCH_2COO^-H^+$
Anion	Strong base	Triethylaminoethyl (TEAE)	$-OCH_2CH_2N^+(CH_2CH_3)_3$
	Intermediate base	Diethylaminoethyl (DEAE)	$-OCH_2CH_2N(CH_2CH_3)_2$
	Weak base	p-Aminobenzyl (PAB)	$-OCH_2(C_6H_4)NH_2$

solutes are retained; cation-exchange chromatography uses columns with immobilized anionic sites, so that cationic species are retained. For biopolymer separations, ion-exchange columns use stationary-phase gels such as cellulose, dextran, agarose, and polyacrylamide. These gel media are derivatized to create immobilized charged groups. Ion-exchangers are classified as either strong or weak, depending on whether the quantity of immobilized charge may easily be controlled through the pH of the mobile phase. Table 14.5 shows representative functional groups generated on the gels by chemical methods, and their classifications.

For any separation by ion-exchange chromatography, the choice of strong or weak exchanger depends on the operating pH and the selectivity required. Weak acids are protonated at $pH \leq 4$, so that they lose their ion-exchange capacity at low pH; similarly, weak and intermediate base ion-exchangers lose their charge at high pH. In addition, high ionic strength mobile phases tend to decrease the strength of interaction between charged solutes and the stationary phase. Therefore, pH and/or ionic strength gradients are used to elute components of a mixture. With cation-exchange chromatography, gradients of high to low pH, or of low to high ionic strength are used (e.g., increasing [HCl] throughout the run). Anion-exchange columns use gradients of low to high pH or of low to high ionic strength (e.g., increasing [NaOH] during the run).

The classical model of selectivity[12] in ion-exchange chromatography is based upon stoichiometric exchange; for example, the competition of Na^+ and Li^+ for sites on a cation-exchanger (R^-):

$$R^-Na^+ + Li^+ \rightleftharpoons R^-Li^+ + Na^+ \tag{14.25}$$

The *selectivity coefficient* for this classical model is the thermodynamic equilibrium constant for this reaction:

$$K = [R^-Li^+][Na^+]/[R^-Na^+][Li^+] \tag{14.26}$$

This classical model works well for small, inorganic ions, but does not adequately describe the interactions of biological macromolecules, such as proteins and nucleic acids, with ion-exchange resins. With these polyelectrolyte species, electrostatic interactions do not follow simple stoichiometric laws, and the multiple charges are not all simultaneously accessible to the ion-exchanger. For this reason, a more elaborate model has been developed to account for the behavior of polyelectrolytes.

14.8.1. Retention Model for Ion-Exchange Chromatography of Polyelectrolytes[13]

An expression relating the capacity factor $k = (V_r - V_o)/V_o$ (and Eq. 14.3) to the Gibbs free energy of retention, ΔG_{ret}, has been developed as a general relation that applies to all retention mechanisms. This expression is shown in Eq. 14.27:

$$\ln(k) = -\Delta G_{ret}/RT + \ln(\phi) \qquad (14.27)$$

where ϕ is the phase ratio of the column, and is equal to the volume occupied by the stationary phase divided by that occupied by the mobile phase (the void volume) of the column:

$$\phi = V_s/V_o \qquad (14.28)$$

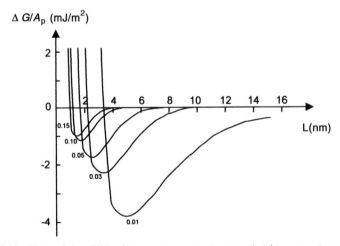

Figure 14.11. Plots of the Gibbs free energy per unit area, $\Delta G/A_p$, as a function of the distance between the two oppositely charged surfaces, L, at different ionic strengths. The results were calculated using $\varepsilon_r = 80$, and by setting -0.16 and $+0.03$ C/m^2 as the charge densities of the stationary and mobile phases, respectively.[13] [Reprinted, with permission, from J. Ståhlberg, B. Jönsson, and C. Horváth, *Anal. Chem.* **63**, 1991, 1867–1874. "Theory for Electrostatic Interaction Chromatography of Proteins". © 1991 by American Chemical Society.]

and the value of φ is considered a constant for a given column. The value of ΔG_{ret} varies with experimental conditions, so that capacity factors vary with pH, ionic strength and temperature, depending on the susceptibility of the retention equilibrium to these parameters.

A model has been developed to describe the dependence of ΔG_{ret} on electrostatic interactions between the stationary phase and a polyelectrolyte molecule. It is based on the assumption of an interaction between two charged, flat surfaces that are in contact with a buffered salt solution. This purely electrostatic treatment yields Eq. 14.29:

$$-\Delta G_{ret} = A_p \sigma_p^2 \{RT\}^{1/2} / (F\{2 I \varepsilon_0 \varepsilon_r\}^{1/2}) \tag{14.29}$$

where A_p is the area of the protein (or polyelectrolyte) surface that interacts with the stationary phase (assumed to be one-half of the total surface area of a sphere), σ_p is the charge density on the protein (the total charge at a given pH divided by total

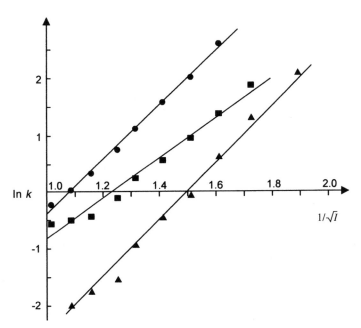

Figure 14.12. Plots of the logarithmic retention factor of proteins as a function of the reciprocal square root of the ionic strength in ion-exchange chromatography of α-chymotrypsinogen (square), lysozyme (circle), and cytochrome c (triangle). A Zorbax Bioseries WCX300 (weak cation-exchanger) was used at pH 6.0, with $(NH_4)_2SO_4$ as the eluting salt.[13] [Reprinted, with permission, from J. Ståhlberg, B. Jönsson, and C. Horváth, *Anal. Chem.* **63**, 1991, 1867–1874. "Theory for Electrostatic Interaction Chromatography of Proteins". © 1991 by American Chemical Society.]

surface area), F is Faraday's constant, I is the ionic strength of the mobile phase, ε_o is the permittivity of a vacuum, and ε_r is the dielectric constant of the mobile phase. Figure 14.11 shows how the free energy varies with distance between the charged stationary phase and the protein surface.

Substituting Eq. 14.29 into Eq. 14.27 and rearranging yields Eq. 14.30:

$$\ln(k) = \{A_p\sigma_p^2/(F\{2RT\varepsilon_o\varepsilon_r\}^{1/2})\}\{1/I\}^{1/2} + \ln(\phi) \qquad (14.30)$$

This equation correctly describes the linear dependence of $\ln(k)$ on $I^{-1/2}$, and furthermore allows the calculation of the net charge on a protein from the slope of this plot, provided that the surface area (or radius) of the protein is known. The constants in the slope can be evaluated, under conditions of an aqueous mobile phase at 25 °C with $0.05 \leq I \leq 0.5\,M$, so that the net charge, q, is given by

$$q = \{(\text{slope})(\text{surface area})/135\}^{1/2} \qquad (14.31)$$

where surface area is in units of ($\mathring{A}^2$).

Figure 14.12 shows the plots of $\ln(k)$ against $I^{1/2}$ obtained for chymotrypsinogen, lysozyme and cytochrome c on a weak cation exchange column at pH 6.0. Except for the nonlinearity of the chymotrypsinogen plot at low ionic strength, the plots show good linearity over a wide range of I, and Eq. 14.31 was used to evaluate the net charges of lysozyme and cytochrome c as $+10.6$ and $+10.7$, respectively.

Similar experiments to those shown in Fig. 14.12 have been performed for a large number of proteins at different pH values and using different eluting salts and stationary phases. The net charge values calculated from the chromatographic plots, q_{chr}, have been compared to values previously obtained from titration data, q_{titr}. The results of these experiments are shown in Table 14.6, where the ratio of q_{chr}/q_{titr} is also tabulated, and should be close to unity if the assumptions inherent in the model are valid.

Some discrepancies may be observed in Table 14.6 between chromatographic and titrimetric values of the net charge, q_{chr} and q_{titr}. If $q_{chr}/q_{titr} > 1$, the protein is believed to have a nonuniform charge distribution, so that the apparent charge density is higher over some areas of the surface that interact more strongly with the stationary phase. If $q_{chr}/q_{titr} < 1$, then the protein is believed to possess a higher charge density than the surface of the stationary phase, so that the stationary-phase properties dominate the electrostatic interactions. In general, there is very good agreement between chromatographic and titrimetric data. Since only fundamental constants are used for the chromatographic estimation of q, and no adjustable parameters are present, as exist in previous models, this is the best quantitative theory in existence that describes the chromatographic retention of polyelectrolytes on ion-exchange columns. Its applicability to other biopolymers, such as highly charged nucleic acids, has yet to be determined.

SUGGESTED REFERENCES

W. S. Hancock, Ed., *High Performance Liquid Chromatography in Biotechnology*, John Wiley & Sons, Inc., New York, 1990.

A. Fallon, R. F. G. Booth, and L. D. Bell, *Applications of HPLC in Biochemistry*, Elsevier, New York, 1987.

J.-C. Janson and L. Ryden, Eds., *Protein Purification: Principles, High-Resolution Methods and Applications*, VCH Publishers, New York, 1989.

P. R. Haddad and P. E. Jackson, *Ion Chromatography: Principles and Applications*, Elsevier, New York, 1987.

S. Yamamoto, *Ion-Exchange Chromatography of Proteins*, Marcel Dekker, New York, 1988.

I. M. Chaiken, *Analytical Affinity Chromatography*, CRC Press, Boca Raton, FL, 1987.

P. D. G. Dean, W. S. Johnson, and F. A. Middle, *Affinity Chromatography: A Practical Approach*, Oxford University Press, Washington, DC, 1985.

D. J. Winzor, From gel filtration to biosensor technology: the development of chromatography for the characterization of protein interactions. Review article in *J. Mol. Recognit.* **13**, 2000, 279–298.

REFERENCES

1. N. A. Parris, *Instrumental Liquid Chromatography*, Elsevier, New York, 1976, Chap. 2, pp. 7–18.

2. S. J. Hawkes, *J. Chem. Educ.* **60**, 1983, 393–398.

3. C. J. O. R. Morris and P. Morris, *Separation Methods in Biochemistry*, John Wiley & Sons, Inc., New York, 1976, Chap. 7, p. 418.

4. H. Ellegren and T. Låås, *J. Chromatogr.* **467**, 1989, 217–226.

5. T. Laurent and J. Killander, *J. Chromatogr.* **14**, 1966, 317.

6. P. L. Dubin, in *Advances in Chromatography*, Vol. 31, J. C. Giddings, E. Grushka, and P. R. Brown, Eds., Marcel Dekker, New York, 1992. Chap. 2, pp. 121–125.

7. C. R. Lowe, *An Introduction to Affinity Chromatography*, Elsevier, New York, 1979, Chap. 5, pp. 428–465.

8. R. R. Walters, *Anal. Chem.* **57**, 1985, 1099A–1101A.

9. P. Brodelius and K. Mosbach, *FEBS Lett.* **35**, 1973, 223–226.

10. A. Jaulmes and C. Vidal-Madjar, in *Advances in Chromatography*, Vol. 28, J. C. Giddings, E. Grushka, and P. R. Brown, Eds., Marcel Dekker, New York, 1989, Chap. 1, pp. 11–39.

11. B. M. Dunn and I. M. Chaiken, *Biochemistry* **14**, 1975, 2343–2349.

12. H. F. Walton and R. D. Rocklin, *Ion-Exchange in Analytical Chemistry*, CRC Press, Boca Raton, FL, 1990. Chap. 3, pp. 40–42.

13. J. Ståhlberg, B. Jönsson, and C. Horváth, *Anal. Chem.* **63**, 1991, 1867–1874.

TABLE 14.6. Chromatographic Conditions and Characteristic Net Charge of Proteins, q_{chr}, from Chromatographic Experiments, and a Comparison between q_{chr} and q_{titr}[a]

Protein	Eluent pH	Stationary Phase (a)	Eluting Salt	r (b)	q_{chr}	q_{titr}	q_{chr}/q_{titr}
Myoglobin (horse muscle)	6.0	Zorbax Bio Series WCX 300	$(NH_4)_2SO_4$	0.996	+6.9	+4.8	1.44
Ribonuclease	6.0	Zorbax Bio Series WCX 300	$(NH_4)_2SO_4$	0.991	+8.8	+6.1	1.44
	6.4	Zorbax Bio Series WCX 300	NaOAc	0.995	+9.5	+5.4	1.76
	6.4	Zorbax Bio Series SCX 300	NaOAc	0.994	+8.3	+5.4	1.54
Cytochrome c	4.9	In-house made WCX	$Ca(OAc)_2$	0.999	+8.6	+8.9	0.97
	6.0	Zorbax Bio Series WCX 300	$(NH_4)_2SO_4$	0.996	+10.7	+7.2	1.49
	6.0	Zorbax Bio Series SCX 300	$(NH_4)_2SO_4$	0.994	+8.6	+7.2	1.19
BSA	7.8	Zorbax Bio Series WAX 300	$(NH_4)_2SO_4$	0.994	-18.5	-15.0	1.23
	7.8	Zorbax Bio Series SAX 300	$(NH_4)_2SO_4$	0.990	-14.8	-15.0	0.99
β-Lactoglobulin A	7.0	SynChropak AX 300	NaCl	0.992	-21.2	-13.5	1.58
	8.0	SynChropak AX 300	NaCl	0.991	-21.2	-18.5	1.15
Human serum albumin	6.5	Mono Q (SAX)	NaCl	0.997	-18.8	-9.8	1.92
	7.5	Mono Q (SAX)	NaCl	0.978	-20.1	-16.2	1.24
	9.6	Mono Q (SAX)	NaCl	0.990	-23.5	-34.8	0.68
Ovalbumin	5.5	Mono Q (SAX)	NaCl	0.991	-11.7	-9.6	1.22
	6.0	SynChropak Q 300	NaCl	0.999	-15.1	-13.1	1.15
	6.5	Mono Q (SAX)	NaCl	0.996	-17.5	-15.3	1.14
	7.0	SynChropak Q 300 (SAX)	NaCl	0.999	-16.6	-17.1	0.97
	7.5	Mono Q (SAX)	NaCl	0.982	-17.4	-18.0	0.97
	7.8	Zorbax Bio Series WAX 300	$(NH_4)_2SO_4$	0.998	-15.3	-18.0	0.81
	7.8	Zorbax Bio Series SAX 300	$(NH_4)_2SO_4$	0.999	-13.3	-18.8	0.71
	8.0	SynChropak Q 300	NaCl	0.999	-19.4	-19.3	1.01
	9.6	Mono Q (SAX)	NaCl	0.965	-19.9	-22.2	0.90
Lysozyme	4.9	In-house made WCX	$Ca(OAc)_2$	0.999	+9.2	+10.6	0.87
	6.0	Zorbax Bio Series WCX 300	$(NH_4)_2SO_4$	0.998	+10.6	+9.0	1.18
	6.0	Zorbax Bio Series SCX 300	$(NH_4)_2SO_4$	0.999	+9.0	+9.0	1.00
	6.4	Zorbax Bio Series WCX 300	NaOAc	0.994	+10.9	+8.6	1.27
	6.4	Zorbax Bio Series SCX 300	NaOAc	0.997	+10.3	+8.6	1.20

[a] (a) SAX = strong anion exchanger, WAX = weak anion exchanger, SCX = strong cation exchanger, WCX = weak cation exchanger. (b) Correlation coefficient for the linearity of ln k' versus $1/\sqrt{I}$ plots.[13] [Reprinted, by permission, from J. Ståhlberg, B. Jönsson, and C. Horváth, *Anal. Chem.* **63**, 1991, 1867–1874. "Theory for Electrostatic Interaction Chromatography of Proteins". © 1991 by American Chemical Society.]

PROBLEMS

1. A mixture consisting of two main protein components was subjected to gel-filtration chromatography, with pH 7.1, 0.05 *M* phosphate as the mobile phase and Sephadex G-100 as the stationary phase. An unretained species, blue dextran, eluted in 2.6 min, while proteins X and Y eluted at times of 6.7 and 9.1 min, respectively.

   (a) Calculate the relative retention, α, of the two protein components.

   (b) Which protein has a higher molecular weight?

2. A gel-filtration column prepared using Bio-Gel P-100 had a void volume of 5.0 mL and a stationary-phase volume of 50.0 mL. The column was calibrated using six proteins of known molecular weight, by measuring the elution volume of each peak (Table.14.7). An unknown protein was then applied to the column, and its elution volume was measured. Given the data in the table below, determine the molecular weight of the unknown protein.

   TABLE 14.7

Protein	MW (kDa)	Elution Volume (mL)
Cytochrome *c*	11.7	42.9
Myoglobin	17.2	37.8
IgG light chain	23.5	33.9
Ovalbumin	43.0	26.1
Aconitase	66.0	20.4
Transferrin	77.0	18.3
Unknown	?	31.2

3. Describe a method by which a dehydrogenase enzyme can be purified from a mixture of proteins (all of similar molecular weight), so that only the free dehydrogenase enzyme is present in the final buffered solution. (*Hint*: Use more than one chromatographic step.)

4. A new receptor protein has been isolated from the sensory tissue found on the barbels (whiskers) of the Louisiana blue catfish. Studies have shown that this protein interacts strongly with glutamate. An affinity column was prepared by

   TABLE 14.8

[Glutamate] (μM)	Elution Volume (mL)
0.1	52.3
0.2	35.6
0.3	27.3
0.4	22.3
0.5	19.0

immobilizing a dipeptide of glutamic acid through its N-terminal amine group onto cyanogen bromide-activated cellulose. The void volume of the final column configuration was determined to be 2.3 mL, by measuring the elution volume of an unretained dye. A series of experiments was performed to determine the elution volume of the receptor protein as a function of mobile phase glutamate concentration. From the data given below (Table 14.8), determine the association constant for the reaction of glutamate with receptor.

CHAPTER 15

Mass Spectrometry of Biomolecules

15.1. INTRODUCTION

Mass spectrometry is widely used for the qualitative analysis of unknowns samples, and in particular, for the identification and characterization of biological macromolecules. Recent decades have seen the introduction and optimization of the so-called "soft" ionization methods that provide intact, vapor-phase biomolecular ions for separation and detection. This chapter considers MS fundamentals, ionization methods, and applications to biological macromolecules. Conventional mass spectrometers used for low volatile molecular weight samples that are introduced in the vapor phase are called single-focusing mass spectrometers, and use an electron-impact ion source.[1] Figure 15.1 shows a diagram of this type of instrument.

With this type of MS, ions are generated by bombarding the vapor-phase sample with electrons at high energy. These electrons are emitted by a hot tungsten or rhenium wire, or filament, and are accelerated toward an electron trap that is held at about $+70$ V. The impact of these electrons with the sample molecules generates ions according to Eq. 15.1:

$$e_{beam}^- + M \rightarrow M^{+\bullet} + e^- + e_{beam}^- \tag{15.1}$$

The sample ions generated in this process are repelled through accelerating plates, which are successive plates held at increasingly negative potentials. The gain in kinetic energy for sample molecules passing through these plates is related to the ion charge (z) and the voltage difference (V) through which the ions pass, according to Eq. 15.2:

$$\text{Kinetic energy} = mv^2/2 = zeV \tag{15.2}$$

where v is the velocity of an ion of mass m and number of charges z, and e is the charge of an electron. The ions then follow a curved flight path as they pass through the mass analyzer, traversing a $60°$ or $90°$ sector of a circumference. As the ion

Bianalytical Chemistry, by Susan R. Mikkelsen and Eduardo Cortón
ISBN 0-471-54447-7   Copyright © 2004 John Wiley & Sons, Inc.

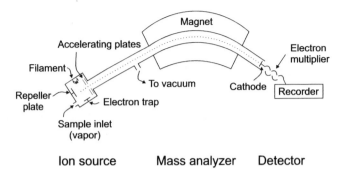

Figure 15.1. Single-focusing MS with electron-impact ionization source.

passes through the magnetic sector, its centrifugal force ($= mv^2/r$, where r is the radius of the sector) and magnetic force ($= Bzv$, where B is the magnetic field strength) are equal. The centrifugal force pushes the ion toward a straight path, while the magnetic force pulls the ion toward the center of the circle.

The ion detector monitors the number of ions colliding with it as a function of their m/z values. The derivation that follows shows how the m/z values of ions are related to the magnetic field strength, the radius of the magnetic sector and the acceleration voltage. We begin by equating the centrifugal and magnetic forces:

$$mv^2/r = Bzev \qquad (15.3)$$

and rearrange this expression to yield:

$$v^2 = (Bzer/m)^2 \qquad (15.4)$$

from Eq. 15.2, we can substitute $v^2 = 2zeV/m$, to obtain

$$2V = (Br)^2 ez/m \qquad (15.5)$$

and this expression readily yields the desired equation for m/z:

$$m/z = (Br)^2 e/2V \qquad (15.6)$$

Therefore, given fixed B, r, and V values, only ions of a particular m/z value will reach the ion detector to generate a signal. Ions that are too heavy (m/z too large) impact with the outer wall of the magnetic sector, while lighter ions (small m/z) have small radii of curvature and impact with the inner wall. Mass spectrometers allow either B or V (or both) to be scanned to detect ions of interest at particular m/z values.

Mass spectrometry is a well-established technique: J.J. Thompson designed the first mass spectrometer in 1912. The methods for ion production in these early times were limited to gaseous samples (or low vapor pressure molecules) for analytical purposes. Organic molecules fragment extensively upon electron impact, but small organic molecules are often identified from their fragmentation patterns.

Soft ionization methods were developed over the last two decades, and allow the formation of gaseous molecular ions from large polymeric biomolecules. These methods allowed MS to be introduced to the biological chemistry area, and are now particularly useful for the study of proteins, glycoproteins, nucleic acids, and their reactions.

Molecular weights are now readily determined for these molecules with very high precision and accuracy. Variation between the experimental and the predicted MW of a protein (predicted from its DNA code) help in the determination of post-translational modifications of the polypeptide chain. Noncovalent interactions between biomolecules and ligands may also be studied using soft ionization methods.

Mass sepetrometry has also been used to analyze nucleic acid mutations and polymorphisms, and for the identification and chemical characterization of microbial cells.

15.2. BASIC DESCRIPTION OF THE INSTRUMENTATION

As shown in Figure 15.1, there are three main components of every mass spectrometer. The ion source is used to produce gas-phase ions by capture or loss of electrons or protons. In the mass analyzer, the ions are separated according to their m/z ratios: ions of a particular m/z value reach the detector, and a current signal is produced. This section describes the soft ionization sources, mass analyzers, and detectors that are used in experiments involving biological macromolecules.

15.2.1. Soft Ionization Sources

The ion source used for the generation of biomolecular parent ions is critical, and only recently have the so-called "soft" ionization methods been developed.[2] Electron-impact ionization sources fall into the category of "hard" sources, whereby the sample must be in the vapor phase initially, and the ionization process produces a very large number of fragments. Soft methods were introduced to overcome the problems associated with the thermal instability and involatility of macromolecular analytes. Soft ionization produces few fragments under relatively mild conditions. In Table 15.1 a comparison is shown between the three main soft ionization methods; some of these values are strongly dependent on individual mass spectrometer configurations and the desired resolution.

15.2.1.1. Fast Atom/Ion Bombardment. This ionization technique utilizes a high-energy beam (keV) of Xe atoms, Cs^+ ions or glycerol–NH_4^+ clusters to sputter the sample and matrix from the probe surface. Matrices commonly used with peptides are glycerol and 3-nitrobenzyl alcohol, which are nonvolatile and do not overload the vacuum pumps. The sample is dissolved in the matrix and inserted into the

TABLE 15.1. Comparison of Soft Ionization Sources Used in Biological MS[a,b]

	FAB[c]	ESI[d]	MALDI[e]
Mass range, typically up to	7000 Da[a]; 25,000 Da[b]	70,000 Da[a]; 200,000 Da[b]	300,000 Da[a,b]
Sensitivity	Low. Analyte at nanomole level	High. Analyte at pico-femtomole level	High. Analyte at pico-femtomole level
Fragmentation	Little	Usually none	Little or none
Background	High (matrix)	Very low	High (matrix)
Not suitable for	Species that are not easily charged	Mixtures. High purity required	Photosensitive analytes
Type of process	Batch	Continuous	Batch
Ionization proceeds	Continuously	Continuously	Short pulses
Matrix	Yes, liquid phase, nonvolatile solvents	Aqueous solution	Yes, solid-phase microcrystaline structure

[a] See Ref. 3.
[b] See Ref. 4.
[c] Fast atom/ion bombardment = FAB.
[d] Electrospray ionization = ESI.
[e] Matrix-assisted laser desorption–ionization = MALDI.

ionization camera over a metallic plate. When a high-energy beam impacts the sample (Fig. 15.2), a fraction of its energy is transferred, mainly to the solvent molecules; in this way excessive analyte fragmentation is avoided. Some analyte and matrix molecules are desorbed to the gas phase, and if they are not already charged, they can be charged in the gas phase by reaction with the surrounding gas-phase ions. Charged molecules are propelled electrostatically to the mass analyzer and, eventually, reach the detector. Fast atom bombardment was the first soft ionization method, and was introduced in 1981. One year later, molecules as large as human insulin were analyzed, and became the benchmark for resolution and sensitivity.[5]

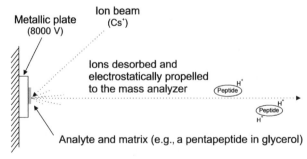

Figure 15.2. FAB ionization source. Total volume of sample and matrix is typically at the microliter (μL) level.

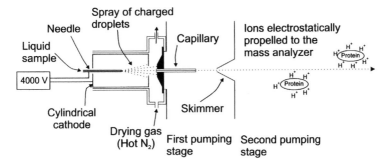

Figure 15.3. ESI ionization source.

15.2.1.2. Electrospray Ionization.
Electrospray is an especially soft ionization method, capable of generating molecular ions (without fragmentation) from biological macromolecules present in aqueous solution. For this reason, it has received increased attention in terms of its applicability to protein and DNA analysis.

Figure 15.3 shows a diagram of this ionization source. The hollow stainless steel needle is maintained at a few kilovolts (kV) relative to the chamber walls and the cylindrical cathode surrounding it. The liquid sample enters at a rate of 1–20 µL/min through the needle, but lower flow rates (10–100 nL/min) are becoming more common in modern instrumentation. The electric field at the needle tip disperses the liquid by Coulombic forces into a fine spray of charged droplets. These droplets migrate in the field toward the glass capillary inlet. A flow of drying gas (typically N_2 at 100 mL/s) evaporates the solvent from the droplets, so that their diameter decreases, and the surface charge density on each droplet increases. When this surface charge density reaches the Rayleigh limit, the Coulombic repulsion is approximately equal to the surface tension, and the droplet explodes into smaller droplets. This process continues until the field due to the surface charge density is strong enough to ionize the macromolecules, which are then propelled through the skimmer and into the magnetic sector. Most biological macromolecules that are ionized in this manner are multiply charged.

15.2.1.3. Matrix-Assisted Laser Desorption/Ionization.
In this ionization method, the analyte is combined with a matrix compound, in a molar ratio of ~1:1000, and evaporated onto a metallic plate. The matrix compound is chosen to absorb strongly at the laser wavelength used. Absorption causes a rapid increase of temperature, allowing vaporization of the sample without extensive fragmentation, which is shown schematically in Figure 15.4.

Short pulses of laser radiation are used, typically at 337 nm, (nitrogen laser) but UV or infrared (IR) lasers can also be employed, depending on the matrix compound selected. Common matrix compounds are 2,5-dihydroxybenzoic acid, nicotinic acid, sinapinic acid, and α-cyanocarboxylic acid.[6]

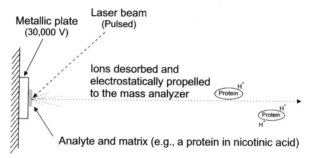

Figure 15.4. MALDI ionization source.

15.2.2. Mass Analyzers[7,8]

Different physical principles are used to analyze the ions produced by the ionization source, and one or more mass analyzer systems may be used in tandem in a given instrument. A brief description of the main mass analyzers will be presented here. The values quoted for the resolution of these analyzers indicate their ability to distinguish closely spaced m/z value ions according to Equation 15.7:

$$\text{Resolution} = \frac{M}{\Delta m} \qquad (15.7)$$

where M is experimentally measured mass of the peak of interest (Da), and the Δm (Da) value is the mass separation between the maxima of two (symmetrical) closely spaced peaks, as shown in Figure 15.5. Usually, the 10% valley definition is used to establish whether a peak is separated from another (Fig. 15.5); by this definition two peaks are considered separated if they coalesce at a maximum intensity of 10% of the peak height. An analogous criterion is the 50% valley definition; by

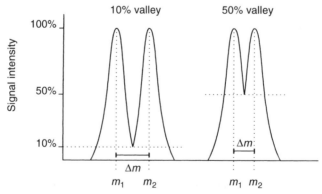

Figure 15.5. Resolution can be calculated using different definitions. The more stringent of these is the 10% valley definition, which is more common in magnetic sector MS.

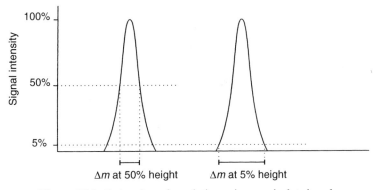

Figure 15.6. Estimation of resolution using one isolated peak.

this definition, the resolution calculated from a given set of data is nominally higher.

A resolution of $R = 1000$ means that the mass spectrometer can separate an ion at m/z 10,000 from another at 10,010. If the resolution is low, two or more peaks may appear as only one.

Alternatively, one isolated symmetrical peak can be used to estimate R. In this case, Δm can be defined as the peak width at 5%, or at the 50% level, which is usually referred to as full with at half-maximum (FWHM) as shown in Figure 15.6.

In quadrupole and ion trap spectrometers, it is customary to use the *unit resolution* definition, which means that each mass can be separated from the next integer mass (500 from 501, 2000 from 2001, etc.). The comparison and conversion between the different resolution definitions is not trivial, considering that they are strongly dependent on peak shape (Gaussian, triangular, trapezoidal, Lorentzian, etc.).[9]

Resolution is also used to compare the theoretical (calculated or predicted) mass value to the experimentally observed value. Resolution defines the peak width, and a parts per million (ppm) error value can be calculated with Eq. 15.8; this allows estimation of the precision of an experimental mass assignment:

$$\text{ppm} = \frac{10^6}{R} = \frac{10^6 \Delta m}{M} \tag{15.8}$$

In Table 15.2, examples of the relationship between these mass spectrometry figures of merit are shown.[10]

The *magnetic sector* basis of separation was explained in the first pages of this chapter; the magnetic field is usually scanned from low to high fields, therefore ions with low m/z ratios reach the detector at the beginning of the scan.

An *electrostatic analyzer* is formed from two curved plates with opposite electric fields that attract and repel ions. Each ion follows a certain path that depends on its momentum; ions with low momentum travel a short path compared to ions with

TABLE 15.2. Resolution as $R = M/\Delta m$ and Its Effect on Precision in MS

m/z	Δm	Resolution	$\pm$ ppm	Mass Range (Da)
2,000	5	400	2,500	1,995–2,005
2,000	0.5	4,000	250	1,999.5–2,000.5
2,000	0.05	40,000	25	1,999.95–2,000.05
20,000	5	4,000	250	19,995–20,005
20,000	0.5	40,000	25	19,999.5–20,000.5
20,000	0.05	400,000	2.5	19,999.95–20,000.05
200,000	5	40,000	25	199,995–200,005
200,000	0.5	400,000	2.5	199,999.5–200,000.5
200,000	0.05	4,000,000	0.25	199,999.95–200,000.05

higher momentum, thus producing a focusing effect. This type of mass analyzer is usually used in combination with a magnetic sector, to increase the final resolution of the instrument.

The *quadrupole analyzer* (or *mass filter*) is formed by four metal rods (usually high-grade steel), positioned precisely in each corner of an imaginary square. Between diagonally opposed rods a direct current (dc) voltage is applied, with a superimposed radio frequency (rf) alternating field. Both dc and rf voltages are scanned. Ion trajectories across the quadrupole are complicated. Under given conditions of dc and rf, ions of only one *m/z* value will reach the detector. All other *m/z* ions possess unstable paths that impact on the rods.

Time of flight analyzers (TOF) allow ions to move across a vacuum tube (usually named *flight tube*). Ions are generally produced by MALDI or by another technique capable of producing discrete bursts of ions. About 100–500 ns after the laser pulse, a strong acceleration field is switched on, and this imparts a fixed kinetic energy to the ions. As the ions travel across the vacuum tube, ions with low *m/z* travel faster and arrive at the detector first.[11] The resolution of MALDI–TOF MS instruments is low (usually <500) but the mass range is practically unlimited. In order to increase resolution, the TOF reflectron has been introduced. The reflectron is an electrostatic mirror used to correct for the distribution in the initial kinetic energy imparted to the ions (the ions do not enter the flight tube precisely at the same time); it is therefore possible to reach higher resolution (Table 15.3) in TOF-reflectron instruments.

MS/MS instruments possess a combination of two or more mass analyzers in tandem. A diversity of different configurations have been assembled, using one or

TABLE 15.3. Comparison of Different Mass Analyzers[a]

Characteristic	Magnetic Sector	Quadrupole	QIT	TOF-Reflectron	FTICR
Mass range, up to (Da)	15,000	4000	100,000	Unlimited	$>10^6$
Resolution	200,000	Unit	30,000	15,000	$>10^6$

[a] See Ref. 12.

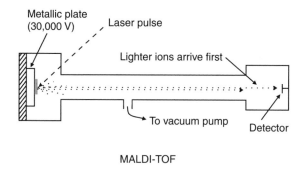

Figure 15.7. Typical configurations used in biological MS. In MALDI–TOF, the ions produced by a short laser pulse travel across a flight tube, arriving at different times at the detector. In ESI–triple quadrupole, the first quadrupole (Q1) is used to separate the sprayed ions, in the second (Q2, also called the fragmentation cell) argon atoms collide with the ions; the resulting ions (daughter ions) are analyzed in Q3, and subsequently detected.

several quadrupole, magnetic, and electrostatic analyzers. A triple quadrupole mass analyzer is composed of two scanning quadrupoles separated by one rf-only quadrupole, where a gas collides with analyte ions in order to produce fragmentation (CID, collision-induced dissociation). Figure 15.7 shows two common combinations of ionization source and mass analyzer.

Relatively new mass analyzers with very high resolution include the quadrupole ion trap (QIT) and the Fourier-transform ion cyclotron resonance (FTICR) instruments. In both analyzers, ions are trapped in a 3D field, and are analyzed once trapped. In Table 15.3, the mass analyzers described here are compared. The combination of MS with other analytical techniques is also very common; MS has been widely used following chromatographic separations, for mass analysis.

15.2.3. Detectors

The ions separated by the mass analyzer are detected through their production of a current signal; this is generally achieved using an electron multiplier or a scintillation counter.

An electron multiplier consists of a series of dynodes held at increasing potential; the ions arrive at the first dynode, producing an emission of secondary electrons. These are accelerated in the electric field and strike a second dynode that produces more electrons. Successive dynodes continue to amplify the current, and the total amplification is high; values of 10^6 are normally obtained. The operation and design are very similar to a photomultiplier tube, but the electron multiplier is not shielded by a glass bulb; for this reason contamination, mainly at the first dynode, is important and the lifetime of the electron multiplier is relatively short.

In the scintillation counter, ions strike a dynode, from which electrons are emitted. These electrons impact a phosphorous-coated screen, causing light emission that is detected by a photomultiplier tube.

15.3. INTERPRETATION OF MASS SPECTRA

High-resolution MS of low molecular weight compounds results in a series of isotopic peaks for the parent ion and each detectable fragment. The intensity of each peak in the series depends on the relative abundance of a given isotope as well as the number of atoms of a given identity in the detected fragment. For example, a mass spectrum of CO_2 shows two main peaks that correspond to $^{12}C^{16}O_2$ and $^{13}C^{16}O_2$ (with the latter showing $\sim 1.1\%$ of the intensity of the former) as well as a number of less intense peaks that result from other isotope combinations. Table 15.4 shows the main isotopes of several elements as well as their relative abundances; from these data, it is clear that molecular and fragment ions containing C and S atoms are expected to produce isotopic peaks with significant intensities.

TABLE 15.4. Isotopic Abundances of Selected Elements

Isotope	Mass (Da)	Abundance (%)
^{12}C	12.0000	98.89
^{13}C	13.003354	1.11
^{1}H	1.007825	99.986
^{2}H	2.014102	0.015
^{14}N	14.003074	99.63
^{15}N	15.000109	0.37
^{16}O	15.994915	99.762
^{17}O	16.999131	0.038
^{18}O	17.999160	0.200
^{32}S	31.972071	95.02
^{33}S	32.971457	0.75
^{34}S	33.967867	4.21
^{36}S	35.967081	0.02
^{31}P	30.973762	100

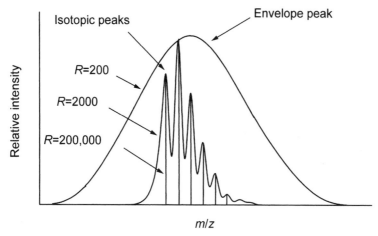

Figure 15.8. Mass spectra for a peptide of nominal mass 2537. Note that at resolution 2000 ($R = 2000$) 6 isotope peaks are clearly observed. The figure represents spectra for a unique molecular ion state, for example $(M + H)^+$, a protonated molecule without fragmentation.

The resolution of the mass spectrometer, R, and the charge state of the detected ion, z, are the most important factors that determine whether individual isotopic peaks will be observed at a given m/z value. At low resolution, these individual species contribute to the observed coalescent "envelope peak", as shown in Figure 15.8. It can be seen in this figure that the position of the envelope peak is offset with respect to that of the most abundant isotope; this offset is predictable based in the natural isotopic abundances of the elements.

The mass of a given chemical species may be calculated in three different ways. The simplest of these uses the integer mass of the most abundant isotope of each element, and results in the *nominal mass* of the species; this value is not particularly useful in MS. The *monoisotopic mass*, calculated using the exact mass of the most abundant isotope of each element, is more useful, especially for high-resolution MS of low molecular weight species. For biomolecules and their fragment ions, the most useful value is the *average mass*. The average mass is calculated using elemental atomic weight values that are averaged over all isotopes; this value allows the position of the envelope peak to be predicted.[13]

Table 15.5 shows examples of nominal, monoisotopic, and average mass values. Although the differences between nominal and average values may appear small, especially for elemental species, they become larger as molecular complexity increases and are important in establishing the identities of proteins. In particular, site-directed mutagenesis and posttranslational modification studies require accurate average mass values to confirm the identities of product proteins.

For purified samples of small molecules studied on low-resolution mass spectrometers, and for singly charged larger molecules, fragment ions are readily identified by straightforward calculations using average mass values. The presence or absence of observable fragments (including the parent ion) are known to depend on the type

TABLE 15.5. Nominal, Monoisotopic, and Average Mass for Some Elements and Molecules[a]

Element or Molecule	Nominal Mass	Monoisotopic Mass	Average Mass
Carbon	12	12.0000	12.0111
Hydrogen	1	1.0078	1.0080
Nitrogen	14	14.0031	14.0067
Oxygen	16	15.9949	15.9994
Sulfur	32	31.9721	32.0660
Phosphorus	31	30.9738	30.9738
Tripeptide $C_{27}H_{29}N_3O_7$	507	507.1998	507.5476
Porcine insulin $C_{256}H_{381}N_{65}O_{76}S_6$	5771	5773.6083	5777.6755

[a] In daltons (Da).

of ion source (e.g., hard vs. soft) as well as ionization source variables (such as ionization and acceleration voltages).

Biomolecules such as proteins and oligonucleotides present mass spectra that are complicated by the presence of multiply charged families of ions. The parent species, for example, generated by a soft ionization method such as FAB of ESI, will yield several m/z peaks in which m is equal to a constant plus (protein) or minus (nucleic acid) a variable number of proton masses, while z is a variable. Egg white lysozyme, for example, yields ESI–mass spectra with fine parent peaks between m/z values of 1194 and 1791; the corresponding z values are 12–8 [see Fig. 15.9(a)].

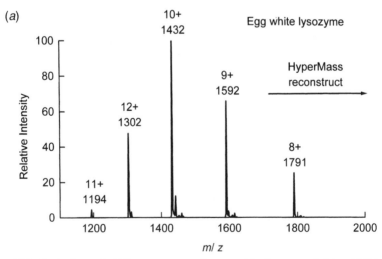

Figure 15.9. Typical protein ESI–mass spectra for egg white lysozyme (a,b) and BSA (c,d), showing an algorithm-based MW calculation (HyperMass reconstruct).[14] [Reprinted, with permission, from From Gary Siuzdak, *"Mass Spectrometry for Biotechnology"*, Academic Press, 1996. Copyright © 1996 by Academic Press, Inc.]

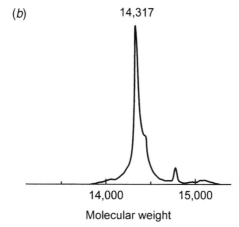

(b)

14,317

14,000   15,000

Molecular weight

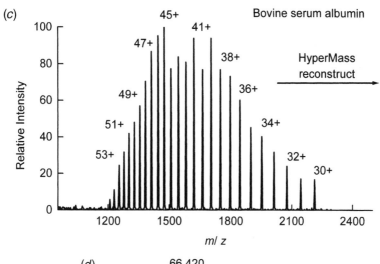

(c)

Bovine serum albumin

HyperMass reconstruct

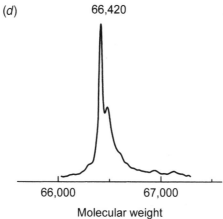

(d)

66,420

66,000   67,000

Molecular weight

Figure 15.9   (*Continued*)

Table 15.6. Bovine Trypsin Autolysis Fragments, Used for *m/z* Calibration

Amino Acid Residues (from–to)	Mass (Da)		Sequence
	Monoisotopic	Average	
110–111	259.19	259.35	LK
157–159	362.20	362.49	CLK
238–243	623.31	632.67	QTIASN
221–228	905.50	906.05	NKPGVYTK
146–156	1152.57	1153.25	SSGTSYPDVLLK
70–89	2162.05	2163.33	LGEDNINVVEGNEQFISASK
21–63	4550.12	4553.14	IVGGYTCGA...VVSAAHCYK

Larger analytes with more protonation–deprotonation sites yield larger families of peaks for each fragment, and the overlap of *m/z* ranges for different fragment families can further complicate the spectra of biomolecules. For this reason, soft ionization methods, that produce few fragments, are particularly useful for biomolecule MS.

Accuracy in MS depends on calibration of the *m/z* axis of the spectra. Calibration is typically performed using separate samples or solutions of the calibrant studied under identical instrumental conditions as the analyte. Calibrants are selected based on the ionization technique and the *m/z* range required for the analyte; typical calibrants used for soft ionization methods with biomolecules include bovine trypsin fragments, bovine insulin (5734 Da), cytochrome *c* (12,361 Da), horse myoglobin (16,951 Da), lysozyme (14,317 Da), and propylene glycols. Separate solutions of the analyte prepared in the absence and presence of a calibrant species may also be used. Table 15.6 shows typical calibration peaks found with bovine trypsin autolysis fragments.[15]

15.4. BIOMOLECULE MOLECULAR WEIGHT DETERMINATION

Electrospray ionization is the most commonly used ionization source for biomolecular studies. In this ionization process, multiply charged ions are formed (without fragmentation), and this has two important consequences. First, large molecules can be detected using MS equipment with low mass range; for example, an ion of mass 10,000 Da with 1 proton would be detected at $m/z \approx 10,001$, whereas the same ion with 9 additional protons will be detected in a different *m/z* position (≈ 1001). Second, the multiplicity of peaks allows molecular weight estimation as a multiple-peak average by means of *ad hoc* computational algorithms, improving the precision of the resulting MW.

There are several computer algorithms, normally incorporated into the instrument software, used to calculate MW. The algorithms are based on the following simple assumptions and calculations. If two adjacent peaks in the spectrum (p_1 and

p_2), are from the same biomolecule, and differ only by the addition of a single proton, we can write the following equations:[16]

$$m_1 = (M + nH)/n \quad \text{and} \quad m_2 = (M + \{n+1\}H)/(n+1) \quad (15.9)$$

where m_1 and m_2 are the m/z values for the peaks p_1 and p_2, respectively, M is the desired experimental MW value, n is the number of charges, and H is the mass of the proton; $m_2 < m_1$.

By using two contiguous peaks, the molecular mass of a biopolymer can be calculated by solving Eqs. 15.10 and 15.11.

$$n_2 = \frac{m_2 - H}{m_1 - m_2} \quad (15.10)$$

$$M = n(m_1 - H) \quad (15.11)$$

These equations use only two peaks; computer algorithms use all identified peaks to obtain more precise MW values. The number and relative abundance of different molecular ions depend on the protein medium (buffer, salt concentration) and instrumental parameters (i.e., voltage, vacuum). Typical MW determination results are shown in Figure 15.9.

Multiply charged molecular ions, as seen in Figure 15.9, allow precise calculation of the molecular weight, as long as the charge states of the molecular ions are known. High-resolution mass spectrometers allow charge states to be determined from the isotopic peaks for a given molecular ion. The peaks corresponding to molecular ions shown in Figure 15.9 are formed by the coalescence of different isotopic peaks. If the molecular ion peak has a charge state of +1, $(M + H)^+$, the differences in mass between the isotopic peaks must be 1 Da; similarly, in the charge state +2, $(M + 2H)^{2+}$, the differences must be 0.5 Da. As shown in Figure 15.10,

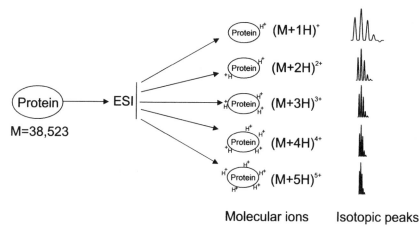

Figure 15.10. Formation of multiply charged protein molecules using ESI methods. For mass and m/z calculations, see Table 15.8.

TABLE 15.7. Calculated Relationship between Molecular Ion Mass and *m/z* Values for Multiple Charged (Protonated) Molecular Ions

Protein Mass (M_r)	Number of Charges	Molecular Ion Mass	*m/z*	Distance (*m/z*) between Isotopic Peaks
38,523.5	1	38,524.5	38,524.5	1
	2	38,525.5	19,262.8	0.5
	3	38,526.5	12,842.2	0.33
	4	38,527.5	9631.88	0.25
	5	38,528.5	7,705.7	0.2
	10	38,533.5	3,853.4	0.1

increased resolution is necessary to observe isotopic peaks when the charge state of the molecular ion is increased; the resolution provided by most instrumentation allows the observation of isotopic peaks only from singly to triply charged molecular ions.

Isotope peaks observed for proteins are mainly produced by the ^{12}C and ^{13}C mass difference of ~1 mass unit. Once one molecular ion charge state is determined, the other charge states are readily assigned by spectral observation. ESI, MALDI, and FAB are usually used in the positive-ion mode, especially for protein MW determination; therefore the proton mass must be used to calculate the ion mass, as shown in Table 15.7.

15.5. PROTEIN IDENTIFICATION

The protein molecular mass is insufficient information for identification, but it is adequate to confirm identity; therefore, MS is one of the preferred techniques for characterization and quality control of recombinant proteins and other biomolecules. In the same way, it has been used to study posttranslational modifications (like glycosylation and disulfide bonding pattern), and other processes that can modify protein mass.[11]

For identification purposes, two basic approaches have been used. In the first, a purified protein is enzymatically digested, the product peptide mixture is analyzed and a mass spectral pattern is obtained. This pattern, called the "MS fingerprint", is used to search in internet-available protein or DNA databases.[17] Information about protein origin and an estimate of its MW are required in order to improve the chances of a correct match. The search algorithm then theoretically digests all appropriate proteins in the database with the specified enzyme, and matches the theoretical and experimental masses. The best matches are ranked, and a confidence parameter is usually calculated.[18]

The second approach uses peptide sequence tags to identify the peptide, and therefore the protein. This method is based on the assumption that in a MS/MS peptide spectrum some consecutive amino acid residues can usually be clearly identified. In a typical protocol, the peptides obtained from a protein tryptic digest are

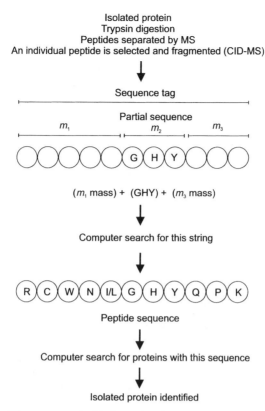

Isolated protein
Trypsin digestion
Peptides separated by MS
An individual peptide is selected and fragmented (CID-MS)

Sequence tag

Partial sequence

m_1 m_2 m_3

G H Y

(m_1 mass) + (GHY) + (m_3 mass)

Computer search for this string

R C W N I/L G H Y Q P K

Peptide sequence

Computer search for proteins with this sequence

Isolated protein identified

Figure 15.11. Peptide sequence elucidation using the sequence tag approach, with posterior protein identification.

ionized using ESI, separated by mass (first quadrupole), selected peptides are fragmented by CID, and the fragments are mass analyzed (Fig. 15.11). The determined amino acid sequence can be very short (two residues), but considering the selectivity of the enzymatic reaction and that the peptides' MW values are known, the probability of random matches is usually low, for example, 3% for a hypothetical situation with only two known amino acid residues.[19] In this example, data presented in Table 15.8 are used to produce a computer search string, and entered into a database.

Figure 15.11 also shows a scheme for the sequence tag approach; in this scheme the definition used is that $m_1 + m_2 + m_3 = M$, where M is the peptide mass. Computer programs that use algorithms designed to search using this tag procedure are accessible on the internet.[20]

Obviously, some knowledge of the protein sequence (or the gene sequence) is necessary in order to use this method for protein identification. Therefore, the probability of a correct match is higher when the protein to be identified comes from an organism possessing a totally sequenced genome, such as *E. coli*, *B. subtilis*, or *H. sapiens*.

TABLE 15.8. Data Used to Search for a Peptide
Sequence Using the Tag Approach[a]

Match Criteria	
Enzyme specificity	Trypsin
Measured peptide molecular mass	2111 ± 0.4
Run of sequence ions	977.4, 1074.5, 1161.5
Type of ion series	b series (see Fig. 15.13)
Partial sequence	PS
Mass of region 1 (m_1)	977.4
Mass of region 3 (m_3)	949.5
Search string	(977.4)PS(949.5)

[a] The masses are monoisotopic and in daltons. The mass difference
between the 997.4 and 1074.5 peaks corresponds to the mass of a
proline residue.

15.6. PROTEIN–PEPTIDE SEQUENCING

There are several instrumental MS methods that can be used to obtain sequence
information from proteins and peptides. ESI-triple-quadrupole is frequently used;
this configuration produces a reasonable number of fragments, the resolution is
usually sufficient and the equipment is relatively inexpensive. With this MS/MS
configuration, peptides up to 2500 Da can be analyzed. A MALDI with a TOF
reflectron and Fourier transform–ion cyclotron resonance mass analyzers are also
beginning to be used.[21]

A typical procedure for protein sequencing begins by digestion with a protease
such as trypsin in order to obtain a collection of tryptic peptides. Other enzymes
and their cleavage sites are described elsewhere.[22] This peptide collection can be
ionized using ESI, separated by the first quadrupole according to m/z ratios, and
then fragmented (CID), with the resulting fragments analyzed in the third quadrupole. The mass spectrum can be analyzed to establish the peptide sequence, or compared directly with a peptide database. Figure 15.12 depicts this typical protocol.

The CID procedure causes more or less random peptide bond cleavage.
Therefore, a number of fragments are obtained that differ by a single amino acid
residue. Several types of fragments are produced, and their nomenclature is
shown in Figure 15.13.

Two main classes of ions are formed by CID, those that contain the C terminus
plus one or more additional residues (ions of types x_n, y_n, and z_n) and those that
contain the N-terminus and one or more additional residues (ions of types a_n, b_n,
and c_n). Ions of types y_n and b_n are formed by the rupture of amide bonds, and the
mass differences between these fragments are limited to the masses of the naturally
occurring 19 different amino acid residues (Table 15.9). Isoleucine and leucine,

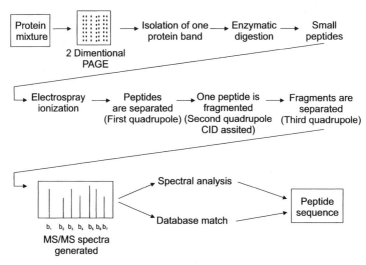

Figure 15.12. Typical procedure used to obtain protein sequence information.

which are structural isomers and therefore have the same molecular weight, cannot be distinguished with MS. Moreover, lysine and glutamine can only be differentiated if resolution is high, since their mass difference is only 0.0432 Da.

The sequencing process consists of the simultaneous measurement of a peptide collection, as shown in Figure 15.14. The sequence is obtained by observation of

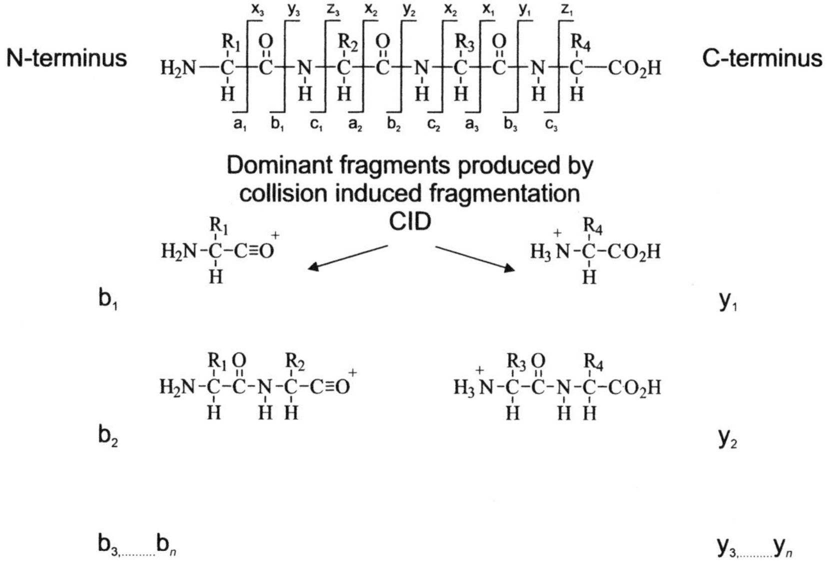

Figure 15.13. Nomenclature used for the more common fragments produced by CID peptide fragmentation. The b and y fragment "families", that occur when the peptide bond is broken, are normally produced at higher concentration.

TABLE 15.9. Average Masses of Amino Acids in Their Free and Residue States[a]

Amino Acid	Codes 3-Letter	M_r		
		1-Letter	Residue	Free
Alanine	Ala	A	71.0786	89.0938
Arginine	Arg	R	156.1870	174.2022
Asparagine	Asn	N	114.1036	132.1188
Aspartic acid	Asp	D	115.0884	133.1036
Cysteine	Cys	C	103.1386	121.1538
Glutamine	Gln	Q	128.1304	146.1456
Glutamic acid	Glu	E	129.1152	147.1304
Glycine	Gly	G	57.0518	75.0670
Histidine	His	H	137.1408	155.1560
Isoleucine	Ile	I	113.1590	131.1742
Leucine	Leu	L	113.1590	131.1742
Lysine	Lys	K	128.1736	146.1888
Methionine	Met	M	131.1922	149.2074
Phenylalanine	Phe	F	147.1762	165.1914
Proline	Pro	P	97.1164	115.1316
Serine	Ser	S	87.0780	105.0932
Threonine	Thr	T	101.1048	119.1200
Tryptophan	Trp	W	186.2128	204.2280
Tyrosine	Tyr	Y	163.1756	181.1908
Valine	Val	V	99.1322	117.1474

[a] One letter codes are customary in the MS area. Reprinted, with permission, from M. Mann and M. Wilm, *Anal. Chem.* **66** (No 24), 1994, 4390-4399. "Error Tolerant Identification of Peptides in Sequence Databases by Peptide Sequence Tags". Copyright © 1994 American Chemical Society.

the mass differences within one family of ions; the "y" family is fully represented in this example. The obtained sequence was then matched to an *E. coli* membrane protein.[23] With this instrumentation, it is possible to analyze peptides with MW up to 2500 Da, although some limitations apply; if only a limited peptide collection is obtained, only limited sequence information can be produced.

Sequence information can also be obtained using Edman degradation to remove amino-terminal residues from a peptide, to produce a collection of peptides. MALDI–TOF can then be used to obtain the peptides masses, and the sequence determined by mass difference between consecutive peptides. This methodology is called protein ladder sequencing, and allows information to be obtained for up to 30 residues. This method is useful for the identification of posttranslational modifications, such as phosphorylated amino acid residues.[24]

Once a peptide family has been sequenced, the next step is to overlap the available sequence information, in order to obtain the protein sequence. However, peptide sequences are more frequently used to identify proteins by searching the peptide sequences in databases. Several databases are available, and their utility

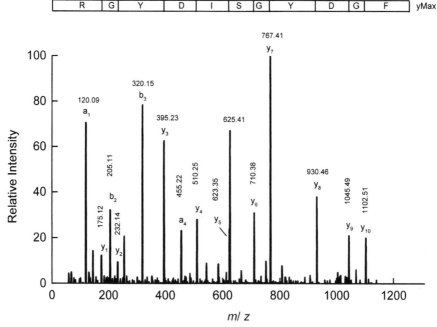

Figure 15.14. Determination of peptide sequence using nanoelectrospray ionization, and a very high-resolution mass analyzer (Q-TOF). In the first quadrupole, a doubly charged peptide ion of $m/z = 625.41$ was selected and later fragmented. The m/z CID spectrum yields the FGDYGSIDYGR sequence, shown at the top.[23] [Reprinted, with permission, from E. Gustafsson, K. Thorén, T. Larsson, P. Davidsson, K. Karlsson, and C. L. Nilsson, Identification of Proteins from *Escherichia coli* Using Two-Dimensional Semi-Preparative Electrophoresis and Mass Spectrometry. *Rapid Communications in Mass Spectrometry* **15**, 2001, 428–432. Copyright © 2001 John Wiley & Sons, Ltd.]

is expected to improve as more experimental data are incorporated. In this regard, it must be noted that two mass spectra are comparable only if they were obtained under very similar instrumental/operational conditions; therefore, several different databases are needed, at least one for each ionization system. MALDI and ESI spectra are quite different, because these two ionization methods produce different types of ions.

15.7. NUCLEIC ACID APPLICATIONS

The application of MS techniques to nucleic acids has been limited in comparison with protein–polypeptide applications, mainly because of the difficulty of ionizing DNA or RNA, since their chemical and structural properties (mainly the negatively charged sugar-phosphate backbone) prevent efficient positive ionization by MALDI or ESI. The use of NH_4^+ counterions is an established method that allows the production of oligonucleotide ions, but the sensitivity and resolution obtained is low. In general, the MALDI ionization process is very inefficient for nucleic acids in comparison with peptides. Generally, 100 times more DNA has to be used to achive similar signal intensity.

Another approach being used to enhance the ionization process is the chemical modification of the nucleic acid molecule in order to improve ionic volatility. For example, the replacement of phosphate protons from native DNA backbones by alkyl groups,[25] or the replacement of phosphate groups by phosphorothioate groups followed by alkylation[26] have been reported.

Another useful possibility is to use ESI in the negative ion mode, which yields a better signal for nucleotides. Multiply charged ions of the type $(M - nH)^{n-}$ are formed. Precise mass measurements for synthetic oligonucleotides with as many as 132 residues have been reported.[27]

Almost all procedures described for nucleic acid analysis (usually DNA) rely on PCR amplification. Short oligonucleotides have been sequenced (usually up to 30 nucleotides), although today MS is used more as a sequence confirmation method than as a sequence determination method. A nomenclature for ion fragments with higher probability of formation has been proposed,[28] and some recent work describes procedures for DNA sequence elucidation using ESI ionization and ion fragmentation (CID spectra).[29] The theory is similar to that used for protein sequencing; when a complete series of fragments is found, the mass difference between two consecutive fragments corresponds to a nucleotide residue mass. In practice, however, MS nucleic acid sequencing methodology is not competitive with alternative sequencing methods.

Several assays have been patented for mutation detection; these are mainly designed for single-nucleotide polymorphism (SNP) analysis and use MALDI–TOF spectrometry (Invader[TM], Sequazyme-PinPoint[TM] assay, MassARRAY[TM], GOOD[TM] assay) all of which use PCR amplification, or require a high DNA concentration in the sample (Invader[TM]).

MAGIChip[TM] is a generic microchip-size device, where spots of polyacrylamide gel are separated by hydrophobic glass. Several gel squares (100 μm) are contained on one glass support, and are separated by 200 μm. Short DNA sequences are immobilized on the gel and are assayed for hybridization; later, the matrix compound is added, allowed to dehydrate, and the spots are analyzed by MALDI–TOF. With this device, different SNP sequences are recognized by their different masses.

Short oligonucleotide mass analysis (SOMA) assay uses PCR amplification followed by restriction endonuclease incubation to digest the amplified DNA; the short oligonucleotides are later separated by HPLC and analyzed by ESI–Triple quadrupole MS (used in the negative ion mode), generating a fingerprint. This method has been used to genotype several variant sites in the human adenomatous polyposis coli gene, for which some variants have been associated with an increased risk of colorectal cancer.[30]

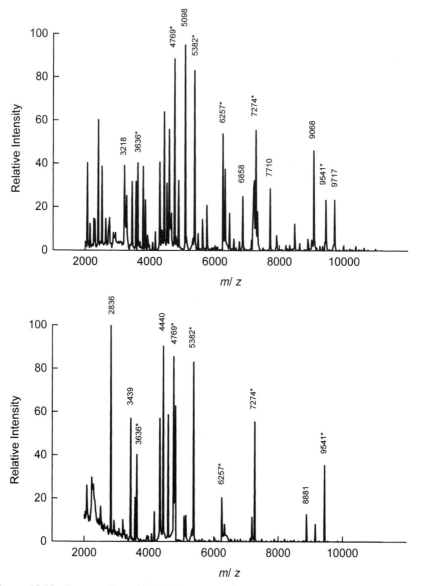

Figure 15.15. Comparasion of MALDI spectra of *E. coli* (strain ATCC 11775) grown in different laboratories, under the same experimental conditions. The asterisks on the bottom trace indicate common peaks in the two spectra.[31,32] [Top trace reprinted, with permission, from Z. Wang, L. Russon, L. Li, D. C. Roser, and S. R. Long, *Rapid Communications in Mass Spectrometry* **12**, 1998, 456–464. "Investigations of Spectral Reproducibility in Direct Analysis of Bacteria Proteins by Matrix-Assisted Laser Desorption/Ionization Time-of-Flight Mass Spectrometry". © 1998 John Wiley & Sons, Ltd. Bottom trace reprinted, with permission, from Catherine Fenselau and Plamen A. Demirev, *Mass Spectrometry Reviews* **20**, 2001, 157–171. "Characterization of Intact Microorganisms by Maldi Mass Spectrometry. Copyright © 2002 by John Wiley & Sons, Inc.]

15.8. BACTERIAL MASS SPECTROMETRY

The classical microbiology approach to microorganism identification or classification relies on differential metabolic pathways. More recently, molecular biology has allowed genetic sequencing methods to be used to identify and/or classify microbiological diversity. Chemical characterization (or chemotaxonomy) of bacteria and other organisms is a third general approach.

Chemical characterization hypothesizes that certain molecules are unique and representative of individual microorganisms, and therefore the mass spectra can be characteristic of given species. A typical bacterial spectrum obtained by MALDI–TOF usually contains between 20 and 40 large peaks, mainly produced by proteins, phospholipids, and cyclic lipopeptides, in the MW range of 4–20 kDa.[31] ESI is less frequently used; samples containing intact cells can clog electrospray devices, and the formation of multiply charged species complicates the spectra.

The cells to be analyzed (mainly bacteria, but fungus or virus particles have also been studied) are typically physically or chemically lysed; exposure to the matrix solution may be sufficient to lyse the cells. Spores are digested (e.g., with strong organic acids), and then the MALDI matrix is added.

Interpretation of bacterial mass spectra is done with difficulty. First, a large portion of cellular compounds are common between different bacterial strains. Second, the spectral reproducibility for a given bacterial strain and protocol/instrument is poor, whether it is tested in the same or different laboratories. There are a number of biological and instrumental sources of variability, including differences in cellular composition for different culture phases, culture media, and stress; minor variations in the sample/matrix ratio cause significant modification in the resulting spectra. Figure 15.15 shows typical bacterial spectra obtained for one strain of *E. coli*; note that many peaks are common to both spectra. Efforts are underway to improve reproducibility, and good results have been obtained by some authors.[33]

Two main approaches have been used to interpret bacterial mass spectra. First, pattern recognition methods have been used, in which reference spectra are obtained under standardized conditions to generate a fingerprint collection. An unknown bacterial spectrum obtained under the same conditions is then compared to the collection and matched if possible.

A second approach exploits the information contained in genome and protein sequencing databases. The masses of a set of proteins from the bacterial strain under study are determined and used to search databases for protein molecular mass matching. This has been used as a method to differentiate *B. subtilis* and *E. coli*, two organisms that have completely sequenced genomes.[34]

SUGGESTED REFERENCES

G. Siuzdak. *Mass Spectrometry for Biotechnology*, Academic Press, New York, 1996.

B. S. Larsen and C. N. McEwen, Eds., *Mass Spectrometry of Biological Materials*, Marcel Dekker, New York, 1998.

A. P. Snyder, *Interpreting Protein Mass Spectra*, Oxford University Press, Oxford, UK, 2000.

C. Dass. *Principles and Practice of Biological Mass Spectrometry*, Wiley-Interscience, New York, 2001.

J. R. Chapman, Ed., *Mass Spectrometry of Proteins and Peptides*, Humana Press, Totowa, New Jersey, 2000, 538 pp.

REFERENCES

1. R. A. W. Johnstone and M. E. Rose, *Mass Spectrometry for Chemists and Biochemists*, Cambridge University Press, New York, 1996, pp. 34–41.

2. G. Siuzdak, *Mass Spectrometry for Biotechnology*, Academic Press, New York, 1996, pp. 9–17.

3. G. Siuzdak, *Mass Spectrometry for Biotechnology*, Academic Press, New York, 1996, pp. 11–18.

4. R. A. W. Johnstone and M. E. Rose, *Mass Spectrometry for Chemists and Biochemists*, Cambridge University Press, New York, 1996, pp. 34, 105–109.

5. C. N. Mc.Ewen and B. S. Larsen, in *Mass Spectrometry of Biological Materials*, B. S. Larsen and C. N. McEwen, Eds., Marcel Dekker, New York, 1998, pp. 10–22.

6. C. N. Mc.Ewen and B. S. Larsen, in *Mass Spectrometry of Biological Materials*, B. S. Larsen and C. N. McEwen, Eds., Marcel Dekker, New York, 1998, pp. 17–19.

7. G. Siuzdak, *Mass Spectrometry for Biotechnology*, Academic Press, New York, 1996, pp. 35–41.

8. R. A. W. Johnstone and M. E. Rose, *Mass Spectrometry for Chemists and Biochemists*, Cambridge University Press, New York, 1996, pp. 37, 54.

9. S. L. Mullen and A. G. Maeshall, *Anal. Chim. Acta* **178**, 1985, 17–26.

10. A. P. Snyder, *Interpreting Protein Mass Spectra*, Oxford University Press, Oxford, UK, 2000, pp. 87–110.

11. M. Mann, R. C. Hendrickson, and A. Pandey. *Annu. Rev. Biochem.* **70**, 2001, 437–473.

12. C. Dass, *Principles and Practice of Biological Mass Spectrometry*, Wiley-Intercience, New York, 2001, pp. 83–84.

13. G. Siuzdak, *Mass Spectrometry for Biotechnology*, Academic Press, New York, 1996, pp. 66–68.

14. G. Siuzdak, *Mass Spectrometry for Biotechnology*, Academic Press, New York, 1996, p. 90.

15. G. Siuzdak, *Mass Spectrometry for Biotechnology*, Academic Press, New York, 1996, pp. 22–24.

16. C. Dass, *Principles and Practice of Biological Mass Spectrometry*, Wiley-Interscience, New York, 2001, pp. 125–128.

17. For example, see http://pepsea.protana.com/ or http://www.proteometrics.com/ or http://prowl.rockefeller.edu/cgi-bin/ProFound.

18. W. J. Griffiths, A. P. Jonsson, S. Liu, D. K. Rai, and Y. Wang, *Biochem. J.* **355**, 2001, 545–561.

19. M. Mann and M. Wilm, *Anal. Chem.* **66**, 1994, 4390–4399.

20. For example, see http://www.mann.embl-heidelberg.de/Services/PeptideSearchIntro.html.

21. G. Siuzdak, *Mass Spectrometry for Biotechnology*, Academic Press, New York, 1996, p. 94.

22. A. P. Snyder, *Interpreting Protein Mass Spectra*, Oxford University Press, Oxford, UK, 2000, pp. 166–167.

23. E. Gustafsson, K. Thorén, T. Larsson, P. Davidsson, K. Karlsson, and C. L. Nilsson, *Rapid Commun. Mass Spectrom.* **15**, 2001, 428–432.

24. B. T. Chait, R. Wang, R. C. Beavis, and S. B. H. Kent. *Science* **262**, 1993, 89–92.

25. T. Keough, T. R. Baker, R. L. Dobson, M. P. Lacey, T. A. Riley, J. A. Hasselfield, and P. E. Hesselberth. *Rapid Commun. Mass Spectrom.* **7**, 1993, 195–200.

26. S. Sauer, D. Lechner, and I. G. Gut, in *Mass Spectrometry and Genomic Analysis*, J. Nicholas Housby, Ed., Kluwer Academic Publishers, Dordrecht, The Netherlands, 2001, pp. 51–53.

27. N. Potier, A. V. Dorsselaer, Y. Cordier, O. Roch, and R. Bischoff, *Nucleic Acid Res.* **22**, 1994, 3895–3903.

28. S. A. McLuckey, G. J. Van Berkel, and G. L. Glish, *J. Am. Soc. Mass Spectrom.* **3**, 1992, 60–70.

29. J. Ni, S. C. Pomerantz, J. Rozenski, Y. Zhang, and J. A. McCloskey, *Anal. Chem.* **68**, 1996, 1989–1999.

30. P. E. Jackson, M. D. Friesen, and J. D. Groopman, in *Mass Spectrometry and Genomic Analysis*, J. Nicholas Housby, Ed., Kluwer Academic Publishers, Dordrecht, 2001, pp. 76–92.

31. C. Fenselau and P. A. Demirev, *Mass Spectrom. Rev.* **20**, 2001, 157–171.

32. Z. Wang, L. Russon, L. Li, D. C. Roser, and S. R. Long, *Rapid Commun. Mass Spectrom.* **12**, 1998, 456–464.

33. A. J. Saenz, C. E. Petersen, N. B. Valentine, S. L. Gantt, K. H. Jarman, M. T. Kingsley, and K. L. Wahl, *Rapid Commun. Mass Spectrom.* **13**, 1999, 1580–1585.

34. J. O. Lay, Jr., *Mass Spectrom. Rev.* **20**, 2001, 172–194.

PROBLEMS

1. Calculate the monoisotopic and average mass (M_r) for the following molecules (Table 15.10).

TABLE 15.10

Compound	Formula	Monoisotopic	Average
Glycine	$C_2H_5NO_2$		
Tryptophan	$C_{11}H_{12}N_2O_2$		
Polipeptide	$C_{84}H_{101}N_{15}O_{14}$		
Porcine insulin	$C_{256}H_{381}N_{65}O_{76}S_6$		
Myoglobin	$C_{769}H_{1212}N_{210}O_{218}S_3$		
Big heterodimer protein	$C_{1802}H_{2933}N_{532}O_{462}S_{12}$		

2. In which case (low or high MW) is the monoisotopic peak intensity higher than isotopic peak intensity?

3. A protein molecule is protonated using an ESI nanoelectrospray system, and analyzed using a mass filter (quadrupole). The M_r (average mass) of this protein is 23,630.36. Use adequate nomenclature to symbolize the molecular ion family, for the charge range +1 to +6. Calculate the mass of each ion, and its position in the mass spectrum obtained.

4. β-Lactoglobulin was purified, and an ESI spectrum was obtained, as shown in the Figure 15.16. Calculate the protein mass for each peak and the average for all the peaks. Present a hypothesis to explain the absence of low-charged species.

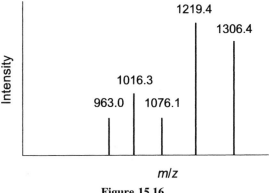

Figure 15.16

5. Calculate the resolution for the following mass spectrum (Fig. 15.17).

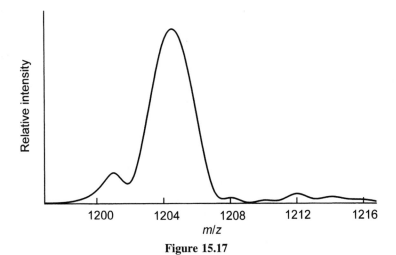

Figure 15.17

CHAPTER 16

Validation of New Bioanalytical Methods

16.1. INTRODUCTION

Validation has the primary goal of selecting the best available method for detection and/or quantitation, based on objective performance data. Regulations require that clinical laboratories show documentation and validation data when new tests are introduced. These data are also important for quality control and troubleshooting. The successful use of an assay depends first on a fundamental understanding of the analytical principles governing the assay. Assay characteristics such as precision, accuracy, and detection limit depend on the properties of the components, the assay format, and the detection system. For example, both manual and automated assays may have problems associated with timing or reaction rates, but the automated method may be expected to have better precision in reagent delivery. Both may have imprecision as a result of low analyte concentration or a slow approach to equilibrium. Even when the assay has been optimized in theory, practical work requires care and technical skill for optimum performance. Completeness of recovery in each step should be checked independently, for example, by using an independent tracer to follow the progress of a reaction. Equilibration rates may be very sensitive to temperature, the order of addition of reagents and stir rates. Assaying standards in each run exactly the same way as samples counterbalances some minor errors; however, poor timing or temperature control often results in increased coefficients of variation for replicate measurements. Incubation times (when necessary) are often shortened to the minimum, and room temperature incubations are common; if equilibration has not occurred, minor technical variations may have exaggerated effects (e.g., room temperature may vary considerably from summer to winter).

Decisions must be made about the method based on the quality of the data produced, and quality criteria must be established. This task is not trivial, since new guidance documents are continually produced to determine whether an analytical

Bianalytical Chemistry, by Susan R. Mikkelsen and Eduardo Cortón
ISBN 0-471-54447-7   Copyright © 2004 John Wiley & Sons, Inc.

method can be used for a quantitative analysis, or whether a new method can replace an established (standard) one. Guidance documents and regulations have been produced by domestic and international agencies for approval of new methods, in order to protect public health, food quality, and other sensitive areas.

Different levels of validation are usually defined. For example, methods proposed to be used worldwide in medically related applications must demonstrate high-quality data related to both precision and accuracy, whereas analytical methods used for research purposes require less stringent validation.[1] High degrees of validation involve several laboratories and the assay of a large number of samples; this tends to be very expensive and impractical for locally or occasionally used methods.

International organizations as AOAC International (International Association of Official Analytical Chemists), FAO (Food and Agriculture Organization of the United Nations), IAEA (International Atomic Energy Agency), IUPAC (International Union of Pure and Applied Chemistry), ISO (International Organization for Standardization), have produced international guidelines that can be used in conjunction with the different local policies in effect around the world. In the specific case of biomedical testing, regulations are contained in CLIA (Clinical Laboratory Improvement Amendments), which are the responsibilities of the FDA. Following established validation procedures is a necessary step to obtain regulatory approval to use a new analytical method, or to use an approved method in a new application.

The definitions, methods, and parameters used to validate analytical and bioanalytical methods are not universal; they vary with the type of assay and the regulatory agency. Here, we introduce the more broadly used figures of merit with their generally accepted definitions. Basic statistical procedures are also presented. More technical sources of further information are offered in the Suggested Reading section.

Bioanalytical methods are classified according to the kind of results they provide. *Qualitative assays* provide information about the presence or absence of the analyte, whereas *quantitative* assays give the analyte concentration and the uncertainty in this value, in the sample. *Semiquantitative* methods give an estimate of concentration and are used to determine whether analyte is present at a level that is above or below a reference concentration; therefore the report can state if the concentration in the sample is higher, similar, or lower than the reference value.

16.2. PRECISION AND ACCURACY

Uncertainty is classified in two major groups: *Random error* is always present in experimental data and can never be completely eliminated. It can result from the random nature of collisions that lead to chemical or biochemical reactions, or may be caused by small voltage fluctuations in measurement instrumentation. Random error causes positive and negative deviations from the true value, and affects the precision of the results. Precision is usually discussed in terms of standard deviation (s) and relative standard deviation (RSD), both defined later in this chapter. *Systematic error* is produced by a more or less constant mistake, and results in a

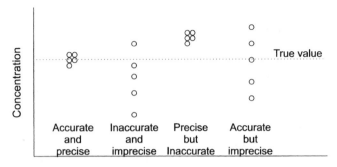

Classification of results

Figure 16.1. Precision and accuracy are not synonymous.

consistent positive or negative deviation of measured values from true values. For example, absorbance measurements that are not corrected for the absorbance of a blank solution that contains no analyte will produce experimental results that are systematically higher than the true concentration values. Systematic errors affect the accuracy of the results, and are studied by comparison of experimental results with "real" concentrations using certified reference materials or with another analytical method that is known to be very accurate. In principle, systematic error can be eliminated from experimental results, because the cause of the error can be found and eliminated. Figure 16.1 shows a graphical comparison of precision and accuracy.

16.3. MEAN AND VARIANCE

Arithmetic mean ($\bar{x}$) and *standard deviation* (s) are defined as follows, where n is the number of individual measurements and x_i are the values of those measurements.

$$\bar{x} = \sum x_i / n \tag{16.1}$$

$$s = \sqrt{\frac{\sum (x_i - \bar{x})^2}{(n-1)}} \tag{16.2}$$

The calculated standard deviation s is a good estimator of the population standard deviation σ if the number of measurements is high enough. An equivalent alternative form of Eq. 16.2 is presented in Eq. 16.3; this form is more useful when a nonprogrammable calculator is used.

$$s = \sqrt{\frac{\sum x_i^2 - \frac{\left(\sum x_i\right)^2}{n}}{n-1}} \tag{16.3}$$

The sample *variance* (s^2) is used instead of s in some calculations.

16.4. RELATIVE STANDARD DEVIATION AND OTHER PRECISION ESTIMATORS

Precision (as discussed previously) is the ability of the assay to give the same result when repeated multiple times either within the same run, or from day to day (i.e., between runs). Precision data are usually collected early in the evaluation at a new method, since the quality of the method depends on the level of precision. If precision is poor, more elaborate validation experiments must be postponed until the sources of imprecision are identified and controlled.

Relative standard deviation and *coefficient of variation* (CV), are used to estimate precision, and are defined by the following equations:

$$\text{RSD} = \frac{s}{\bar{x}} \tag{16.4}$$

$$\text{CV} = \frac{s}{\bar{x}} \times 100 \tag{16.5}$$

Other parameters frequently used are the *mean absolute deviation* (MAD), as an estimator of the dispersion of groups of single observations, and the *standard error of the mean* (SEM), which is used to calculate the confidence limits of a mean value. Their formulae are presented in Eqs. 16.6 and 16.7.

$$\text{MAD} = \frac{\sum |x_i - \bar{x}|}{n} \tag{16.6}$$

$$\text{SEM} = \frac{s}{\sqrt{n}} \tag{16.7}$$

Precision varies with concentration; because of this, precision should be evaluated at the low, middle, and high concentration regions of the standard curve, and should be evaluated in the different matrices that will be encountered in real assays. The precision profile (Section 16.4.3) is used to establish the working range of the assay.

16.4.1. Distribution of Errors and Confidence Limits

When the concentration values obtained from repetitive measurements of replicate samples are plotted, a normal (or Gaussian) distribution of results is often obtained. The normal curve is symmetrical about $\bar{x}$, and the greater the s value, the wider the spread of the curve.

Another important characteristic of the normal distribution is that $\sim 95\%$ of the data values lie within $\pm 2s$, and 99.7% in the range $\pm 3s$, as shown in Figure 16.2. This distribution of error in a normal distribution allows the calculation of confidence intervals for $\bar{x}$. The confidence interval is the range of concentration within which the real sample concentration is expected to occur, for a given degree of confidence. Therefore, the interval size depends of the degree of confidence (for greater

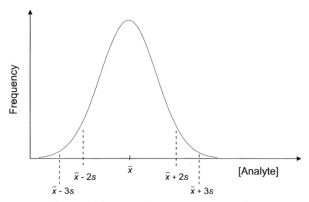

Figure 16.2. Gaussian (normal) distribution.

confidence, a larger interval is required), the value of s and the sample size n, as shown in Eq. 16.8, for a confidence interval of 95%

$$x = \bar{x} \pm \frac{2s}{\sqrt{n}} \tag{16.8}$$

Other commonly used limits are 99.0 and 99.7%, as shown in Eqs. 16.9 and 16.10, respectively.

$$x = \bar{x} \pm \frac{2.58s}{\sqrt{n}} \tag{16.9}$$

$$x = \bar{x} \pm \frac{2.97s}{\sqrt{n}} \tag{16.10}$$

When the sample size (n) is small, the values of the t-distribution are used, as shown in the following equation:

$$x = \bar{x} \pm \frac{ts}{\sqrt{n}} \tag{16.11}$$

Values of the t-statistic are tabulated for commonly used confidence intervals and degrees of freedom $(n - 1)$.

16.4.2. Linear Regression and Calibration

When a linear relationship is observed between two variables, the correlation is quantified by a method such as linear least-squares regression. This method determines the equation for the best straight line that fits the experimental data.

The line has the form of the Eq. 16.12, were x and y are variables, a is the slope and b is the intercept on the y axis. Linear regression determines the best values of a

and b, along with their uncertainties.

$$y = ax + b \tag{16.12}$$

A linear relationship between a measurable parameter (like absorbance) and concentration is observable in many analytical methods, and the least-squares method is used to fit the best calibration curve.

Normally, a minimum of 6–8 points are necessary to ensure the linearity of the calibration curve and to calculate the regression parameters. In this case, the following equations are used to calculate a and b values and their confidence limits. We assume that $\sigma \approx s$ and that σ is independent of x.

$$a = \frac{n \sum x_i y_i - \sum x_i \sum y_i}{D} \pm \sqrt{\frac{n s_y^2}{D}} \tag{16.13}$$

$$a = \bar{a} \pm s_a \tag{16.14}$$

$$b = \frac{\sum y_i \sum (x_i)^2 - \sum x_i \sum x_i y_i}{D} \pm \sqrt{\frac{s_y^2 \sum (x_i)^2}{D}} \tag{16.15}$$

$$b = \bar{b} \pm s_b \tag{16.16}$$

where

$$D = n \sum (x_i)^2 - \left(\sum x_i \right)^2 \tag{16.17}$$

and

$$s_y^2 = \frac{\sum (y_i - y_{\text{pred}})^2}{n - 2} \tag{16.18}$$

To determine $\bar{x}$ from a measured $\bar{y} \pm s_y$ value:

$$x = \frac{y - b}{a} \pm \left(\frac{y - b}{a} \right) \sqrt{\left(\frac{\sqrt{s_y^2 + s_b^2}}{y - b} \right)^2 + \left(\frac{s_a}{a} \right)^2} \tag{16.19}$$

$$x = \bar{x} \pm s_x \tag{16.20}$$

where y_{pred} is the y predicted by the calibration curve, s_y is the standard error of the estimate [sometimes denoted $s_{(y/x)}$], and n the number of points on the calibration curve. Confidence limits for x are calculate using Eq. 16.11.

The correlation coefficient r, or more frequently r^2, is used to calculate the quality of the calibration line obtained (Eq. 16.21). A value of 1 indicates a perfect positive correlation; values > 0.95 are usually considered acceptable.

$$r^2 = \frac{\left[\sum (x_i - \bar{x})(y_i - \bar{y})\right]^2}{\sum (x_i - \bar{x})^2 \sum (y_i - \bar{y})^2} \tag{16.21}$$

16.4.3. Precision Profiles[2]

Precision profiles are useful to establish the concentration range (for a given level of precision) for which a method can be used. Precision is generally worst at the lowest concentration of this range. Figure 16.3 demonstrates how the dynamic (or

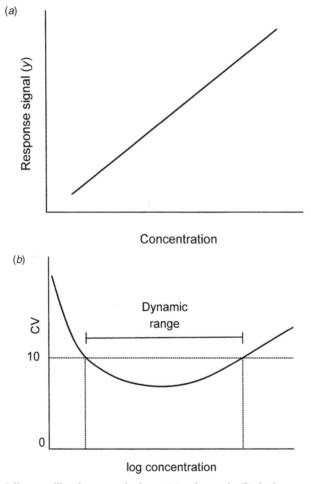

Figure 16.3. A linear calibration curve is shown (*a*), whereas in (*b*) the log concentration has been plotted to recognize the poor precision at both ends of the dynamic range. If lower precision is acceptable, the dynamic range becomes wider.

working) concentration range of an assay is determined. In Figure 16.3(*b*), the dynamic range of the assay is shown as the range over which the CV is <10%. Precision profiles are a very useful means of comparing two different assay method for the same analyte.

Precision studies can be performed under different conditions, and are strongly influenced by variables such as temperature, source and quality of reagents, reproducibility of reagent delivery, and instrumental noise. Therefore, if all precision studies are done in the same laboratory (*intralaboratory* study) higher precision is expected in comparison with *interlaboratory* studies, where several laboratories produce the data used to prepare the method precision profile.

Intralaboratory precision studies are classified as *intraassay*, where the profile is obtained doing the replication in only one run, using the same batch of reagents, or *interassay*, where the precision profile is obtained by comparison of runs done on different days. Poorer precision is generally obtained in interassay studies.

16.4.4. Limit of Quantitiation and Detection

Limit of detection (LOD) is defined as that analyte concentration that gives a signal significantly different from the blank or background (b) signal. For practical use, the meaning of "significantly different" is usually defined as the blank signal plus two standard deviations of a well-characterized blank (s_b). Well characterized means at least 10 blank measurements.

$$y_{LOD} = b + 2\,s_b \tag{16.22}$$

The LOD value is then calculated from linear regression parameters obtained from calibration data,

$$(b + 2\,s_b) = ax + b \tag{16.23}$$

where x is the concentration at the detection limit of the assay.

The limit of quantitation (LOQ) is defined by the maximum acceptable level of uncertainty in the measured values, and is generally considered to be the analyte concentration that yields a signal equal to the blank signal plus 10 standard deviations, as shown below. These and alternative definitions of LOD and LOQ are presented elsewhere.[3]

$$y_{LOQ} = b + 10\,s_b \tag{16.24}$$

16.4.5. Linearizing Sigmoidal Curves (Four-Parameter Log–Logit Model)

When an assay presents a nonlinear calibration curve (Fig. 16.4), the data can be linearized using standard functions.[4] The log–logit function transforms a sigmoid curve with a single point of inflection into a straight line, and is used extensively with data from competitive immunoassays.

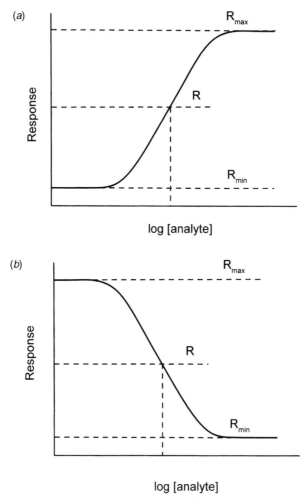

Figure 16.4. Sigmoid response curves: positive (*a*) and negative (*b*) sigmoid responses are shown.

The first step to linearizing a sigmoidal curve is to normalize the response values so that $0 < y < 1$. Normalized values can be calculated using Eq. 16.25.

$$y = \frac{(R - R_{min})}{(R_{max} - R_{min})} \tag{16.25}$$

The data are then linearized using the logit function:

$$\text{logit}(y) = \ln \frac{(1 - y)}{y} \tag{16.26}$$

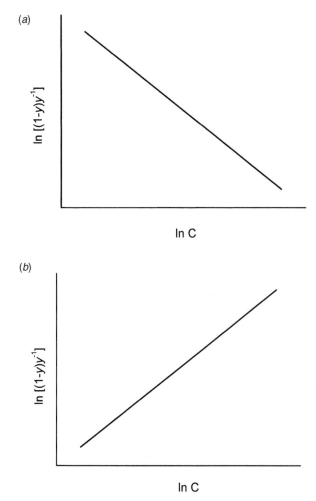

Figure 16.5. Linear plots obtained using the logit function. (*a*) When response increases with the concentration and (*b*) when response decreases with concentration.

Values obtained for logit(y) are then plotted against log[analyte], and a linear plot is obtained as shown in Figure 16.5.

16.4.6. Effective Dose Method

This method is used for estimating analyte concentrations by immunoassay with maximum precision. Serial dilutions of a standard analyte solution are prepared and assayed; serial dilutions of the unknown are also prepared and assayed. Responses from both dilution series are plotted against $\log_{10}$ of the dilution factor, as shown in Figure 16.6.

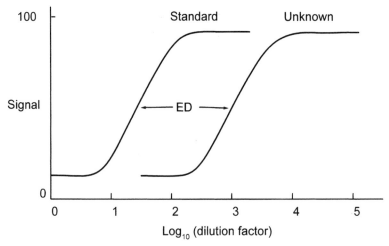

Figure 16.6. By plotting standard and unknown signals, the effective dose (ED) can be calculated graphically.

The best precision (lowest RSD) occurs at the steepest point on the curves. The difference between these points (unknown minus standard) is the *effective dose* (ED):

$$ED = \log_{10}(\text{unknown dilution}) - \log_{10}(\text{standard dilution}) \tag{16.27}$$

$$ED = \log_{10} \frac{[\text{unknown}]}{[\text{standard}]} \tag{16.28}$$

therefore, the unknown concentration can be calculated from Eq. 16.29:

$$[\text{unknown}] = [\text{standard}] \times 10^{ED} \tag{16.29}$$

16.5. ESTIMATION OF ACCURACY

Accuracy is the ability of any assay to provide the "correct" result. Ideally, the assay should detect all of the analyte (100% recovery) and nothing else (no interference or cross-reactivity). To estimate the method accuracy, a comparison of method results with true sample concentrations must be completed. A straightforward procedure involves the use of a *standard reference material*, in which the analyte concentration is known with high accuracy and precision. Standard reference materials are not generally available for biochemical analytes, however. When a reference material is not available, accuracy can be established by comparison with alternative previously validated analytical techniques, or currently accepted methods. Intralaboratory tests of matrix effects and interferences are also conducted in order to establish the accuracy of a new method.

16.5.1. Standardization

Standard solutions are the fundamental basis for the accuracy of assays. In preparing standards, it is usually necessary to make a number of assumptions, related to the purity and behavior of the standards. Standards and samples are assumed to react with assay reagents in the same way, and matrix effects are assumed absent. Standard solutions should be dilutions of a stock solution of highly purified material. Often, in the case of a hormone or other complex molecule, the standard is assigned a concentration or unit value based on a comparison with a master standard curve prepared from an international standard.

16.5.2. Matrix Effects

16.5.2.1. Recovery. Recovery experiments are designed to show whether a method measures all the analyte present in the sample. Interactions with matrix components (e.g., formation of complexes) can sometimes prevent a fraction of the analyte present from be detected.

The recovery of a known amount of analyte from a sample matrix has long been taken to demonstrate accuracy. For this to hold true, the standard material must be pure and the matrix must not interfere with the reactions. When the experimental conditions have been set, the recovery experiments involve spiking a base matrix with analyte. Recovery (R) is calculated from the signal difference between the spiked base and the base itself, and is equal to (measured concentration-base concentration)/(added concentration). The volume of the spike should be small ($< 10\%$ of the total volume), so that the matrix is essentially unchanged by the spike. The recovery (%) can be calculated using Eq. 16.30.

$$\% \, \text{Recovery} = \frac{\text{Measured concentration} \times 100}{\text{Calculated concentration}} \tag{16.30}$$

Recoveries at different concentrations are used to construct the recovery plot, as shown in Figure 16.7, where linearity with slope $= 1$ demonstrates 100% recovery. Recoveries $> 90\%$ are considered acceptable for bioassays. In the same figure, expected deviations from the ideal behavior at low and high concentration are shown.

16.5.2.2. Parallelism. The parallelism experiment is used to determine whether an assay's accuracy depends on the analyte concentration in the sample used. This characteristic is tested by diluting a standard sample, and determining whether results agree after correction for the dilution. Alternatively, results can be plotted against expected values, ideally to yield a straight (flat) line in which the concentration is independent of the dilution factor. The diluent must be carefully selected to ensure a consistent matrix among the dilutions. Failure of the parallelism test may indicate extreme sensitivity of the assay to matrix composition, lack of sensitivity at the extremes of the standard curve or inaccurate standards.

Nonideal behavior at low dilution (as shown in Fig. 16.8) can result from association of analyte with matrix components, association of assay reagents with

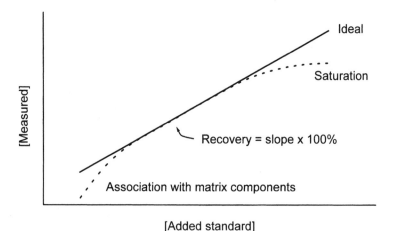

[Added standard]

Figure 16.7. Recovery plot. Low recovery can be explained by association of the analyte with matrix components, losses of analyte during sample manipulation, or inadequate concentration of assay reagents.

matrix components, the presence of enzyme activators/inhibitors in the matrix and/or nonspecific contributions to the signal, for example, from turbidity or background absorbance.

16.5.3. Interferences

Interferences are substances that can be present in the sample, that cause overestimation or underestimation of the true analyte concentration. These species are often structurally related to the analyte, and interact with assay reagents in similar ways to produce positive interference or overestimation.

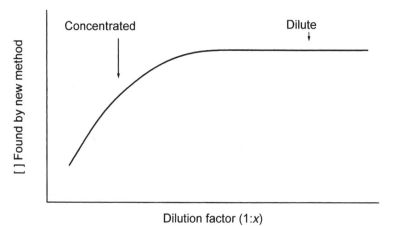

Dilution factor (1:*x*)

Figure 16.8. Parallelism plot. At high dilution factors matrix effect are minimized. The ideal curve is a flat horizontal line, where the measured concentration, after correction for dilution, is independent of the dilution factor.

This kind of interference is called cross-reactivity, or specific interference, and is usually quantitated by assaying for the interferent in the absence of analyte. Validation procedures involve screening of a large number of potential interferents at concentrations higher than their expected levels in real samples. Cross-reactivity methods for immunoassays have been described in Chapter 6.

16.6. QUALITATIVE (SCREENING) ASSAYS

As was stated at the beginning of this chapter, qualitative assays report whether the analyte is present or not in the sample. A qualitative "yes/no" result is important in many clinical assays, where the presence of a certain antibody, for example, indicates a disease state regardless of its concentration. Automated instrumental methods, however, provide numerical results, and it is the conversion of these numerical results into a binary "yes/no" format that is a large component of the validation process. Conversion to the binary format requires a *cutoff value*. This value may be selected as the signal corresponding to the LOD or LOQ; it may also be chosen as the signal produced by a legal or accepted cutoff concentration. A large sample population is then screened by the new assay and by an accepted (validated) quantitative method. Each sample is then characterized as

True positive: Both methods yield positive results.

True negative: Both methods yield negative results.

False positive: New assay positive, accepted assay negative.

False negative: New assay negative, accepted assay positive.

For this categorization, the important assumption has been made that the quantitative method always yields the correct results. The predetermined cutoff value is then adjusted to minimize or eliminate false negatives. Screening assays are used on large sample populations, often to determine which samples require further investigation. For this reason, false positives are not as problematic as false negatives, since they will be further examined. False negatives, however, can result in misdiagnoses, with severe consequences.

16.6.1. Figures of Merit for Qualitative (Screening) Assays

$$\text{Sensitivity} = \text{percent positives identified}$$

$$= \frac{\text{True positives}}{\text{True positives} + \text{false negatives}} \times 100\% \qquad (16.31)$$

$$\text{Specificity} = \text{percent negatives identified}$$

$$= \frac{\text{True negatives}}{\text{True negatives} + \text{false positives}} \times 100\% \qquad (16.32)$$

$$\text{Efficiency} = \text{percent correct results}$$

$$= \frac{\text{True positives} + \text{True negatives}}{\text{Total number of samples}} \times 100\% \qquad (16.33)$$

Sensitivity is generally considered the most important of these figures of merit, because it is an indication of the likelihood of false negative results.

16.7. EXAMPLES OF VALIDATION PROCEDURES

16.7.1. Validation of a Qualitative Antibiotic Susceptibility Test[5]

Medical laboratories are frequently asked to determine which antibiotic compounds are effective against clinically isolated microorganisms. An accepted method for determining antibiotic susceptibilities involves culturing the microorganism on an agar plate in the absence or presence of antibiotic on a small filter-paper disk placed on the agar surface. Growth of the culture (reproduction) is then allowed to occur during overnight incubation at 37 °C. A ring with no growth around a filter-paper disk indicates that the microorganism is susceptible to that antibiotic. This method requires at least 8 h and may require several days, depending on the growth rate of the organism.

In the proposed new method, microorganisms suspended in a liquid culture are incubated at 37 °C for 20 min in the absence or presence of an antibiotic, using the same drug concentration as is present in the accepted agar plate method. Respiratory activity (breathing) is then measured by a new electrochemical method. If respiratory activity is $< 90\%$ of the control measurement, made in the absence of antibiotic, the microorganism is susceptible to the antibiotic. The new method requires only 25 min, and can therefore provide results (and effective treatment regimes) much more rapidly than the accepted method.

Validation was performed using a common laboratory strain of *E. coli* with 13 antibiotics possessing different mechanisms of action, using both the accepted and the new method. Comparison of results for each antibiotic allowed classification of the results as follows: A *true positive* shows no growth by the agar plate method and

TABLE 16.1. Validation Results for the New Qualitative Antibiotic Susceptibility Test

Antibiotic	Agar Plate Result	Respiration Result	Validation Result
Penicillin G	No growth	Decrease	True positive
D-Cycloserine	No growth	Decrease	True positive
Vancomycin	No growth	Decrease	True positive
Bacitracin	Growth	Increase	True negative
Cephalosporin C	No growth	Decrease	True positive
Tetracycline	No growth	Decrease	True positive
Erythromycin	Growth	No change	True negative
Chloramphenicol	No growth	Decrease	True positive
Streptomycin	Growth	No change	True negative
Nalidixic acid	No growth	Decrease	True positive
Rifampicin	No growth	Decrease	True positive
Trimethoprim	No growth	Decrease	True positive
Nystatin	Growth	No change	True negative

decreased respiratory activity by the new method. A false positive involves growth on the agar plate but decreased respiration in the new method. A true negative involves growth on agar, and no charge or an increase in respiration. Finally, a false negative result occurs if there is no observable growth around the filter paper disk by the agar plate method, while the new method shows no charge or an increase in respiratory activity. Table 16.1 summarizes the results obtained in this validation experiment.

From the results shown in Table 16.1, three parameters may be calculated: selectivity, sensitivity and efficiency, as defined in Eqs. 16.31–16.33. In this example, all three parameters are equal to 100%.

Prior to use in a clinical laboratory, this kind of test must undergo regulatory approval. Data are required from several hospital labs, from numerous control and sample organisms, as well as a range of antibiotics.

16.7.2. Measurement of Plasma Homocysteine by Fluorescence Polarization Immunoassay (FPIA) Methodology[6]

Elevated concentrations of plasma homocysteine (HCY) are related to an increased risk of cardiovascular disease, which exists in numerous forms in plasma, with the main form existing as a disulfide with itself, cysteine, or albumin. Therefore, the first step in the measurement involves treatment with a reducing agent, in this case dithiothreitol (DTT), to obtain HCY in its free form (Eq. 16.34). Some amino acids (e.g., L-cysteine and L-methionine) are present in human plasma at higher molar concentrations than HCY and may interfere with this assay. To avoid this possible interference, the highly selective enzymatic conversion of HCY to S-adenosyl-L-homocysteine (SAH), as shown in Eq. 16.34, is used. Both reactions (reduction and conjugation) are accomplished in 30 min at 34 °C.

$$(16.34)$$

A monoclonal antibody was raised against SAH, and used in a competitive reaction where the antibody binds either SAH or a fluoresceinated SAH analogue (tracer), during 20 min at room temperature, as shown in Eq. 16.35.

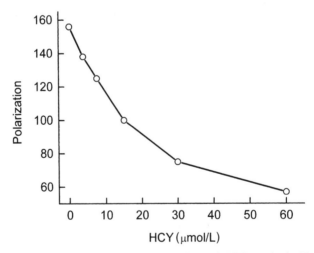

$$(16.35)$$

The fluorescence polarization (P) was calculated from measured fluorescence values according to Eq. 16.36, where $F_\parallel$ is the fluorescence intensity parallel to the excitation plane, and $F_\perp$ is the fluorescence intensity perpendicular to the excitation plane.

$$P = \frac{(F_\parallel - F_\perp)}{(F_\parallel + F_\perp)} \tag{16.36}$$

A calibration curve was constructed using five HCY concentration and a blank solution, and the nonlinear curve obtained (Fig. 16.9) was used to calculate the unknown concentrations.

Figure 16.9. Calibration curve based on determinations of HCY standards. [Reprinted, with permission, from Mohammed T. Shipchandler, and Ewin G. Moore, *Clinical Chem.* **41**:7, 1995, 991–994. "Rapid, Fully Automated Measurement of Plasma Homocyst(e)ine with the Abbot Imx Analyzer". © Amer. Assos. for Clinical Chemistry.]

TABLE 16.2. Intraassay and Interassay Variation Coefficients for Calibration Data from the Plasma Homocysteine Assay[a]

Standard (μmol/L)	CV	
	Intraassay ($n = 3$)	Interassay ($n = 3$)
3.75	8.0	9.1
7.50	3.6	4.5
15.00	3.4	6.5
30.00	2.7	3.0
60.00	1.8	2.6

[a] Reprinted, with permission, from Mohammed T. Shipchandler and Ewin G. Moore. *Clinical Chem.* **41**:7, 1995, 991–994. "Rapid, Fully Automated Measurement of Plasma Homocyst(e)ine with the Abbot Imx Analyzer". © Amer. Assos. for Clinical Chemistry.

Coefficients of variation for intraassay and interassay calibration data are shown in Table 16.2. Matrix effects were studied by dilution with buffer of two real plasma samples. Both parallelism tests (Table 16.3) show that matrix effects are not significant for this method, since the observed–calculated results show apparently random scatter $\sim 100\%$

Another method used to study matrix effects is the recovery experiment. Table 16.4 shown that recoveries are within acceptable limits, since the calculated values are again scattered $\sim 100\%$.

Cross-reactivity was study for similar molecules expected to be present in plasma. L-Cysteine (hydrochloride) and L-methionine at 5 and 4.5 mmol/L, respectively, were assayed in buffer. The observed polarization values correspond to HCY concentrations of 0.0 and 0.1 μmol/L, respectively. Thus, the antibody-binding reaction is sufficiently selective in the presence of these potential interferents.

Accuracy was assessed by comparison with established methods. Regression equations were determined for clinical samples analyzed by this new method and four established HCY quantitation methods. Results from this new method were plotted as *x* values and the established method results were plotted as *y*. Table 16.5 shows the regression equations and their statistical quality parameters.

The correlation coefficients (*r*) obtained between the tested methods are good, although the differences in slopes are large, indicating either overestimation of HCY concentration by methods B–D, or underestimation of HCY concentrations by method A and the new method. The authors believe that these differences can be attributed to different purities of commercial standards (HCY) used in the different laboratories.

The use of standard reference materials, as indicated previously in this chapter, would allow a better assessment of the accuracy of the new method. However, for many biological analytes, these materials are not available. Under these conditions,

TABLE 16.3. Parallelism, Tested in Two Real Plasma Samples (*Plasma 1 and Plasma 2*) Diluted with Buffer[a]

Buffer:Plasma (vol. ratio)	HCY (μmol/L) Observed	HCY (μmol/L) Calculated	Obs. × 100% Calculated
	Plasma 1		
0:1	47.51	47.51	100.0
1:3	33.78	35.63	94.8
1:2	32.74	31.64	103.5
1:1	25.29	23.76	106.5
3:1	12.79	11.88	107.7
7:1	6.82	5.94	114.8
		Mean (SD)	104.5 (6.9)
	Plasma 2		
0:1	40.40	40.40	100.0
1:3	29.14	30.30	96.2
1:2			
1:1	19.48	20.20	96.4
3:1	10.09	10.10	99.9
7:1	5.13	5.00	102.6
		Mean (SD)	99.0 (2.7)

[a] Reprinted, with permission, from Mohammed T. Shipchandler and Ewin G. Moore. *Clinical Chem.* **41**:7, 1995, 991–994. "Rapid, Fully Automated Measurement of Plasma Homocyst(e)ine with the Abbot Imx Analyzer". © Amer. Assos. for Clinical Chemistry.

a comparison of results obtained using different methods, as shown in Table 16.5, is appropriate.

16.7.3. Determination of Enzymatic Activity of β-Galactosidase[7]

Neutral lactase (β-Galactosidase) is an enzyme produced on an industrial scale, which catalyzes the hydrolysis of lactose. Industry utilizes this enzyme in the production of low-lactose dairy products, whey treatment, and fermented lactic products. The goal of this study is to demonstrate that a new colorimetric method to measure the enzymatic activity is adequate for use as a standard method.

The method is proposed to be applicable to solutions of lactase between 2000 and 5000 units/g, where one enzymatic unit is defined as the quantity of enzyme that liberates 1.30 μmol *o*-nitrophenol/min under assay conditions. Lactase decomposes the artificial substrate *o*-nitrophenyl-β-D-galactopyranoside into *o*-nitrophenol and galactose. Absorbance of *o*-nitrophenol (yellow in alkaline medium) is measured to estimate the enzymatic activity, at 420 nm.

TABLE 16.4. Recovery of L-Homocysteine Added to Plasma Samples[a]

HCY (μmol/L) Added	Calculated	Observed	Recovery (%)
Plasma 1			
None		7.29	
0.00	6.07	6.05	99.7
3.75	6.70	5.77	86.1
7.50	7.32	7.20	98.4
15.00	8.57	7.93	92.5
30.00	11.07	10.56	95.4
60.00	16.07	16.36	101.8
120.00	26.07	27.56	105.7
		Mean (SD)	97.7 (6.4)
Plasma 2			
None		16.2	
0.00	13.5	13.56	100.4
3.75	14.13	14.50	102.6
7.50	14.75	14.77	100.1
15.00	16.00	15.35	95.9
30.00	18.50	18.55	100.3
		Mean (SD)	99.9 (2.4)

[a] Reprinted, with permission, from Mohammed T. Shipchandler and Ewin G. Moore. *Clinical Chem.* **41**:7, 1995, 991–994. "Rapid, Fully Automated Measurement of Plasma Homocyst(e)ine with the Abbot Imx Analyzer". © Amer. Assos. for Clinical Chemistry.

TABLE 16.5. Accuracy of the HCY Assay, Determined by Comparison of Results from Clinical Samples with Results from Established Methods (A to D), Performed in Different Laboratories

Method	Equation	r	s_y	n
A	$y = 1.030x + 0.184$	0.980	1.183	42
B	$y = 1.212x - 0.319$	0.995	0.457	10
C	$y = 1.119x + 0.167$	0.996	0.557	8
D	$y = 1.493x - 1.145$	0.997	0.877	21

Thirteen laboratories participated in this study; each one received the complete protocol including precise instructions about the preparation and storage of each reagent, test samples to calibrate the procedure before beginning the interlaboratory study, and five blind duplicate samples. The relatively low number of samples is due to the fact that only two concentrations (2000 and 5000 units/g) were tested in this study.

TABLE 16.6. Interlaboratory Comparison of Results for the Determination of β-Galactosidase Activity Expressed in NLU/g[a]

	Samples				
Laboratory	1 + 10	2 + 7	3 + 8	4 + 6	5 + 9
A	4401, 4343	1680, 1655	1716, 1690	4583, 4385	1586, 1542
B	5278, 5591	2232, 2139	2280, 2347	5386, 5340	2004, 2024
C	4522, 4569	1821, 1825	1870, 1874	4879, 4638	1718, 1649
D	5040, 5099	1903, 1856	1955, 2059	5113, 5037	1780, 1706
E	5320, 5097	2086, 2036	2107, 2088	5405, 5340	1924, 1884
F	5404, <u>5533</u>	2125, 2108	2189, 2146	5470, 5472	1959, 1940
G	5462, 5246	2249, 2099	1980, 2114	5329, 5250	2012, 1916
H	4925, 4765	1693, 1838	<u>1356, 1569</u>	<u>4022</u>, 5066	1834, 1578
Mean	5031	1959	1959	5044	1816
RSD_r (%)	3.20	3.75	8.62	7.53	5.01
RSD_R (%)	8.77	10.83	16.35	10.88	10.26

[a] In samples 1, 4, 6, and 10, the expected activity is 5000 NLU/g, whereas in the rest of the samples the expected activity is 2000 NLU/g. The data presented are averages of two replicates samples. The underlined data were later identified as outliers and were not used in further calculations. Intralaboratory (RSD_r) and interlaboratory (RSD_R) precision values are also shown. [Reprinted, with permission, from A. J. Engelen and P.H.G Randsdorp, *Journal of AOAC International* **82** No. 1, 1999, 112–118. "Determination of Neutral Lactase Activity in Industrial Enzyme Preparations by a Colorimetric Enzymatic Method: Collaborative Study". © 1999 by Association of Official Analytical Chemists.]

Data received from three laboratories were excluded from the statistical analysis because all three reported low enzymatic activity ($< 50\%$ of the expected values), due to procedural errors. For example, laboratory "J" performed sample pretreatment over a period of 4 days and assayed for activity during the fifth day, this caused significant loss of enzymatic activity. Two laboratories did not return results. Statistical analyses were conducted using the data produced by the remaining participants; results are shown in Table 16.6.

Analysis of the data presented in Table 16.6 allows the rejection of anomalous data (outliers) using statistical tests, producing better precision, from 2.94 to 5.01 (RSD_r) and from 7.50 to 13.84 (RSD_R). These precision parameters are considered acceptable, and this preliminary validation of the colorimetric method was therefore considered successful.

16.7.4. Establishment of a Cutoff Value for Semi-Quantitative Assays for Cannabinoids[8]

When drugs of abuse or other banned substances are determined, a two-step procedure is often used. In the first step, specimens are tested by a semiquantitative assay; specimens determined positive in this step are examined by a confirmatory, quantitative assay in a second step. The second method is generally based on a

TABLE 16.7. Performance Characteristics of Immunoassays for the Detection of Cannabinoids in Urine[a]

| Assay | Number of Results | | | | | | | | Figures of Merit (%) | | | | | |
| | True Positives | | True Negatives | | False Positives | | False Negatives | | Sensitivity | | Specificity | | Efficiency | |
	100[a]	50[b]	100[a]	50[b]	100[a]	50[b]	100[a]	50[b]	100[a]	50[b]	100[a]	50[b]	100[a]	50[b]
Emit d.a.u. 100	76	120	790	796	16	37	75	31	50.3	79.5	98.0	95.4	90.5	92.9
Emit II 100	51	94	800	788	6	18	100	57	33.8	62.3	99.3	97.8	88.9	92.2
Abuscreen online	47	112	802	794	4	12	104	39	31.1	74.2	99.5	98.5	88.7	94.7
Abuscreen RIA	37	91	806	794	0	12	114	60	24.5	60.3	100.0	98.5	88.1	92.5
DRI	40	86	802	789	4	17	111	65	26.5	57.0	99.5	97.9	88.0	91.4
ADx	31	112	802	787	4	19	120	39	20.5	74.2	99.5	97.6	87.0	93.9
Emit d.a.u. 50		111		780		26		40		73.5		96.8		93.1
Emit II 50		101		787		19		50		66.9		97.6		92.8
Mean	47	103.4	800.3	786.0	5.7	20.0	104.0	47.6	31.1	68.5	99.3	97.5	88.5	92.9
SD	15.9	13.9	5.4	9.2	5.4	9.2	15.9	13.9	10.5	9.2	0.7	1.1	1.2	1.2

[a] Cutoff concentration 100 µg/L for immunoassay, 15 µg/L for GC–MS.
[b] Cutoff concentration 50 µg/L for immunoassay, 15 µg/L for GC–MS. [Reprinted, with permission, from M. A. Huestis, J. M. Mitchell, and E. J. Cone, *Clinical Chemistry* **40**, No. 5, 1994, 729–733. "Lowering the Federally Mandated Cannabinoid Immunoassay Cutoff Increases True-Positive Results". © 1994 by American Asso. of Clinical Chem.]

fundamentally different analytical principle, to reduce the potential impact of interfering species (such as cross-reactants in immunoassays). The first assay must be highly sensitive to identify presumptive positive samples and eliminate false negatives, while the second, quantitative method must be highly specific to eliminate false positive results.

In this example, urine specimens from six healthy male subjects with a history of marijuana use, who were undergoing medical treatment, were initially tested using eight semiquantitative cannabinoid immunoassays (Table 16.7). In these assays, the metabolite 11-nor-9-carboxy-Δ9-tetrahydrocannabinol (THCCOOH) was detected. The effect of two different cutoff values (100- and 50-µg/L THCCOOH) were investigated with respect to sensitivity, specificity and efficiency, as defined previously in Eq. 16.31–16.33. In this study, all the specimens were also assayed by GC–MS, as a quantitative confirmatory method, using a cutoff value of 15 µg/L; this value was used to define positive and negative test results ($\geq$15 and <15 µg/L, respectively).

True positive specimens were defined as having immunoassay results equal to or greater than the specified cutoff concentration (100 or 50 µg/L) and $\geq$15 µg/L THCCOOH by GC–MS. True negative specimens had results less than the cutoff concentrations for the immunoassays and the GC–MS test. False positive samples had immunoassay results greater than or equal to the immunoassay cutoff concentration and <15 µg/L THCCOH by GC–MS. False negative samples had immunoassay results lower than the specified immunoassay cutoff concentration and $\geq$15 µg/L THCCOOH by GC–MS.

In this study, from 957 specimens analyzed, GC–MS results indicate 151 positive and 806 negative results. These are compared with results of the immunoassay methods in Table 16.7. This study concluded that lowering the cannabinoid cutoff concentration from 100 to 50 µg/L THCCOOH increased the percent correct identification of true-positive specimens by all of the commercial immunoassays tested. Therefore, the sensitivities of all of the immunoassays were enhanced, resulting in improved efficiency for these assays. As expected, the specificity decreased slightly when the lower (50 µg/L) cutoff value was used. It is also evident from the results shown in Table 16.7 that there are discrepancies between the results of the eight commercial immunoassays. Preliminary tests suggest that these discrepancies can be attributed to differences in antibody selectivities, since different antibodies are used in the different assays; these antibodies may be expected to show different cross-reactivities with other cannabinoid metabolites, as well as other interfering species.

SUGGESTED REFERENCES

J. C. Miller and J. N. Miller, *Statistics for Analytical Chemistry*, Elis Horwood PTR Prentice Hall, New York, 1993.

G. T. Wernimont, *Use of Statistics to Develop and Evaluate Analytical Methods*, W. Spendley Ed., Association of Official Analytical Chemists, Arlington, VA, 1985.

R. Caulcutt and R. Boddy, *Statistics for Analytical Chemists*, Chapman and Hall, London, 1983.

A. Fajgelj and A. Ambrus, Eds., *Principles and Practices of Method Validation*, The Royal Society of Chemistry, Cambridge, UK, 2000.

M. L. Bishop, J. L. Duben-Engelkirk, and E. P. Fody, *Clinical Chemistry*, Lippincott Williams & Wilkins, Philadelphia, 2000.

D. C. Harris, *Quantitative Chemical Analysis*, W. H. Freeman and Co., New York, 2002.

REFERENCES

1. V. P. Shah et al., *Pharmaceut. Res.* **17**, 2000, 1551–1557.

2. W. Horwitz, *Anal. Chem.* **54**, 1982, 67A–76A.

3. P. Willetts and R. Wood, in *Principles and Practices of Method Validation*, A. Fajgelj and A. Ambrus, Eds., Royal society of Chemistry, Cambridge, UK, 2000, pp. 271–288

4. B. Nix and D. Wild, in *The Immunoassay Handbook*, D. Wild, Ed., 2nd Ed., Nature Publishing Group, New York, 2001, pp. 198–210.

5. P. Ertl and S. R. Mikkelsen, U.S. Pat. 6,391,577 BI, granted May 21, 2002, and P. Ertl, E. Robello, F. Battaglini, and S. R. Mikkelsen, *Anal. Chem.* **72**, 2000, 4957–4964.

6. M. T. Shipchandler and E. G. More, *Clin. Chem.* **41**, 1995, 991–994.

7. A. J. Engelen and P. H. G. Randsdorp, *J. AOAC Inter.* **82**, 1999, 112–118.

8. M. A. Huestis, J. M. Mitchell, and E. J. Cone, *Clin. Chem.* **40**, 1994, 729–733.

Answers to Problems

CHAPTER 1

1. It is likely that interferences (Cu^{2+} reducing agents, possibly reducing sugars) are present in the first Lowry assay; thorough oxygenation of the solution oxidizes these species, so that they do not interfere in the second assay.

2. The sensitivities of these methods are related to the amino acid compositions of the standard and unknown proteins, among other factors including the exposure of particular residues, which affects their reactivities. In addition, glycoproteins often react differently with the assay reagents than do proteins lacking a polysaccharide component. Therefore, when calibration curves are made using different standard proteins, their slopes are different. The choice of a standard protein will determine whether total protein is over- or underestimated in the unknown sample.

3. After correction for dilution, the ferricyanide method yields 18 ± 3 µg/mL. This concentration of carbohydrate will produce a signal in the DABA assay that is equivalent to 33 µg/mL of DNA ($18 \times 330/180$). The remaining signal in the DABA assay is due to DNA; its concentration is therefore ≈ 11 µg/mL. A dye-binding assay (e.g., ethidium bromide) can be used as a confirmative method.

CHAPTER 2

1. (a) Two substrates, since water is 55 M and does not change.

   (b) The [NAD^+] must be $>10K_m$ for NAD^+, that is, >2.2 mM; linearity should be observed from 0 to 1.7 mM ($0.1K_m$ for alanopine).

   (c) The NAD^+ is a competitive inhibitor. The apparent K_m for NADH is higher in the presence of added NAD^+ than in its absence; V_{max} is not affected.

Bianalytical Chemistry, by Susan R. Mikkelsen and Eduardo Cortón
ISBN 0-471-54447-7   Copyright © 2004 John Wiley & Sons, Inc.

2. (a) Requires an enzyme solution of 0.5 I.U./L concentration, so for 100 mL we need 0.05 I.U.; from enzyme's specific activity we calculate 0.05 I.U./700 I.U./mg = 70-ng enzyme.

(b) From specific activity and MW we find $(700 \text{ I.U./mg})(1000 \text{ mg/g})(45000 \text{ g/mol}) = 3.15 \times 10^{10}$ I.U./mol $= 3.15 \times 10^4$ mol substrate/(min $\times$ mol enzyme); converting minutes to seconds gives $k_{cat} = 525 \text{ s}^{-1}$.

(c) 0.32 I.U./L $\times$ (2 mL/0.5 mL) = 1.28 I.U./L; 1.28 I.U./L $\times$ 0.1 L = 0.128 I.U.; 0.128 I.U./1.6 mg = 0.080 I.U.mg.

3. Enzyme preparation has 20 ng/mL $(2.0 \times 10^{-5} \text{ mg/mL})$ of solid having 16 I.U./mg, so the activity of this solution is $2.0 \times 10^{-5} \times 16 = 3.2 \times 10^{-4}$ I.U./mL, or 0.32 I.U./L. This solution converts 0.32-μM substrate/min. From Beer's law, we calculate $A = (2.15 \times 10^4 \, M^{-1} \text{ cm}^{-1})(1 \text{ cm})(3.2 \times 10^{-7} \text{ M/min}) = 0.0069 \text{ AU/min}$.

4. Eadie–Hofstee method: plot v versus v/[S], slope $= -K_m$, y intercept is V_{max}. Values obtained by linear regression are $K_m = 0.62$ mM and $V_{max} = 0.74$ µmol/min, with R $= -0.996$, using the data shown below:

v	0.648	0.488	0.418	0.353	0.310	0.275
v/[S]	0.162	0.365	0.523	0.621	0.692	0.747

Cornish–Bowden–Eisenthal method: each of the v, v[S] points shown above gives the equation for a straight line, according to $V_{max} = v + (v/[S])K_m$. The six equations are thus

$$V_{max} = 0.648 + 0.162 \, K_m \tag{1}$$

$$V_{max} = 0.488 + 0.365 \, K_m \tag{2}$$

$$V_{max} = 0.418 + 0.523 \, K_m \tag{3}$$

$$V_{max} = 0.353 + 0.621 \, K_m \tag{4}$$

$$V_{max} = 0.310 + 0.692 \, K_m \tag{5}$$

$$V_{max} = 0.275 + 0.747 \, K_m \tag{6}$$

Each pair of equations is then used to solve for V_{max} and K_m (2 equations / 2 unknowns), so that $5 + 4 + 3 + 2 + 1 = 15$ values are obtained for each. These values are

V_{max} : 0.776, 0.751, 0.753, 0.753, 0.752, 0.652, 0.683,

0.691, 0.694, 0.770, 0.763, 0.756, 0.750,

0.741, 0.727 (avg $\pm$ SD $= 0.73 \pm 0.04 \, \mu M/min$)

K_m : 0.788, 0.637, 0.643, 0.638, 0.638, 0.443, 0.527, 0.554,

0.558, 0.663, 0.639, 0.638, 0.606,

0.619, 0.636 (avg $\pm$ SD $= 0.62 \pm 0.08$ mM)

The CB–E method is likely to yield better estimates of K_m and V_{max}.

To find k_{cat}, we use the V_{max} value determined by the CB–E method and the enzyme concentration: $(0.73\text{-}\mu M$ substrate/min)/0.050-μM enzyme) $= 15$ min$^{-1} = 0.25$ s^{-1}.

5. Hanes plots yield slopes of $1/V_{max}$ and y intercepts of K_m/V_{max}. So, from the values given, $V_{max} = 0.0100 \, \mu M/\text{min}$ and $K_m = 2.0 \times 10^{-5} \, \mu M$, or 20 p$M$. Specific activity is calculated from $V_{max} = 0.0100 \, \text{I.U.}/\text{L}$, since enzyme concentration is known to be 10 ng/mL (i.e., 0.010 mg/L): (0.0100 I.U./L)/(0.010 mg/L) $= 1.0$ I.U./mg.

CHAPTER 3

1. The kinetic method requires $[S] < 0.1 \, K_m$, since initial rates are measured, and these depend linearly on $[S]$ only under these conditions.

2. For a substrate assay, $[S] < 0.1 \, K_m$, and for an activity assay, $[S] > 10 \, K_m$.

3. Detection limits would be considerably improved (lower) if the indicator reaction generates and inhibitor of a fluorophore-producing enzymatic reaction, because of the additional catalytic amplification provided by this reaction.

4. (a) Only the first reaction is critical for the AP assay. The other three reactions are indicator reactions that could be used with any primary reaction producing FAD.

   (b) 0.1 μmol substrate/L converted in 10 min $= 1 \times 10^{-5} \, \mu$mol/(min mL) $= 1 \times 10^{-5}$ I.U./mL.

   (c) Calculate $V_{max} = (3.2 \, \text{s}^{-1})1.2 \times 10^{-8} \, M$ enzyme) $= 4.32 \times 10^{-8} \, M/\text{s}$; after 10 min (600 s), $4.32 \times 10^{-8} \, M/\text{s} \times 600 \, \text{s} = 2.59 \times 10^{-5} \, M$ substrate has been converted. Since 1:1 stoichiometry prevails, the absorbance change is $(2.3 \times 10^4 \, M^{-1}\text{cm}^{-1})$ (1 cm)(2.59 $\times 10^{-5} \, M) = 0.60$ AU.

   (d) In the activity assay, $1.2 \times 10^{-8} \, M$ enzyme would yield and absorbance change of 0.3 AU. Assuming a linear relationship between ΔA and $[S]$, $0.4 \times 10^{-8} \, M$ enzyme would yield 0.1 AU, and $8.0 \times 10^{-8} \, M$ would give 2.0 AU. The approximate dynamic range of the alkaline phosphatase assay is thus $0.004 - 0.080 \, \mu M$. Since $k_{cat} = 3.6 \, \text{s}^{-1}$ (or 216 min^{-1}), and $V_{max} = k_{cat}[E]$, the V_{max} values range from 0.86 $\mu M/\text{min}$ to 17 $\mu M/\text{min}$. These values convert to 8.6×10^{-4} and 1.7×10^{-2} I.U./mL.

5. From Beer's law, $[\text{ABTS}]_{ox} = 0.174 \, \text{AU}/(3.48 \times 10^4 \, M^{-1} \, \text{cm}^{-1} \times 1 \, \text{cm}) = 5.00 \times 10^{-6} \, M$. The AG/ABTS stoichiometry is 1:1, so taking dilution into account, the original [AG] was $5.00 \times 10^{-5} \, M$. Converting units yields [AG] $= 8.20$ mg/mL, indicating that the patient is diabetic.

CHAPTER 4

1. Steps consist of support activation, enzyme coupling, and column packing. Polystyrene must first be activated to generate suitable functional groups for

immobilization. This requires nitration, reduction, and diozotization in the activation step. The subsequent coupling step to immobilize the enzyme then consists of exposing the support to a buffered aqueous solution containing cholesterol oxidase. Following this mild coupling step, the polystyrene particles are filtered, rinsed, and packed into a column. Cholesterol oxidase, like other oxidase enzymes, generates hydrogen peroxide which can be detected ampero-metrically at +0.7 V versus SCE. To use an absorbance detector, peroxidase must be coimmobilized with the cholesterol oxidase, and a dye reagent must be included in the mobile phase so that a colored reaction product is generated.

2. (a) This copolymer yields residual carboxylate groups after immobilization. The negative charge attracts counterions including H^+, so that local pH at the particle surface is lower than bulk pH. This means that the apparent pH optimum will be higher for the immobilized enzyme than for the dissolved, homogeneous enzyme. This applies to any enzyme immobilized on ethylene/malic anhydride copolymer.

   (b) Due to the stagnant (diffusion) layer at the particle surface, the apparent K_m for urea will be larger than the homogeneous K_m. The same is true for any immobilized enzyme.

3. (a) This is a physical immobilization method. The enzyme is trapped between membranes, but is not chemically bound. The effect of this immobilization on K_m could be to increase it slightly (K_m is never lower for immobilized enzymes), but the pH optimum is not expected to change. In effect, the enzyme exists in solution that is trapped between membranes, but it is not chemically bound to the surface. Physical immobilization methods have less effect on K_m and pH optima than do chemical methods.

   (b) Response increases linearly with [acetylcholine], then levels off at high concentrations. K'_m occurs at half-maximal response.

   (c) Same as Figure 2.12.

4. (a) Silanol groups are derivatized using APTS, giving primary amine groups. These can be reacted directly with protein carboxylate groups, using a dehydrating reagent such as a carbodiimine/NHS reaction; alternately, the primary amines can be diazotized, then coupled to protein residues.

   (b) From Eq. 4.23, capacity $C = kE_T\beta = \{(1.45 \times 10^3 \text{ s}^{-1})(0.3\text{-g enzyme})/(5 \times 10^4 \text{ g/mol}\} \times (3 \text{ mL/5mL}) = 3.5 \times 10^{-3} \text{ mol/s} = 0.21 \text{ mol min}^{-1}$.

   (c) From Eq. 4.24, substitute $C = 0.21$ mol/min, $P = 0.95$ and given K'_m values. Using the homogeneous K_m value, maximum flow rates of 1.2×10^4, 1.2×10^4 and 9.2×10^3 L/min are obtained with the values given (this column has an extremely large capacity).

CHAPTER 5

1. Preparation A is more concentrated, since it must be diluted to a much greater extent to reach the dilute end of the equivalence–agglutination zone.

2. Monoclonal antibody preparations contain a homogeneous population of identical antibody molecules with identical binding sites. They have a well-defined interactions between this site and a single antigen epitope. This results in a linear Scarchard plot. Polyclonal preparations consist of a mixture of antibody molecules, some reacting with antigen. Those that bind antigen possess a number of different paratopes, so that a variety of antigen epitopes bind, each with a different affinity. This results in a curved Scatchard plot.

3. (a) No precipitin forms. Anti-B, -C, or -D must also be present to obtain precipitin.

(b) Precipitin is present since antigen is multivalent.

(c) No precipitin forms because only the primary Ag:Ab reaction can occur.

(d) Precipitin forms [same argument as for (b)].

4. Yes, the bacterium was present, since an agglutination zone was observed. A prozone is only observed if the antigen is present in excess, at the low dilution (concentrated) end of the dilution series; since no prozone was observed, the bacterium was initially present at too dilute a concentration for the observation of a prozone.

5. The precipitin line would curve toward the antigen well.

CHAPTER 6

1. (a) Estrogen, (b) human α-fetoprotein, (c) mouse antibody, and (d) acetylcholine.

2. (a) Antibody, (b) antibody, and (c) labeled ligand and ligand/unknown.

3.

Fluorophore	Continuous
Chemiluminescent	Single event (e.g., 1 photon/reaction)
Enzyme	Catalytic amplification
Cofactor	Continuous (like substrate)
Lysing agent	Catalytic amplification (trapped E)
Secondary label	Catalytic amplification (binds E)
Prosthetic group	Catalytic amplification (creates active E)
Electroactive	Single event (e.g., 1 electron/ferrocene)
Spin label	Continuous
Substrate	Continuous (e.g. A of product)

4. (a) Ab is present at limited concentration, while H, H—H, and Ab* are all present in excess.

(b) Only one other species will be present, Ab—H.

(c) Of these potential interferents, E is expected to cross-react to the greatest extent, since it is the most identical to C, especially in the region most likely to comprise the epitope. The least likely to interfere is D, since it is different in the Ab-binding region.

(d) Plot is sigmoid, with signals high at low [testosterone] and decreasing as concentration increases.

5. (a) Plot is sigmoid, with higher signals at low [free biotin] values, decreasing as [biotin] increases.

(b) [Substrate] > 1.0 mM should be used ($10\ K_m$).

(c) Ideally, the plot of light intensity, or photons detected, as a function of time will be flat, that is, photons are emitted and detected at a constant rate. In contrast, absorbing products accumulate in an enzymatic reaction, so that absorbance increases with time.

6. This assay is noncompetitive, so that linear emission–concentration (rather than sigmoid emission–log[Ag]) plots are observed. Linear log–log plots will also be observed, and as shown in the Figure 6.6, F1 emission decreases with increasing concentration while F2 emission increases with increasing Ag.

CHAPTER 7

1. (a) $i = 26$ μA; do not forget to convert cubic centimeter (cm^3) units to liters (L).

(b) Same as (a)—current does not change.

2. The assumption is invalid when very small sample volumes are used. Then analyte is consumed, and the concentration of analyte in the reaction layer decreases with time.

3. We want NADPH to be produced very close to the surface of the optical fiber, so that the evanescent wave will detect it. Option (a) is not the best, since adsorption would lead to a loss of activity, and the monolayer would have a low total activity. Option (c) also produces a monolayer, and may not fully convert glucose (i.e., [S]$_o$ does not equal 0). Option (b) looks the best, since complete conversion of glucose would occur in the outer layers, and the production of NADPH occur in the outer layers, and the production of NADPH occurs at the surface of the fiber.

4. Thermal sensors rely on heat retention, to measure heat generated by the enzymatic reaction. This will result in denaturation over the longer term. Amperometric sensors are not insulated, so that heat dissipation is more efficient.

CHAPTER 8

1. False. Biological and analytical "figures of merit" are not always the same.

2. The main criterion is the availability of knowledge to guide a rational design process. Secondary criteria would include the lack of an available method to screen a large library, and the nature of the desired improvement.

3. In the first case, the diversity generated will be low; the PCR products will consist mainly of perfect copies of the template, with no mutations. In the second case, it is unlikely that a functional protein will be produced due to the very high error rate.

4. The number of original DNA molecules in 5 ng of template is $\sim 6.72 \times 10^9$. The mass of the template molecules was not included in the total DNA mass calculation.

Number of Cycles	Number of DNA Copies (for each template molecule)	Molecules Copied	DNA Mass (in g)
5	31	2.08×10^{11}	1.55×10^{-7}
15	32,767	2.20×10^{14}	1.64×10^{-4}
30	1.07×10^9	7.19×10^{18}	5.35
60	1.15×10^{18}	7.72×10^{27}	5.75×10^9
500	3.27×10^{150}	2.20×10^{160}	1.64×10^{142}

CHAPTER 9

1. (a) The pH 6: Both proteins are positively charged, and will migrate to the cathode.

   (b) The pH 8.5: Myoglobin has a negative charge and will migrate to the anode, while cytochrome c has a positive charge and will migrate to the cathode.

2. (a) Sodium dodecyl sulfate gives a uniformly negative charge/mass ratio, so samples are applied at the cathodic end of the gel and will migrate to the anode.

   (b) The smallest protein migrates farthest (i.e., protein X).

   (c) Distance migrated is proportional to log(MW). Using this relationship with the two points given, the MW of the unknown is 26.3 kDa.

3. Proteins stained with Coomassie Blue show as visible blue bands. The DNA stained with ethidium show bands only under UV illumination.

4. If pH > 3, DNA is always negatively charged, and proteins will all be positively charged, so that DNA and proteins will transfer in different directions.

5. Ethidium will stain all DNA present, while the DNA probe will bind only to its complementary sequence.

CHAPTER 10

1. Regression of distance (cm) against log(MW) gives the line $y = (-1.56)x + 11.6$; from this, $y = 3.03$ cm gives log(MW) $= 5.49$, so MW $= 309$ kDa.

2. The protein has a subunit ratio of 2 (75 kDa): 1 (50 kDa), with a minimum MW of 200 kDa.

3. Regression of distance (cm) against log(length, BP) yields $y = (-55.45)x + 224.3$; from this, $y = 15.27$ cm gives log (length) $= 3.77$, so the length of the unknown is 5884 BP.

4. (a) Cathode = bottom; Anode = top; samples are applied at the bottom.

   (b) 5'-GCGATCACCG-3'.

CHAPTER 11

1. Zone EP causes a continuous migration in one direction, as long as the voltage is applied; bands can migrate right off the gel. The IEF causes initial migration, but reaches a steady-state when all proteins are at their pI values and have zero net charge.

2. The pH range of the gel is 6.5 (anode) $-$ 7.5 (cathode), over the 10-cm length. The three HB variants will be 0.8 cm (pI 7.42), 2.9 cm (pI 7.21), and 4.5 cm (pI 7.05) from the cathodic end of the gel.

3. After the IEF run, the gel is exposed to a solution containing the E—Ab conjugate. The enzyme is chosen to convert a colorless substrate into a highly absorbing product. A colored band appears where the Ab binds its Ag. Reaction of proteins with SDS, needed in the second dimension of the 2D separation, results in a loose of the tertiary–quaternary structure of the antigen, so that Ab probably will not bind.

4. IEF is performed on proteins in their native state, and is based on migration to the pI value in the gel. The SDS disturbs the protein structure, and gives a uniformly negative charge/mass ratio, so that separation is based only on size (sieving).

5. Nucleic acids inherently have a uniformly negative ratio charge/mass m/z from the sugar–phosphate backbone. An IEF run over the pH 6–8 range would cause migration of nucleic acids off the anodic end of the gel.

6. The IPG–DALT gel shows the presence of two subunits. One has pI 8.2 and MW 50 kDa, while the other has pI 7.1 and MW 80 kDa. Since the 80 kDa band is darker (larger), the intact protein probably contains two 80- and one 50-kDa subunits.

7. Molecular weight markers are added at the end of the IEF run, to one end of the IEF gel, so that they do not interfere with (or obscure) the focussing of analyte proteins, and provide a distinct series of bands on one side of the 2D gel.

CHAPTER 12

1. By using Eq. 12.7, retention times (a) increase by a factor of 4, (b) decrease by a factor of 2, and (c) increase by a factor of 2.

2. The APTES modification will cause an anodic drift, so the detector should be placed at the anodic end of the capillary.

3. The lowest m/z corresponds to the highest charge (+8), and the highest m/z corresponds to the lowest charge (+5). The four calculated MW values are thus 7504, 7497, and 7500, so MW = 7500.

4. All amino acids have the reactive primary amino group, which can be derivatized, for example, with an aldehyde group present on a redox-active species, such as ferrocenecarboxaldehyde, $(C_5H_5)Fe(C_5H_4CHO)$.

5. The coincidence detector eliminates background noise by requiring both PMTs to simultaneously record a count.

6. The G:A:C:T ratio is 8:4:2:1. Thus the sequence can be read as 5'-CTGCAG-CATCATG-3'. The complementary (analyte) sequence is therefore 5'-CAT-GATGCTGCAG-3'.

7. To elute the pattern at the anodic (acid) end of the capillary, the focused proteins should develop a negative charge. This is done by shifting to basic conditions by adding a salt (e.g., NaCl) to the anodic buffer reservoir.

CHAPTER 13

1. (a) $RCF_{min} = 97,890$ g, $RCF_{max} = 179,465$ g, $RCF_{av} = 138,678$ g.
   (b) No; the relationship between rpm and g force is not linear.

2. (a) $k = 114$.
   (b) The vertical rotor with $k = 8$.

3. 48,036 rpm.

4. The k values are 469 (40,000 rpm); 833 (30,000 rpm); 1875 (20,000 rpm) and 7500 (10,000 rpm).

5. Influenza virus particles ≈ 16 min; ribosome subunits, 4.5 and 3 h for small and large subunits, respectively.

6. 5.52 S (Svedberg)

CHAPTER 14

1. (a) 1.6
   (b) Protein X has higher MW because it elutes earlier than does protein Y.

2. A plot of $(V_E - V_o)$ versus log(MW) yields the regression equation $y = (-30)x + 160$; if $y = 26.2$ mL, $x = 4.46$, corresponding to MW = 28.8 kDa.

3. If the mixture contains other dehydrogenases, an affinity column (e.g., cibachron blue) will separate all dehydrogenases. Then ion exchange chromatography can be used to isolate the dehydrogenase of interest based on net charge and charge distribution characteristics in the dehydrogenase of interest.

4. A plot of $1/(V_R - V_o)$ (in L units) against [glutamate] (in μM units) yield $y = 100x + 10$. The association constant is calculated as slope/intercept, and is 10 (μM^{-1}), or $1.0 \times 10^7 \ M^{-1}$.

CHAPTER 15

1.

Compound	Formula	Monoisotopic	Average
Glycine (Gly or G)	$C_2H_5NO_2$	75.0319	75.0677
Tryptophan (Trp or W)	$C_{11}H_{12}N_2O_2$	204.0896	204.2303
Polypeptide	$C_{84}H_{101}N_{15}O_{14}$	1543.7629	1544.8325
Porcin insulin	$C_{256}H_{381}N_{65}O_{76}S_6$	5773.6183	5777.6755
Myoglobin	$C_{769}H_{1212}N_{210}O_{218}S_3$	16,972.9091	16,983.7061
Big heterodimer protein	$C_{1802}H_{2933}N_{532}O_{462}S_{12}$	39,770.8635	39,796.4794

2. Monoisotopic peak is the more intense in low molecular weight compounds. When the number of carbon atoms increase, molecules bearing one or more ^{13}C atoms becomes common, and therefore these peaks are more intense. At high MW, there is a negligible probability of not bearing at least one ^{13}C atom; therefore the monoisotopic peak is not observable in these spectra.

3.

Ion Charge	Symbol	Ion Mass	Position (m/z)
+1	$(M+1H)^+$	23,631.36	23,631.36
+2	$(M+2H)^{+2}$	23,632.36	11,816.1800
+3	$(M+3H)^{+3}$	23,633.36	7877.7866
+4	$(M+4H)^{+4}$	23,634.36	5908.5900
+5	$(M+5H)^{+5}$	23,635.36	4727.0720
+6	$(M+6H)^{+6}$	23,636.36	3939.3933

4.

m_2	m_1	Δm	n_2 (calculated)	n_2 (integral units)	M
1306.4	1219.4	87.0	14.00	14	18,275.6
1219.4	1143.3	76.1	15.01	15	18,276.0
1143.3	1076.1	67.2	16.00	16	18,276.8
1076.1	1016.3	59.8	16.98	17	18,276.7
1016.3	963	53.3	18.05	18	18,275.4

(a) The m/z range of spectrometer. (b) Concentration negligible.

5. The parameter Δm at 50% level is equal to ≈ 2.9, and $M \approx 1204.5$, therefore the calculated resolution is equal to ≈ 415 using FWHM definition.

Bianalytical Chemistry, by Susan R. Mikkelsen and Eduardo Cortón
ISBN 0-471-54447-7   Copyright © 2004 John Wiley & Sons, Inc.